THE ART AND SCIENCE OF PROTECTIVE RELAYING

THE ART AND SCIENCE OF PROTECTIVE RELAYING

C. RUSSELL MASON

Engineering Planning and Development Section
General Electric Company
Schenectady, N.Y.

One of a series written by General Electric authors for the advancement of engineering practice

JOHN WILEY & SONS, INC.

New York · London · Sydney

17 16 15

Library of Congress Catalog Card Number: 56–8694

Printed in the United States of America

ISBN 0 471 57552 6

TO MY WIFE, DOT WITHOUT WHOSE PATIENCE AND UNDER-
STANDING THIS BOOK WOULD NOT HAVE BEEN WRITTEN

PREFACE

"Science is systematized knowledge. Art is knowledge made efficient by skill."

Webster's Collegiate Dictionary, Fifth Edition, G. and C. Merriam Co., Springfield, Mass., 1942

This is a textbook on protective relaying. Much of the material has been used for several years as notes for teaching the subject in the Power Systems Engineering Course given by the General Electric Company. These notes were written because there was no suitable reference book available that properly presented the subject to the novice.

From the experience gained, the notes were revised to improve their clarity and to add necessary explanatory material. Therefore, the material has been tested in the classroom and it should prove useful both as a reference for teaching the subject and for the purpose of self-education.

As already intimated, the book assumes no prior knowledge of protective relaying. In fact, as the student will quickly discover, one needs to know only the fundamental principles of electrical engineering to become well acquainted with protective relaying. He will not be able to master the subject without practical experience in the field, but the subject will lose most of the mystery generally associated with it.

In spite of the elementary nature of the book, it should also be useful to the practicing relay engineer. The book contains new material, it treats many subjects in a different manner from that found elsewhere, and it may help many to understand better what they already know. At least, that has been my experience in writing the material and presenting it to the student. Also, the book contains references to basic source material that has been found to be most authoritative and useful.

Theoretical considerations that have no practical utility have been studiously avoided. Only material is included that I know is useful on the basis of over 25 years of experience. This is not to say that the book will answer every question that may arise, but it will at least help one to find the answer. Neither is the book an historical reference or a reference to foreign practices.

So far as it is possible to make it so, the book is timeless. It contains fundamental information that is applicable to present-day North American practice, but the nature of the material is such that it will be applicable whenever or wherever protective relays are involved. It follows that the book tries to be impartial to all manufacturers of protective relays. Naturally, I have had the most ready access to data, etc., pertaining to relays manufactured by the company with which I am associated, but I have conscientiously tried to avoid any taint of commercialism.

In general, proofs are avoided. References are given to source material containing proofs where such is considered necessary. There are too many worth-while things to be covered to burden the book with proofs unless they are an actual aid to a better understanding of the subject.

The references are not always to prove or confirm a point in the text, but are sometimes other opinions on the subject. With few exceptions, the references are not intended to give credit for original work or contributions, but to present the most up-to-date treatments of the subjects. In general, references are made only to publications that are easily available to the student; many other valuable contributions to the literature have been made, but they consist of conference papers and publications of individual companies or associations that are not readily available in most libraries; in this respect, it is strongly recommended that authors of such material seek eventual publication in the technical press.

This book studiously avoids photographs and nomenclature of actual relays. Such things would "date" the book, take up valuable space, and would add but little to its value. One does not need to know what a relay looks like to learn how it operates and how to apply it. A relay photograph, especially as reproducible in a book, shows practically nothing of real value. The student does not have to go far to see actual relays. Also, the manufacturers will provide well-illustrated publications.

C. Russell Mason

Schenectady, N. Y.
January, 1956

ACKNOWLEDGMENTS

The experiences and knowledge of many people are contained in this book; most of these contributions I can only humbly acknowledge in a general way. Specifically, I am greatly indebted to my immediate associates for their contributions to this book, both directly and indirectly, and particularly to Messrs. R. E. Cordray, H. T. Seeley, W. C. New, C. G. Dewey, R. H. Macpherson, J. R. McGlynn, W. C. Morris, H. Bany, L. F. Kennedy, A. J. McConnell, and D. B. Brandt. The sharing of their knowledge and philosophies, and their criticisms and suggestions for improving the book, have been most valuable.

For their painstaking stenographic help in the various stages of preparing the manuscript, I am indebted to the Misses Alicia Corrado, Charlotte Carpenter, and Rose Cuda.

CONTENTS

Chapter 1 The Philosophy of Protective Relaying 1
What is Protective Relaying? The Function of Protective Relaying. Fundamental Principles of Protective Relaying: Primary Relaying, Back-up Relaying, Protection against Other Abnormal Conditions. Functional Characteristics of Protective Relaying: Sensitivity, Selectivity, and Speed, Reliability. Are Protective Practices Based on the Probability of Failure? Protective Relaying Versus a Station Operator. Undesired Tripping Versus Failure to Trip When Desired. The Evaluation of Protective Relaying. How Do Protective Relays Operate?

Chapter 2 Fundamental Relay-Operating Principles and Characteristics 16
General Considerations: Operating Principles, Definitions of Operation, Operation Indicators, Seal-In and Holding Coils, and Seal-In Relays, Adjustment of Pickup or Reset, Time Delay and Its Definitions. Single-Quantity Relays of the Electromagnetic-Attraction Type: Operating Principle, Ratio of Reset to Pickup, Tendency toward Vibration, Directional Control, Effect of Transients, Time Characteristics. Directional Relays of the Electromagnetic-Attraction Type: Operating Principle, Efficiency, Ratio of Continuous Thermal Capacity to Pickup, Time Characteristics. Induction-Type Relays—General Operating Principles: The Production of Actuating Force, Types of Actuating Structure, Accuracy. Single-Quantity Induction Relays: Torque Control, Effect of Frequency, Effect of D-C Offset, Ratio of Reset to Pickup, Reset Time, Time Characteristics. Directional Induction Relays: Torque Relations in Terms of Actuating Quantities, The Significance of the Term "Directional," The Polarizing Quantity of a Directional Relay, The Operating Characteristic of a Directional Relay, The "Constant-Product" Characteristic, Effect of D-C Offset and Other Transients, The Effect of Frequency, Time Characteristics. The Universal Relay-Torque Equation.

Chapter 3 Current, Voltage, Directional, Current (or Voltage)-Balance, and Differential Relays 42
General Protective-Relay Features: Continuous and Short-Time Ratings, Contact Ratings, Holding-Coil or Seal-In-Relay and Target Ratings, Burdens. Overcurrent, Undercurrent, Overvoltage, and Undervoltage Relays: Adjustment, Time Characteristics, Overtravel, Reset Time, Compensation for Frequency or Temperature Changes, Combination of Instantaneous and Inverse-Time Relays. D-C Directional Relays: Current-Directional Relays, Voltage-Directional

Relays, Voltage-and-Current-Directional Relays, Voltage-Balance Directional Relays, Current-Balance Directional Relays, Directional Relays for Vacuum-Tube or Rectified A-C Circuits, Polarizing Magnet versus Field Coil, Use of Shunts, Time Delay. A-C Directional Relays: Power Relays, Directional Relays for Short-Circuit Protection, Directional-Overcurrent Relays. Current (or Voltage)-Balance Relays: Overcurrent Type, Directional Type. Differential Relays.

Chapter 4 Distance Relays 70

The Impedance-Type Distance Relay. The Modified-Impedance-Type Distance Relay. The Reactance-Type Distance Relay. The Mho-Type Distance Relay. General Considerations Applicable to All Distance Relays: Overreach, Memory Action, The Versatility of Distance Relays, The Significance of *Z*

Chapter 5 Wire-Pilot Relays 86

Why Current-Differential Relaying Is Not Used. Purpose of a Pilot. Tripping and Blocking Pilots. D-C Wire-Pilot Relaying. Additional Fundamental Considerations. A-C Wire-Pilot Relaying: Circulating-Current Type, Opposed-Voltage Type, Advantages of A-C over D-C Wire-Pilot Equipments, Limitations of A-C Wire-Pilot Equipments, Supervision of Pilot-Wire Circuits, Remote Tripping over the Pilot Wires, Pilot-Wire Requirements, Pilot Wires and Their Protection against Overvoltage.

Chapter 6 Carrier-Current-Pilot and Microwave-Pilot Relays . . . 100

The Carrier-Current Pilot. The Micro-Wave Pilot. Phase-Comparison Relaying. Directional-Comparison Relaying. Looking Ahead.

Chapter 7 Current Transformers 112

Types of Current Transformers. Calculation of CT Accuracy: Current-Transformer Burden, Ratio-Correction-Factor Curves, Calculation of CT Accuracy Using a Secondary-Excitation Curve, ASA Accuracy Classification, Series Connection of Low-Ratio Bushing CT's, The Transient or Steady-State Errors of Saturated CT's, Overvoltage in Saturated CT Secondaries, Proximity Effects, Polarity and Connections: Wye Connection, Delta Connection, The Zero-Phase-Sequence-Current Shunt.

Chapter 8 Voltage Transformers 133

Accuracy of Potential Transformers. Capacitance Potential Devices: Standard Rated Burdens of Class A Potential Devices, Standard Accuracy of Class A Potential Devices, Effect of Overloading, Non-Linear Burdens, The Broken-Delta Burden and the Winding Burden, Coupling-Capacitor Insulation Coordination and Its Effect on the Rated Burden, Comparison of Instrument Potential Transformers and Capacitance Potential Devices. The Use of Low-Tension Voltage. Polarity and Connections: Low-Tension Voltage for Distance Relays, Connections for Obtaining Polarizing Voltage for Directional-Ground Relays.

Chapter 9 Methods for Analyzing, Generalizing, and Visualizing Relay Response 155

The *R-X* Diagram: Principle of the *R-X* Diagram, Conventions for Superimposing Relay and System Characteristics. Short Circuits: Three-Phase Short Circuits, Phase-to-Phase Short Circuits, Discussion of Assumptions, Determination of Distance-Relay Operation,

Effect of a Wye-Delta or a Delta-Wye Power Transformer between Distance Relays and a Fault. Power Swings and Loss of Synchronism. Effect on Distance Relays of Power Swings or Loss of Synchronism. Response of Polyphase Directional Relays to Positive- and Negative-Phase-Sequence Volt-Amperes. Response of Single-Phase Directional Relays to Short Circuits. Phase-Sequence Filters.

Chapter 10 A-C Generator and Motor Protection 193

Generator Protection: Short-Circuit Protection of Stator Windings by Percentage-Differential Relays, The Variable-Percentage-Differential Relay, Protection against Turn-to-Turn Faults in Stator Windings, Combined Split-Phase and Over-All Differential Relaying, Sensitive Stator Ground-Fault Relaying, Stator Ground-Fault Protection of Unit Generators, Short-Circuit Protection of Stator Windings by Overcurrent Relays, Protection against Stator Open Circuits, Stator-Overheating Protection, Overvoltage Protection, Loss-of-Synchronism Protection, Field Ground-Fault Protection, Protection against Rotor Overheating Because of Unbalanced Three-Phase Stator Currents, Loss-of-Excitation Protection, Protection against Rotor Overheating Because of Overexcitation, Protection against Vibration, Protection against Motoring, Overspeed Protection, External-Fault Back-Up Protection, Bearing-Overheating Protection, Miscellaneous Other Forms of Protection, Generator Potential-Transformer Fusing and Fuse Blowing, Station Auxiliary Protection. Motor Protection: Short-Circuit Protection of Stator Windings, Stator-Overheating Protection, Rotor-Overheating Protection, Loss-of-Synchronism Protection, Undervoltage Protection, Loss-of-Excitation Protection, Field Ground-Fault Protection.

Chapter 11 Transformer Protection 241

Power Transformers and Power Autotransformers: The Choice of Percentage-Differential Relaying for Short-Circuit Protection, Current-Transformer Connections for Differential Relays, The Zero-Phase-Sequence-Current Shunt, Current-Transformer Ratios for Differential Relays, Current-Transformer Accuracy Requirements for Differential Relays, Choice of Percent Slope for Differential Relays, Protecting a Three-Winding Transformer with a Two-Winding Percentage-Differential Relay, Effect of Magnetizing-Current Inrush on Differential Relays, Protection of Parallel Transformer Banks, Short-Circuit Protection with Overcurrent Relays, Gas-Accumulator and Pressure Relays, Grounding Protective Relay, Remote Tripping, External-Fault Back-Up Protection. Regulating Transformers: Protection of In-Phase Type, Protection of Phase-Shifting Type, External-Fault Back-Up Protection. Step Voltage Regulators. Grounding Transformers. Electric-Arc-Furnace Transformers. Power-Rectifier Transformers.

Chapter 12 Bus Protection 275

Protection by Back-Up Relays. The Fault Bus. Directional-Comparison Relaying. Current-Differential Relaying with Overcurrent Relays. Partial-Differential Relaying. Current-Differential Relaying with Percentage-Differential Relays. Voltage-Differential Relaying with "Linear Couplers." Current-Differential Relaying with Overvoltage Relays. Combined Power-Transformer and Bus Protection. Ring-Bus Protection. The Value of Bus Sectionalizing. Back-Up Protection for Bus Faults. Grounding the Secondaries of Differentially Connected CT's. Automatic Reclosing of Bus Breakers. Practices with Regard to Circuit-Breaker By-Passing. Once-a-Shift Testing of Differential-Relaying Equipment.

Chapter 13 Line Protection with Overcurrent Relays 296

How to Adjust Inverse-Time-Overcurrent Relays for Coordination. Arc and Ground Resistance. Effect of Loop Circuits on Overcurrent-Relay Adjustments. Effect of System on Choice of Inverseness of Relay Characteristic. The Use of Instantaneous Overcurrent Relays. An Incidental Advantage of Instantaneous Overcurrent Relaying. Overreach of Instantaneous Overcurrent Relays. The Directional Feature. Use of Two vs. Three Relays for Phase-Fault Protection. Single-Phase vs. Polyphase Directional-Overcurrent Relays. How to Prevent Single-Phase Directional-Overcurrent-Relay Misoperation during Ground Faults. Adjustment of Ground vs. Phase Relays. Effect of Limiting the Magnitude of Ground-Fault Current. Transient CT Errors. Detection of Ground Faults in Ungrounded Systems. Effect of Ground-Fault Neutralizers on Line Relaying. The Effect of Open Phases Not Accompanied by a Short Circuit. The Effect of Open Phases Accompanied by Short Circuits. Polarizing the Directional Units of Ground Relays. Negative-Phase-Sequence Directional Units for Ground-Fault Relaying. Current-Balance and Power-Balance Relaying. Automatic Reclosing. Restoration of Service to Distribution Feeders After Prolonged Outages. Coordinating with Fuses. A-C and Capacitor Tripping.

Chapter 14 Line Protection with Distance Relays 340

The Choice between Impedance, Reactance, or Mho. The Adjustment of Distance Relays. The Effect of Arcs on Distance-Relay Operation. The Effect of Intermediate Current Sources on Distance-Relay Operation. Overreach Because of Offset Current Waves. Overreach of Ground Relays for Phase Faults. Use of Low-Tension Voltage. Use of Low-Tension Current. Effect of Power-Transformer Magnetizing-Current Inrush on Distance-Relay Operation. The Connections of Ground Distance Relays. Operation When PT Fuses Blow. Purposeful Tripping on Loss of Synchronism. Blocking Tripping on Loss of Synchronism. Automatic Reclosing. Effect of Presence of Expulsion Protective Gaps. Effect of a Series Capacitor. Cost-Reduction Schemes for Distance Relaying. Electronic Distance Relays.

Chapter 15 Line Protection with Pilot Relays 373

Wire-Pilot Relaying: Obtaining Adequate Sensitivity, The Protection of Multiterminal Lines, Current-Transformer Requirements, Back-Up Protection. Carrier-Current-Pilot Relaying: Automatic Supervision of Carrier-Current Channel, Carrier-Current Attenuation, Use of Carrier Current to Detect Sleet Accumulation, Types of Relaying Equipment. Phase Comparison: Obtaining Adequate Sensitivity, The Protection of Multiterminal Lines, Back-Up Protection. Directional Comparison: Relation between Sensitivities of Tripping and Blocking Units for Two-Terminal Lines, The Protection of Multiterminal Lines, Effect of Transients. Combined Phase and Directional Comparison: The Effect of Mutual Induction on Directional-Ground Relays. All-Electronic Directional-Comparison Equipment. Microwave: The Microwave Channel, Remote Tripping. High-Speed Reclosing.

Index 403

1 THE PHILOSOPHY OF PROTECTIVE RELAYING

What is Protective Relaying?

We usually think of an electric power system in terms of its more impressive parts—the big generating stations, transformers, high-voltage lines, etc. While these are some of the basic elements, there are many other necessary and fascinating components. Protective relaying is one of these.

The role of protective relaying in electric-power-system design and operation is explained by a brief examination of the over-all background. There are three aspects of a power system that will serve the purposes of this examination. These aspects are as follows:

A. Normal operation.
B. Prevention of electrical failure.
C. Mitigation of the effects of electrical failure.

The term "normal operation" assumes no failures of equipment, no mistakes of personnel, nor "acts of God." It involves the minimum requirements for supplying the existing load and a certain amount of anticipated future load. Some of the considerations are:

A. Choice between hydro, steam, or other sources of power.
B. Location of generating stations.
C. Transmission of power to the load.
D. Study of the load characteristics and planning for its future growth.
E. Metering.
F. Voltage and frequency regulation.
G. System operation.
H. Normal maintenance.

The provisions for normal operation involve the major expense for equipment and operation, but a system designed according to this aspect alone could not possibly meet present-day requirements. Electrical equipment failures would cause intolerable outages. There must be additional provisions to minimize damage to equipment and interruptions to the service when failures occur.

Two recourses are open: (1) to incorporate features of design aimed at preventing failures, and (2) to include provisions for mitigating the effects of failure when it occurs. Modern power-system design employs varying degrees of both recourses, as dictated by the economics of any particular situation. Notable advances continue to be made toward greater reliability. But also, increasingly greater reliance is being placed on electric power. Consequently, even though the probability of failure is decreased, the tolerance of the possible harm to the service is also decreased. But it is futile—or at least not economically justifiable—to try to prevent failures completely. Sooner or later the law of diminishing returns makes itself felt. Where this occurs will vary between systems and between parts of a system, but, when this point is reached, further expenditure for failure prevention is discouraged. It is much more profitable, then, to let some failures occur and to provide for mitigating their effects.

The type of electrical failure that causes greatest concern is the short circuit, or "fault" as it is usually called, but there are other abnormal operating conditions peculiar to certain elements of the system that also require attention. Some of the features of design and operation aimed at preventing electrical failure are:

A. Provision of adequate insulation.
B. Coordination of insulation strength with the capabilities of lightning arresters.
C. Use of overhead ground wires and low tower-footing resistance.
D. Design for mechanical strength to reduce exposure, and to minimize the likelihood of failure causable by animals, birds, insects, dirt, sleet, etc.
E. Proper operation and maintenance practices.

Some of the features of design and operation for mitigating the effects of failure are:

A. Features that mitigate the immediate effects of an electrical failure.
   1. Design to limit the magnitude of short-circuit current.[1]
      a. By avoiding too large concentrations of generating capacity.
      b. By using current-limiting impedance.
   2. Design to withstand mechanical stresses and heating owing to short-circuit currents.
   3. Time-delay undervoltage devices on circuit breakers to prevent dropping loads during momentary voltage dips.
   4. Ground-fault neutralizers (Petersen coils).
B. Features for promptly disconnecting the faulty element.
   1. Protective relaying.
   2. Circuit breakers with sufficient interrupting capacity.
   3. Fuses.

C. Features that mitigate the loss of the faulty element.
   1. Alternate circuits.
   2. Reserve generator and transformer capacity.
   3. Automatic reclosing.
D. Features that operate throughout the period from the inception of the fault until after its removal, to maintain voltage and stability.
   1. Automatic voltage regulation.
   2. Stability characteristics of generators.
E. Means for observing the effectiveness of the foregoing features.
   1. Automatic oscillographs.
   2. Efficient human observation and record keeping.
F. Frequent surveys as system changes or additions are made, to be sure that the foregoing features are still adequate.

Thus, protective relaying is one of several features of system design concerned with minimizing damage to equipment and interruptions to service when electrical failures occur. When we say that relays "protect," we mean that, together with other equipment, the relays help to minimize damage and improve service. It will be evident that all the mitigation features are dependent on one another for successfully minimizing the effects of failure. *Therefore, the capabilities and the application requirements of protective-relaying equipments should be considered concurrently with the other features.*[2] This statement is emphasized because there is sometimes a tendency to think of the protective-relaying equipment after all other design considerations are irrevocably settled. Within economic limits, an electric power system should be designed so that it can be adequately protected.

The Function of Protective Relaying

The function of protective relaying is to cause the prompt removal from service of any element of a power system when it suffers a short circuit, or when it starts to operate in any abnormal manner that might cause damage or otherwise interfere with the effective operation of the rest of the system. The relaying equipment is aided in this task by circuit breakers that are capable of disconnecting the faulty element when they are called upon to do so by the relaying equipment.

Circuit breakers are generally located so that each generator, transformer, bus, transmission line, etc., can be completely disconnected from the rest of the system. These circuit breakers must have sufficient capacity so that they can carry momentarily the maximum short-circuit current that can flow through them, and then interrupt this current; they must also withstand closing in on such a short circuit and then interrupting it according to certain prescribed standards.[3]

Fusing is employed where protective relays and circuit breakers are not economically justifiable.

Although the principal function of protective relaying is to mitigate the effects of short circuits, other abnormal operating conditions arise that also require the services of protective relaying. This is particularly true of generators and motors.

A secondary function of protective relaying is to provide indication of the location and type of failure. Such data not only assist in expediting repair but also, by comparison with human observation and automatic oscillograph records, they provide means for analyzing the effectiveness of the fault-prevention and mitigation features including the protective relaying itself.

Fundamental Principles of Protective Relaying

Let us consider for the moment only the relaying equipment for the protection against short circuits. There are two groups of such equipment—one which we shall call "primary" relaying, and the other "back-up" relaying. Primary relaying is the first line of defense, whereas back-up relaying functions only when primary relaying fails.

PRIMARY RELAYING

Figure 1 illustrates primary relaying. The first observation is that circuit breakers are located in the connections to each power-system element. This provision makes it possible to disconnect only a faulty element. Occasionally, a breaker between two adjacent elements may be omitted, in which event both elements must be disconnected for a failure in either one.

The second observation is that, without at this time knowing how it is accomplished, a separate zone of protection is established around each system element. The significance of this is that any failure occurring within a given zone will cause the "tripping" (i.e., opening) of all circuit breakers within that zone, and only those breakers.

It will become evident that, for failures within the region where two adjacent protective zones overlap, more breakers will be tripped than the minimum necessary to disconnect the faulty element. But, if there were no overlap, a failure in a region between zones would not lie in either zone, and therefore no breakers would be tripped. The overlap is the lesser of the two evils. The extent of the overlap is relatively small, and the probability of failure in this region is low; consequently, the tripping of too many breakers will be quite infrequent.

Finally, it will be observed that adjacent protective zones of Fig. 1 overlap *around a circuit breaker.* This is the preferred practice

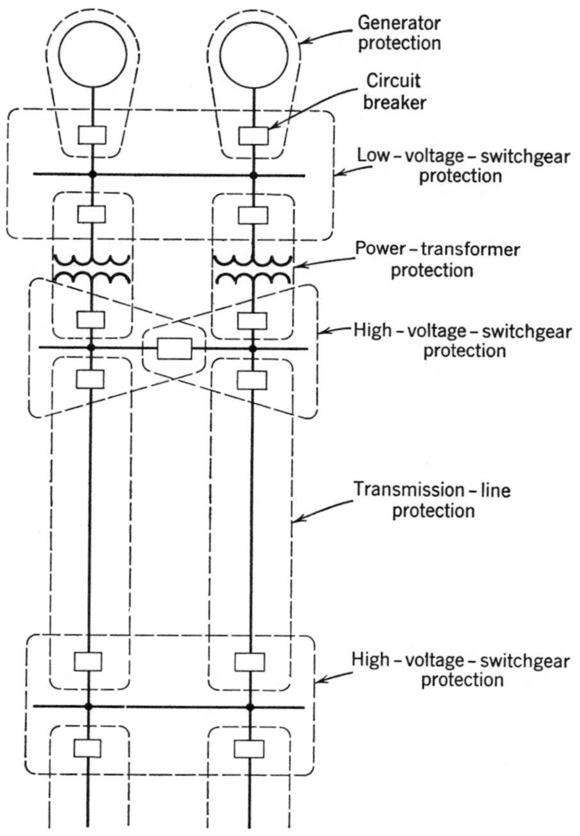

Fig. 1. One-line diagram of a portion of an electric power system illustrating primary relaying.

because, for failures anywhere except in the overlap region, the minimum number of circuit breakers need to be tripped. When it becomes desirable for economic or space-saving reasons to overlap on one side of a breaker, as is frequently true in metal-clad switchgear, the relaying equipment of the zone that overlaps the breaker must be arranged to trip not only the breakers within its zone but also one or more breakers of the adjacent zone, in order to completely disconnect certain faults. This is illustrated in Fig. 2, where it can be seen that, for a short circuit at X, the circuit breakers of zone B, including breaker C, will be tripped; but, since the short circuit is outside zone A, the relaying equipment of zone B must also trip certain breakers in zone A if that is necessary to interrupt the

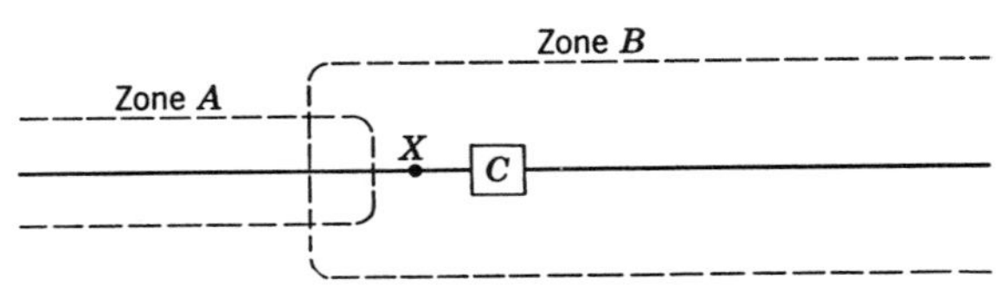

Fig. 2. Overlapping adjacent protective zones on one side of a circuit breaker.

flow of short-circuit current from zone A to the fault. This is not a disadvantage for a fault at X, but the same breakers in zone A will be tripped unnecessarily for other faults in zone B to the right of breaker C. Whether this unnecessary tripping is objectionable will depend on the particular application.

BACK-UP RELAYING

Back-up relaying is employed only for protection against short circuits. Because short circuits are the preponderant type of power-system failure, there are more opportunities for failure in short-circuit primary relaying. Experience has shown that back-up relaying for other than short circuits is not economically justifiable.

A clear understanding of the possible causes of primary-relaying failure is necessary for a better appreciation of the practices involved in back-up relaying. When we say that primary relaying may fail, we mean that any of several things may happen to prevent primary relaying from causing the disconnection of a power-system fault. Primary relaying may fail because of failure in any of the following:

A. Current or voltage supply to the relays.
B. D-c tripping-voltage supply.
C. Protective relays.
D. Tripping circuit or breaker mechanism.
E. Circuit breaker.

It is highly desirable that back-up relaying be arranged so that anything that might cause primary relaying to fail will not also cause failure of back-up relaying. It will be evident that this requirement is completely satisfied only if the back-up relays are located so that they do not employ or control anything in common with the primary relays that are to be backed up. So far as possible, the practice is to locate the back-up relays at a different station. Consider, for example, the back-up relaying for the transmission line section *EF* of Fig. 3. The back-up relays for this line section are normally arranged to trip breakers *A*, *B*, *I*, and *J*. Should breaker *E* fail to trip for a fault on the line section *EF*, breakers *A* and *B* are tripped; breakers *A* and *B* and their associated back-up-relaying equipment, being physically apart from the equipment that has failed, are not likely to be simultaneously affected as might be the case if breakers *C* and *D* were chosen instead.

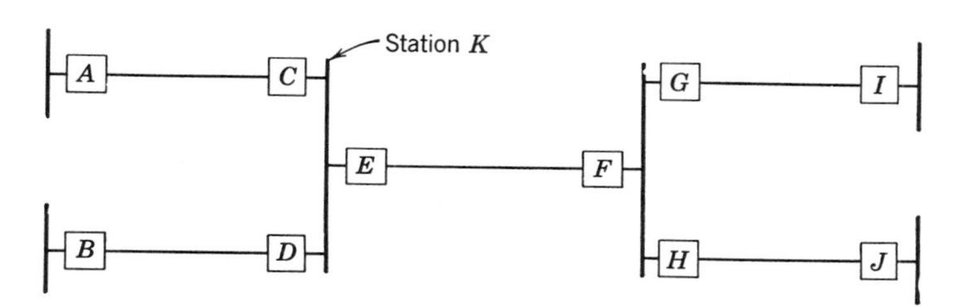

Fig. 3. Illustration for back-up protection of transmission line section *EF*.

The back-up relays at locations *A*, *B*, and *F* provide back-up protection if bus faults occur at station *K*. Also, the back-up relays at *A* and *F* provide back-up protection for faults in the line *DB*. In other words, the zone of protection of back-up relaying extends in one direction from the location of any back-up relay and at least overlaps each adjacent system element. Where adjacent line sections are of different length, the back-up relays must overreach some line sections more than others in order to provide back-up protection for the longest line.

A given set of back-up relays will provide incidental back-up protection of sorts for faults in the circuit whose breaker the back-up relays control. For example, the back-up relays that trip breaker *A* of Fig. 3 may also act as back-up for faults in the line section *AC*. However, this duplication of protection is only an incidental benefit and is not to be relied on to the exclusion of a conventional back-up arrangement when such arrangement is possible; to differentiate between the two, this type might be called "duplicate primary relaying."

A second function of back-up relaying is often to provide primary protection when the primary-relaying equipment is out of service for maintenance or repair.

It is perhaps evident that, when back-up relaying functions, a larger part of the system is disconnected than when primary relaying operates correctly. This is inevitable if back-up relaying is to be made independent of those factors that might cause primary relaying to fail. However, it emphasizes the importance of the second requirement of back-up relaying, that it must operate with sufficient time delay so that primary relaying will be given enough time to function if it is able to. In other words, when a short circuit occurs, both primary relaying and back-up relaying will normally start to operate, but primary relaying is expected to trip the necessary breakers to remove the short-circuited element from the system, and back-up relaying will then reset without having had time to complete its function. When a given set of relays provides back-up protection for several adjacent system elements, the slowest primary relaying of any of those adjacent elements will determine the necessary time delay of the given back-up relays.

For many applications, it is impossible to abide by the principle of complete segregation of the back-up relays. Then one tries to supply the back-up relays from sources other than those that supply the primary relays of the system element in question, and to trip other breakers. This can usually be accomplished; however, the same tripping battery may be employed in common, to save money and because it is considered only a minor risk. This subject will be treated in more detail in Chapter 14.

In extreme cases, it may even be impossible to provide any back-up protection; in such cases, greater emphasis is placed on the need for better maintenance. In fact, even with complete back-up relaying, there is still much to be gained by proper maintenance. When primary relaying fails, even though back-up relaying functions properly, the service will generally suffer more or less. Consequently, back-up relaying is not a proper substitute for good maintenance.

PROTECTION AGAINST OTHER ABNORMAL CONDITIONS

Protective relaying for other than short circuits is included in the category of primary relaying. However, since the abnormal conditions requiring protection are different for each system element, no universal overlapping arrangement of relaying is used as in short-circuit protection. Instead, each system element is independently provided with whatever relaying is required, and this relaying is arranged to trip the necessary circuit breakers which may in some cases be different from those tripped by the short-circuit relaying. As previously mentioned, back-up relaying is not employed because experience has not shown it to be economically justifiable. Frequently, however, back-up relaying for short circuits will function when other abnormal conditions occur that produce abnormal currents or voltages, and back-up protection of sorts is thereby incidentally provided.

Functional Characteristics of Protective Relaying

SENSITIVITY, SELECTIVITY, AND SPEED

"Sensitivity," "selectivity," and "speed" are terms commonly used to describe the functional characteristics of any protective-relaying equipment. All of them are implied in the foregoing considerations of primary and back-up relaying. Any relaying equipment must be sufficiently *sensitive* so that it will operate reliably, when required, under the actual condition that produces the least operating tendency. It must be able to *select* between those conditions for which prompt operation is required and those for which no operation, or time-delay operation, is required. And it must operate at the required *speed*. How well any protective-relaying equipment fulfills each of these requirements must be known for each application.

The ultimate goal of protective relaying is to disconnect a faulty system element as quickly as possible. Sensitivity and selectivity are essential to assure that the proper circuit breakers will be tripped, but speed is the "pay-off." The benefits to be gained from speed will be considered later.

RELIABILITY

That protective-relaying equipment must be reliable is a basic requirement. When protective relaying fails to function properly, the allied mitigation features are largely ineffective. Therefore, it is essential that protective-relaying equipment be inherently reliable, and that its application, installation, and maintenance be such as to assure that its maximum capabilities will be realized.

Inherent reliability is a matter of design based on long experience, and is much too extensive and detailed a subject to do justice to here. Other things being equal, simplicity and robustness contribute to reliability, but they are not of themselves the complete solution. Workmanship must be taken into account also. Contact pressure is an important measure of reliability, but the contact materials and the provisions for preventing contact contamination are fully as important. These are but a few of the many design considerations that could be mentioned.

The proper application of protective-relaying equipment involves the proper choice not only of relay equipment but also of the associated apparatus. For example, lack of suitable sources of current and voltage for energizing the relays may compromise, if not jeopardize, the protection.

Contrasted with most of the other elements of an electric power system, protective relaying stands idle most of the time. Some types of relaying equipment may have to function only once in several years. Transmission-line relays have to operate most frequently, but even they may operate only several times per year. This lack of frequent exercising of the relays and their associated equipment must be compensated for in other ways to be sure that the relaying equipment will be operable when its turn comes.

Many electric utilities provide their test and maintenance personnel with a manual that experienced people in the organization have prepared and that is kept up to date as new types of relays are purchased. Such a manual specifies minimum test and maintenance procedure that experience has shown to be desirable. The manual is prepared in part from manufacturers' publications and in part from the utility's experience. As a consequence of standardized techniques, the results of periodic tests can be compared to detect changes or deterioration in the relays and their associated devices. Testers are encouraged to make other tests as they see fit so long as they make the tests required by the manual. If a better testing technique is devised, it is incorporated into the manual. Some organizations include information on the purpose of the relays, to give their people better appreciation of the importance of their work. Courses may be given, also. Such activity is highly recommended. Unless a person is thoroughly acquainted with relay testing and maintenance, he can do more harm than good, and he might better leave the equipment alone.

In some cases, actual field tests are made after installation and after careful preliminary testing of the individual relays. These field tests provide an excellent means for checking the over-all operation of all equipment involved.

Careful maintenance and record keeping, not only of tests during maintenance but also of relay operation during actual service, are the best assurance that the relaying equipment is in proper condition. Field testing is the best-known way of checking the equipment prior to putting it in service, but conditions may arise in actual service that were not anticipated in the tests. The best assurance that the relays are properly applied and adjusted is a record of correct operation through a sufficiently long period to include the various operating conditions that can exist. It is assuring not only when a particular relaying equipment trips the proper breakers when it should for a given fault but also when other relaying equipments properly refrain from tripping.

Are Protective Practices Based on the Probability of Failure?

Protective practices are based on the probability of failure to the extent that present-day practices are the result of years of experience in which the frequency of failure undoubtedly has played a part. However, the probability of failure seldom if ever enters directly into the choice of a particular type of relaying equipment except when, for one reason or another, one finds it most difficult to apply the type that otherwise would be used. In any event, the probability of failure should be considered only together with the consequences of failure should it occur. It has been said that the justification for a given practice equals the likelihood of trouble times the cost of the trouble. Regardless of the probability of failure, no portion of a system should be entirely without protection, even if it is only back-up relaying.

Protective Relaying versus a Station Operator

Protective relaying sometimes finds itself in competition with station operators or attendants. This is the case for protection against abnormal conditions that develop slowly enough for an operator to have time to correct the situation before any harmful consequences develop. Sometimes, an alert and skillful operator can thereby avoid having to remove from service an important piece of equipment when its removal might be embarrassing; if protective relaying is used in such a situation, it is merely to sound an alarm. To some extent, the preference of relying on an operator has a background of some unfortunate experience with protective relaying whereby improper relay operation caused embarrassment; such an attitude is understandable, but it cannot be supported logically. Where quick and accurate action is required for the protection of important equipment, it is unwise to rely on an operator. Moreover, when trouble occurs, the operator usually has other things to do for which he is better fitted.

Undesired Tripping versus Failure to Trip When Desired

Regardless of the rules of good relaying practice, one will occasionally have to choose which rule may be broken with the least embarrassment. When one must choose between the chance of undesired or unnecessary tripping and failure to trip when tripping is desired, the best practice is generally to choose the former. Experience has shown that, where major system shutdowns have resulted from one or the other, the failure to trip—or excessive delay in tripping—has been by far the worse offender.

The Evaluation of Protective Relaying

Although a modern power system could not operate without protective relaying, this does not make it priceless. As in all good engineering, economics plays a large part. Although the protection engineer can usually justify expenditures for protective relaying on the basis of standard practice, circumstances may alter such concepts, and it often becomes necessary to evaluate the benefits to be gained. It is generally not a question of whether protective relaying can be justified, but of how far one should go toward investing in the best relaying available.

Like all other parts of a power system, protective relaying should be evaluated on the basis of its contribution to the best economically possible service to the customers. The contribution of protective relaying is to help the rest of the power system to function as efficiently and as effectively as possible in the face of trouble.[2] How protective relaying does this is as follows. By minimizing damage when failures occur, protective relaying minimizes:

A. The cost of repairing the damage.
B. The likelihood that the trouble may spread and involve other equipment.
C. The time that the equipment is out of service.
D. The loss in revenue and the strained public relations while the equipment is out of service.

By expediting the equipment's return to service, protective relaying helps to minimize the amount of equipment reserve required, since there is less likelihood of another failure before the first failure can be repaired.

The ability of protective relaying to permit fuller use of the system capacity is forcefully illustrated by system stability. Figure 4 shows how the speed of protective relaying influences the amount of power that can be transmitted without loss of synchronism when short circuits occur.[4] More load can be carried over an existing system by speeding up the protective relaying. This has been shown to be a relatively inexpensive way to increase the transient stability limit.[5] Where stability is a problem, protective relaying can often be evaluated against the cost of constructing additional transmission lines or switching stations.

Other circumstances will be shown later in which certain types of protective-relaying equipment can permit savings in circuit breakers and transmission lines.

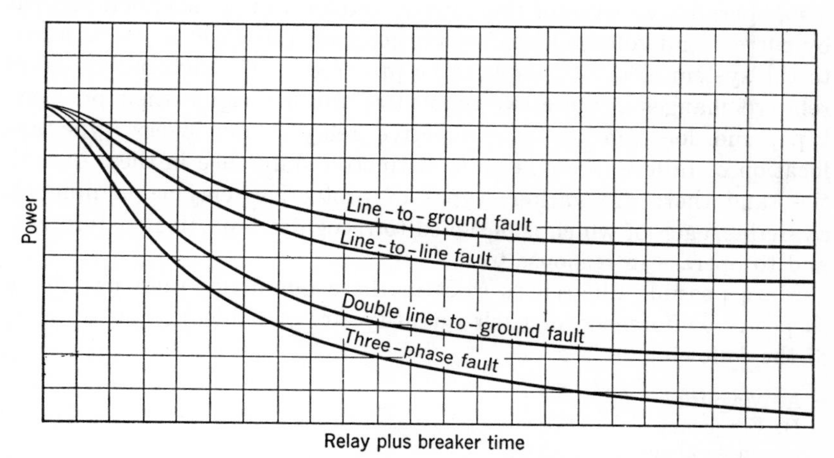

Fig. 4. Curves illustrating the relation between relay-plus-breaker time and the maximum amount of power that can be transmitted over one particular system without loss of synchronism when various faults occur.

The quality of the protective-relaying equipment can affect engineering expense in applying the relaying equipment itself. Equipment that can still operate properly when future changes are made in a system or its operation will save much future engineering and other related expense.

One should not conclude that the justifiable expense for a given protective-relaying equipment is necessarily proportional to the value or importance of the system element to be directly protected. A failure in that system element may affect the ability of the entire system to render service, and therefore that relaying equipment is actually protecting the service of the entire system. Some of the most serious shutdowns have been caused by consequential effects growing out of an original failure in relatively unimportant equipment that was not properly protected.

How Do Protective Relays Operate?

Thus far, we have treated the relays themselves in a most impersonal manner, telling what they do without any regard to how they do it.

This fascinating part of the story of protective relaying will be told in much more detail later. But, in order to round out this general consideration of relaying and to prepare for what is yet to come, some explanation is in order here.

All relays used for short-circuit protection, and many other types also, operate by virtue of the current and/or voltage supplied to them by current and voltage transformers connected in various combinations to the system element that is to be protected. Through individual or relative changes in these two quantities, failures signal their presence, type, and location to the protective relays. For every type and location of failure, there is some distinctive difference in these quantities, and there are various types of protective-relaying equipments available, each of which is designed to recognize a particular difference and to operate in response to it.[6]

More possible differences exist in these quantities than one might suspect. Differences in each quantity are possible in one or more of the following:

A. Magnitude.
B. Frequency.
C. Phase angle.
D. Duration.
E. Rate of change.
F. Direction or order of change.
G. Harmonics or wave shape.

Then, when both voltage and current are considered in combination, or relative to similar quantities at different locations, one can begin to realize the resources available for discriminatory purposes. It is a fortunate circumstance that, although Nature in her contrary way has imposed the burden of electric-power-system failure, she has at the same time provided us with a means for combat.

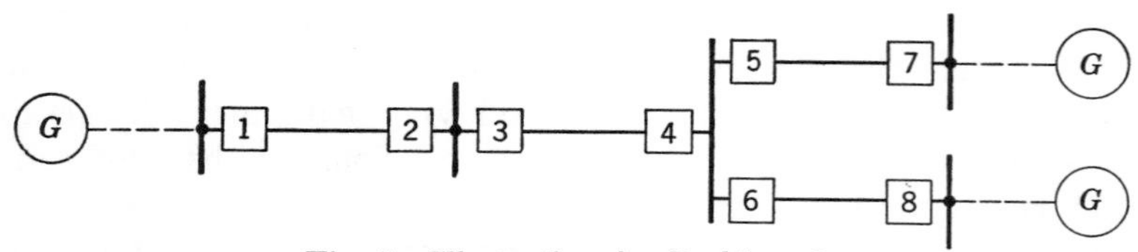

Fig. 5. Illustration for Problem 2.

Problems

1. Compare protective relaying with insurance.

2. The portion of a power system shown by the one-line diagram of Fig. 5, with generating sources back of all three ends, has conventional primary and back-up

relaying. In each of the listed cases, a short circuit has occurred and certain circuit breakers have tripped as stated. Assume that the tripping of these breakers was correct under the circumstances. Where was the short circuit? Was there any failure of the protective relaying, including breakers, and if so, what failed? Assume only one failure at a time. Draw a sketch showing the overlapping of primary protective zones and the exact locations of the various faults.

Case	Breakers Tripped
a	4, 5, 8
b	3, 7, 8
c	3, 4, 5, 6
d	1, 4, 5, 6
e	4, 5, 7, 8
f	4, 5, 6

Bibliography

1. "Power System Fault Control," AIEE Committee Report, *AIEE Trans.*, *70* (1951), pp. 410–417.

2. "Protective Relay Modernization Program Releases Latent Transmission Capacity," by M. F. Hebb, Jr., and J. T. Logan, *AIEE District Conference Paper 55-354*.

"Plan System and Relaying Together," *Elec. World*, July 25, 1955, p. 86.

3. "Standards for Power Circuit Breakers," *Publ. SG4-1954*, National Electrical Manufacturers Association, 155 East 44th St., New York 17, N. Y.

"Interrupting Rating Factors for Reclosing Service on Power Circuit Breakers," *Publ. C37.7-1952*, American Standards Association, Inc., 70 East 45th St., New York 17, N. Y.

4. *Power System Stability*, Vol. II, by S. B. Crary, John Wiley & Sons, New York, 1947.

5. "Costs Study of 69- to 345-Kv Overhead Power-Transmission Systems," by J. G. Holm, *AIEE Trans.*, *63* (1944), pp. 406–422.

6. "A Condensation of the Theory of Relays," by A. R. van C. Warrington, *Gen. Elec. Rev.*, *43*, No. 9 (Sept., 1940), pp. 370–373.

"Principles and Practices of Relaying in the United States," by E. L. Harder and W. E. Marter, *AIEE Trans.*, *67*, Part II (1948), pp. 1005–1022. Discussions, pp. 1022–1023.

"Principles of High-Speed Relaying," by W. A. Lewis, *Westinghouse Engineer*, *3* (Aug., 1943), pp. 131–134.

2 FUNDAMENTAL RELAY-OPERATING PRINCIPLES AND CHARACTERISTICS

Protective relays are the "tools" of the protection engineer. As in any craft, an intimate knowledge of the characteristics and capabilities of the available tools is essential to their most effective use. Therefore, we shall spend some time learning about these tools without too much regard to their eventual use.

General Considerations

All the relays that we shall consider operate in response to one or more electrical quantities either to close or to open contacts. We shall not bother with the details of actual mechanical construction except where it may be necessary for a clear understanding of the operation. One of the things that tend to dismay the novice is the great variation in appearance and types of relays, but actually there are surprisingly few fundamental differences. Our attention will be directed to the response of the few basic types to the electrical quantities that actuate them.

OPERATING PRINCIPLES

There are really only two fundamentally different operating principles: (1) electromagnetic attraction, and (2) electromagnetic induction. Electromagnetic-attraction relays operate by virtue of a plunger being drawn into a solenoid, or an armature being attracted to the poles of an electromagnet. Such relays may be actuated by d-c or by a-c quantities. Electromagnetic-induction relays use the principle of the induction motor whereby torque is developed by induction in a rotor; this operating principle applies only to relays actuated by alternating current, and in dealing with those relays we shall call them simply "induction-type" relays.

DEFINITIONS OF OPERATION

Mechanical movement of the operating mechanism is imparted to a contact structure to close or to open contacts. When we say that a relay "operates," we mean that it either closes or opens its contacts —whichever is the required action under the circumstances. Most relays have a "control spring," or are restrained by gravity, so that they assume a given position when completely de-energized; a contact that is closed under this condition is called a "closed" contact, and one that is open is called an "open" contact. This is standardized nomenclature, but it can be quite confusing and awkward to use. A much better nomenclature in rather extensive use is the designation *"a"* for an "open" contact, and *"b"* for a "closed" contact. This nomenclature will be used in this book. The present standard method for showing *"a"* and *"b"* contacts on connection diagrams is illustrated in Fig. 1. Even though an *"a"* contact may be closed under normal operating conditions, it should be shown open as in Fig. 1; and similarly, even though a *"b"* contact may normally be open, it should be shown closed.

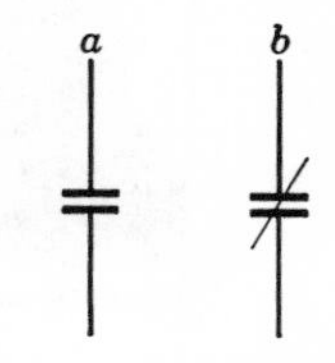

Fig. 1. Contact symbols and designations.

When a relay operates to open a *"b"* contact or to close an *"a"* contact, we say that it "picks up," and the smallest value of the actuating quantity that will cause such operation, as the quantity is slowly increased from zero, is called the "pickup" value. When a relay operates to close a *"b"* contact, or to move to a stop in place of a *"b"* contact, we say that it "resets"; and the largest value of the actuating quantity at which this occurs, as the quantity is slowly decreased from above the pickup value, is called the "reset" value. When a relay operates to open its *"a"* contact, but does not reset, we say that it "drops out," and the largest value of the actuating quantity at which this occurs is called the "drop-out" value.

OPERATION INDICATORS

Generally, a protective relay is provided with an indicator that shows when the relay has operated to trip a circuit breaker. Such "operation indicators" or "targets" are distinctively colored elements that are actuated either mechanically by movement of the relay's operating mechanism, or electrically by the flow of contact current, and come into view when the relay operates. They are arranged to be reset manually after their indication has been noted, so as to be ready for the next operation. One type of indicator is shown in

Fig. 2. Electrically operated targets are generally preferred because they give definite assurance that there was a current flow in the contact circuit. Mechanically operated targets may be used when the closing of a relay contact always completes the trip circuit where

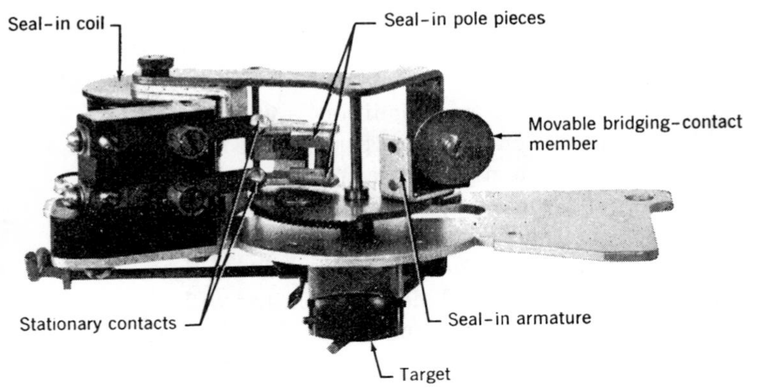

Fig. 2. One type of contact mechanism showing target and seal-in elements.

tripping is not dependent on the closing of some other series contact. A mechanical target may be used with a series circuit comprising contacts of other relays when it is desired to have indication that a particular relay has operated, even though the circuit may not have been completed through the other contacts.

SEAL-IN AND HOLDING COILS, AND SEAL-IN RELAYS

In order to protect the contacts against damage resulting from a possible inadvertent attempt to interrupt the flow of the circuit-breaker trip-coil current, some relays are provided with a holding mechanism comprising a small coil in series with the contacts; this coil is on a small electromagnet that acts on a small armature on the moving-contact assembly to hold the contacts tightly closed once they have established the flow of trip-coil current. This coil is called a "seal-in" or "holding" coil. Figure 2 shows such a structure. Other relays use a small auxiliary relay whose contacts by-pass the protective-relay contacts and seal the circuit closed while tripping current flows. This seal-in relay may also display the target. In either case, the circuit is arranged so that, once the trip-coil current starts to flow, it can be interrupted only by a circuit-breaker auxiliary switch that is connected in series with the trip-coil circuit and that opens when the breaker opens. This auxiliary switch is defined as an *"a"* contact. The circuits of both alternatives are shown in Fig. 3.

Figure 3 also shows the preferred polarity to which the circuit-breaker trip coil (or any other coil) should be connected to avoid corrosion because of electrolytic action. No coil should be connected

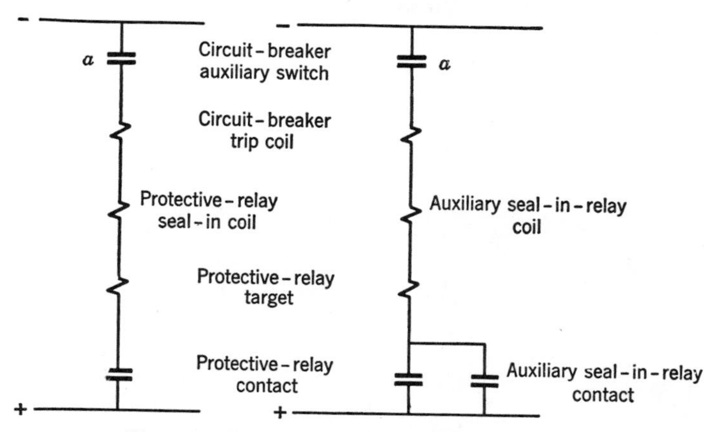

Fig. 3. Alternative contact seal-in methods.

only to positive polarity for long periods of time; and, since here the circuit breaker and its auxiliary switch will be closed normally while the protective-relay contacts will be open, the trip-coil end of the circuit should be at negative polarity.

ADJUSTMENT OF PICKUP OR RESET

Adjustment of pickup or reset is provided electrically by tapped current coils or by tapped auxiliary potential transformers or resistors; or adjustment is provided mechanically by adjustable spring tension or by varying the initial air gap of the operating element with respect to its solenoid or electromagnet.

TIME DELAY AND ITS DEFINITIONS

Some relays have adjustable time delay, and others are "instantaneous" or "high speed." The term "instantaneous" means "having no intentional time delay" and is applied to relays that operate in a minimum time of approximately 0.1 second. The term "high speed" connotes operation in less than approximately 0.1 second and usually in 0.05 second or less. The operating time of high-speed relays is usually expressed in cycles based on the power-system frequency; for example, "one cycle" would be $\frac{1}{60}$ second in a 60-cycle system. Originally, only the term "instantaneous" was used, but, as relay speed was increased, the term "high speed" was felt to be necessary in order to differentiate such relays from the earlier, slower types. This book will use the term "instantaneous" for general reference to either instantaneous or high-speed relays, reserving the term "high-speed" for use only when the terminology is significant.

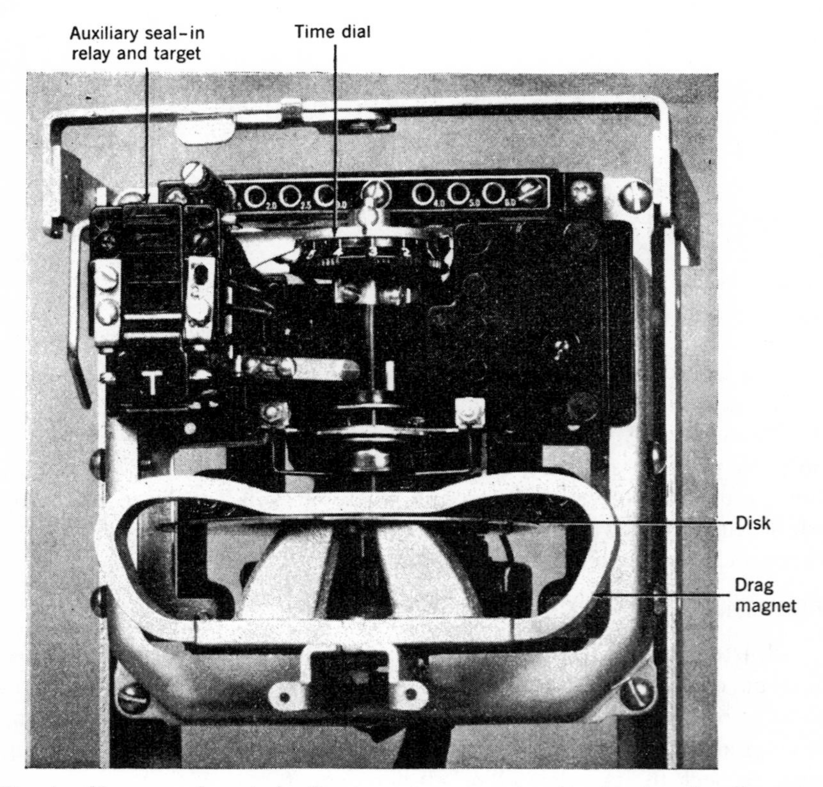

Fig. 4. Close-up of an induction-type overcurrent unit, showing the disc rotor and the drag magnet.

Occasionally, a supplementary auxiliary relay having fixed time delay may be used when a certain delay is required that is entirely independent of the magnitude of the actuating quantity in the protective relay.

Time delay is obtained in induction-type relays by a "drag magnet," which is a permanent magnet arranged so that the relay rotor cuts the flux between the poles of the magnet, as shown in Fig. 4. This produces a retarding effect on motion of the rotor in either direction. In other relays, various mechanical devices have been used, including dash pots, bellows, and escapement mechanisms.

The terminology for expressing the shape of the curve of operating time versus the actuating quantity has also been affected by developments throughout the years. Originally, only the terms "definite time" and "inverse time" were used. An inverse-time curve is one in which the operating time becomes less as the magnitude of the actuating quantity is increased, as shown in Fig. 5. The more pro-

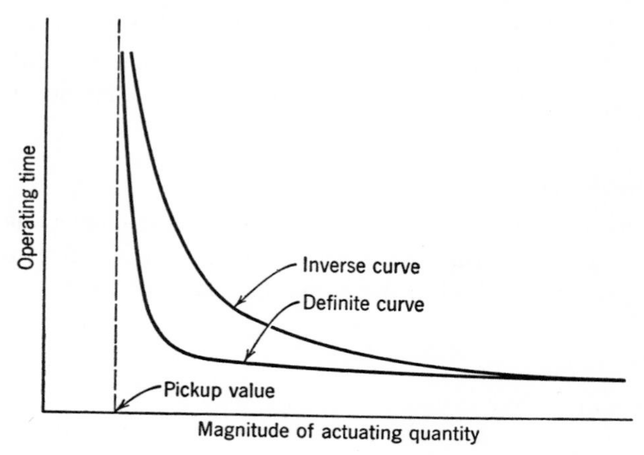

Fig. 5. Curves of operating time versus the magnitude of the actuating quantity.

nounced the effect is, the more inverse is the curve said to be. Actually, all time curves are inverse to a greater or lesser degree. They are most inverse near the pickup value and become less inverse as the actuating quantity is increased. A definite-time curve would strictly be one in which the operating time was unaffected by the magnitude of the actuating quantity, but actually the terminology is applied to a curve that becomes substantially definite slightly above the pickup value of the relay, as shown in Fig. 5.

As a consequence of trying to give names to curves of different degrees of inverseness, we now have "inverse," "very inverse," and "extremely inverse." Although the terminology may be somewhat confusing, each curve has its field of usefulness, and one skilled in the use of these relays has only to compare the shapes of the curves to know which is best for a given application. This book will use the term "inverse" for general reference to any of the inverse curves, reserving the other terms for use only when the terminology is significant.

Thus far, we have gained a rough picture of protective relays in general and have learned some of the language of the profession.

References to complete standards pertaining to circuit elements and terminology are given in the bibliography at the end of this chapter.[1] With this preparation, we shall now consider the fundamental relay types.

Single-Quantity Relays of the Electromagnetic-Attraction Type

Here we shall consider plunger-type and attracted-armature-type a-c or d-c relays that are actuated from either a single current or voltage source.

OPERATING PRINCIPLE

The electromagnetic force exerted on the moving element is proportional to the square of the flux in the air gap. If we neglect the effect of saturation, the total actuating force may be expressed:

$$F = K_1 I^2 - K_2,$$

where F = net force.
K_1 = a force-conversion constant.
I = the rms magnitude of the current in the actuating coil.
K_2 = the restraining force (including friction).

When the relay is on the verge of picking up, the net force is zero, and the operating characteristic is:

$$K_1 I^2 = K_2,$$

or

$$I = \sqrt{\frac{K_2}{K_1}} = \text{constant}$$

RATIO OF RESET TO PICKUP

One characteristic that affects the application of some of these relays is the relatively large difference between their pickup and reset values. As such a relay picks up, it shortens its air gap, which permits a smaller magnitude of coil current to keep the relay picked up than was required to pick it up. This effect is less pronounced in a-c than in d-c relays. By special design, the reset can be made as high as 90% to 95% of pickup for a-c relays, and 60% to 90% of pickup for d-c relays. Where the pickup is adjusted by adjusting the initial air gap, a higher pickup calibration will have a lower ratio of reset to pickup. For overcurrent applications where such relays are often used, the relay trips a circuit breaker which reduces the current to zero, and hence the reset value is of no consequence. However, if a low-reset relay is used in conjuction with other relays in such a way that a breaker is not always tripped when the low-reset relay operates, the application should be carefully examined. When the reset value is a low percentage of the pickup value, there is the possibility that an abnormal condition might cause the relay to pick up (or to reset), but that a return to normal conditions might not return the relay to its normal operating position, and undesired operation might result.

TENDENCY TOWARD VIBRATION

Unless the pole pieces of such relays have "shading rings" to split the air-gap flux into two out-of-phase components, such relays are not suitable for continuous operation on alternating current in the picked-up position. This is because there would be excessive vibration that would produce objectionable noise and would cause excessive wear. This tendency to vibrate is related to the fact that a-c relays have higher reset than d-c relays; an a-c relay without shading rings has a tendency to reset every half cycle when the flux passes through zero.

DIRECTIONAL CONTROL

Relays of this group are used mostly when "directional" operation is not required. More will be said later about "directional control" of relays; suffice it to say here that plunger or attracted-armature relays do not lend themselves to directional control nearly as well as induction-type relays, which will be considered later.

EFFECT OF TRANSIENTS

Because these relays operate so quickly and with almost equal facility on either alternating current or direct current, they are affected by transients, and particularly by d-c offset in a-c waves. This tendency must be taken into consideration when the proper adjustment for any application is being determined. Even though the steady-state value of an offset wave is less than the relay's pickup value, the relay may pick up during such a transient, depending on the amount of offset, its time constant, and the operating speed of the relay. This tendency is called "overreach" for reasons that will be given later.

TIME CHARACTERISTICS

This type of relay is inherently fast and is used generally where time delay is not required. Time delay can be obtained, as previously stated, by delaying mechanisms such as bellows, dash pots, or escapements. Very short time delays are obtainable in d-c relays by encircling the magnetic circuit with a low-resistance ring, or "slug" as it is sometimes called. This ring delays changes in flux, and it can be positioned either to have more effect on air-gap-flux increase if time-delay pickup is desired, or to have more effect on air-gap-flux decrease if time-delay reset is required.

Directional Relays of the Electromagnetic-Attraction Type

Directional relays of the electromagnetic-attraction type are actuated by d-c or by rectified a-c quantities. The most common use of such relays is for protection of d-c circuits where the actuating quantity is obtained either from a shunt or directly from the circuit.

OPERATING PRINCIPLE

Figure 6 illustrates schematically the operating principle of this type of relay. A movable armature is shown magnetized by current flowing in an actuating coil encircling the armature, and with such

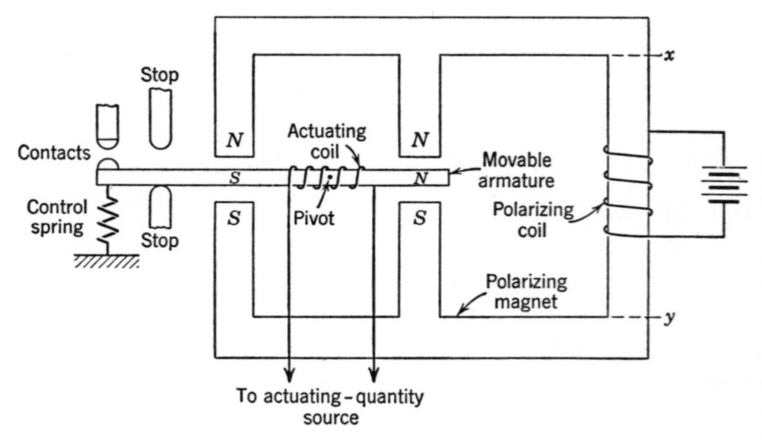

Fig. 6. Directional relay of the electromagnetic-attraction type.

polarity as to close the contacts. A reversal of the polarity of the actuating quantity will reverse the magnetic polarities of the ends of the armature and cause the contacts to stay open. Although a "polarizing," or "field," coil is shown for magnetizing the polarizing magnet, this coil may be replaced by a permanent magnet in the section between x and y. There are many physical variations possible in carrying out this principle, one of them being a construction similar to that of a d-c motor.

The force tending to move the armature may be expressed as follows, if we neglect saturation:

$$F = K_1 I_p I_a - K_2,$$

$$F = \text{net force}$$

where K_1 = a force-conversion constant.
I_p = the magnitude of the current in the polarizing coil.
I_a = the magnitude of the current in the armature coil.
K_2 = the restraining force (including friction).

At the balance point when $F = 0$, the relay is on the verge of operating, and the operating characteristic is:

$$I_p I_a = \frac{K_2}{K_1} = \text{constant}$$

I_p and I_a are assumed to flow through the coils in such directions that a pickup force is produced, as in Fig. 6. It will be evident that, if the direction of either I_p or I_a (but not of both) is reversed, the direction of the force will be reversed. Therefore, this relay gets its name from its ability to distinguish between opposite directions of actuating-coil current flow, or opposite polarities. If the relative directions are correct for operation, the relay will pick up at a constant magnitude of the product of the two currents.

If permanent-magnet polarization is used, or if the polarizing coil is connected to a source that will cause a constant magnitude of current to flow, the operating characteristic becomes:

$$I_a = \frac{K_2}{K_1 I_p} = \text{constant}$$

I_a still must have the correct polarity, as well as the correct magnitude, for the relay to pick up.

EFFICIENCY

This type of relay is much more efficient than hinged-armature or plunger relays, from the standpoint of the energy required from the actuating-coil circuit. For this reason, such directional relays are used when a d-c shunt is the actuating source, whether directional action is required or not. Occasionally, such a relay may be actuated from an a-c quantity through a full-wave rectifier when a low-energy a-c relay is required.

RATIO OF CONTINUOUS THERMAL CAPACITY TO PICKUP

As a consequence of greater efficiency, the actuating coil of this type of relay has a high ratio of continuous current or voltage capacity to the pickup value, from the thermal standpoint.

TIME CHARACTERISTICS

Relays of this type are instantaneous in operation, although a slug may be placed around the armature to get a short delay.

Induction-Type Relays—General Operating Principles

Induction-type relays are the most widely used for protective-relaying purposes involving a-c quantities. They are not usable with d-c quantities, owing to the principle of operation. An induction-type relay is a split-phase induction motor with contacts. Actuating force is developed in a movable element, that may be a disc or other form of rotor of non-magnetic current-conducting material, by the interaction of electromagnetic fluxes with eddy currents that are induced in the rotor by these fluxes.

THE PRODUCTION OF ACTUATING FORCE

Figure 7 shows how force is produced in a section of a rotor that is pierced by two adjacent a-c fluxes. Various quantities are shown

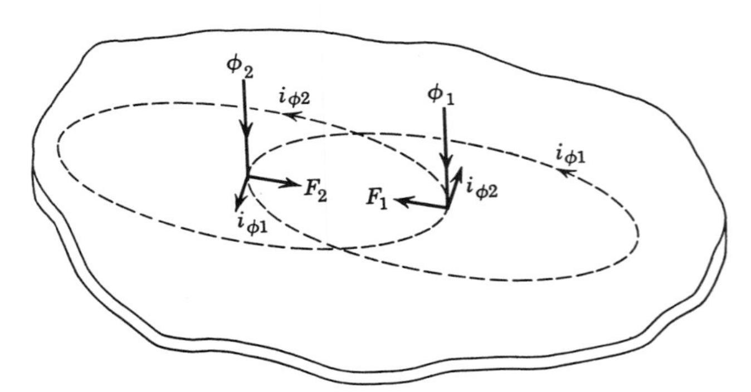

Fig. 7. Torque production in an induction relay.

at an instant when both fluxes are directed downward and are increasing in magnitude. Each flux induces voltage around itself in the rotor, and currents flow in the rotor under the influence of the two voltages. The current produced by one flux reacts with the other flux, and vice versa, to produce forces that act on the rotor.

The quantities involved in Fig. 7 may be expressed as follows:

$$\phi_1 = \Phi_1 \sin \omega t$$

$$\phi_2 = \Phi_2 \sin (\omega t + \theta),$$

where θ is the phase angle by which ϕ_2 leads ϕ_1. It may be assumed with negligible error that the paths in which the rotor currents flow have negligible self-inductance, and hence that the rotor currents are in phase with their voltages:

$$i_{\phi 1} \propto \frac{d\phi_1}{dt} \propto \Phi_1 \cos \omega t$$

$$i_{\phi 2} \propto \frac{d\phi_2}{dt} \propto \Phi_2 \cos (\omega t + \theta)$$

We note that Fig. 7 shows the two forces in opposition, and consequently we may write the equation for the net force (F) as follows:

$$F = (F_2 - F_1) \propto (\phi_2 i_{\phi 1} - \phi_1 i_{\phi 2}) \tag{1}$$

Substituting the values of the quantities into equation 1, we get:

$$F \propto \Phi_1 \Phi_2 [\sin (\omega t + \theta) \cos \omega t - \sin \omega t \cos (\omega t + \theta)] \tag{2}$$

which reduces to:

$$F \propto \Phi_1 \Phi_2 \sin \theta \tag{3}$$

Since sinusoidal flux waves were assumed, we may substitute the rms values of the fluxes for the crest values in equation 3.

Apart from the fundamental relation expressed by equation 3, it is most significant that the net force is the same *at every instant*. This fact does not depend on the simplifying assumptions that were made in arriving at equation 3. The action of a relay under the influence of such a force is positive and free from vibration. Also, although it may not be immediately apparent, the net force is directed from the point where the leading flux pierces the rotor toward the point where the lagging flux pierces the rotor. It is as though the flux moved across the rotor, dragging the rotor along.

In other words, actuating force is produced in the presence of out-of-phase fluxes. One flux alone would produce no net force. There must be at least two out-of-phase fluxes to produce any net force, and the maximum force is produced when the two fluxes are 90° out of phase. Also, the direction of the force—and hence the direction of motion of the relay's movable member—depends on which flux is leading the other.

A better insight into the production of actuating force in the induction relay can be obtained by plotting the two components of the expression inside the brackets of equation 2, which we may call the "per-unit net force." Figure 8 shows such a plot when θ is assumed to be 90°. It will be observed that each expression is a double-frequency sinusoidal wave completely offset from the zero-force axis.

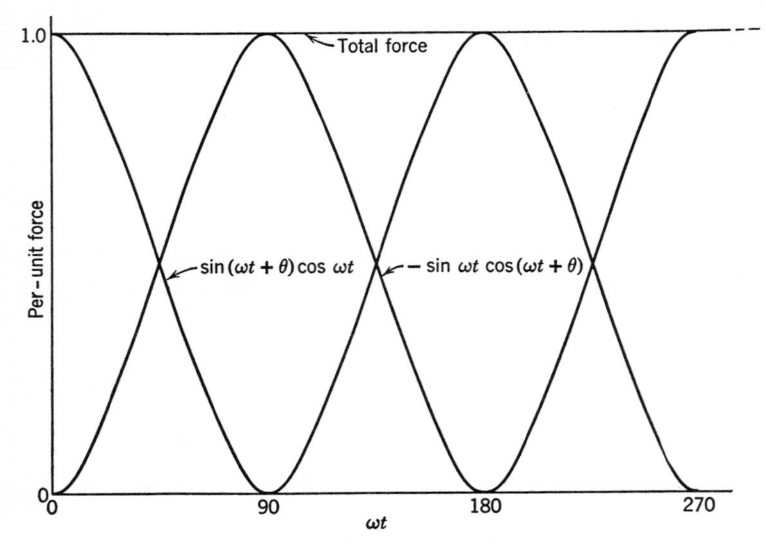

Fig. 8. Per-unit net force.

The two waves are displaced from one another by 90° in terms of fundamental frequency, or by 180° in terms of double frequency. The sum of the instantaneous values of the two waves is 1.0 at every instant. If θ were assumed to be less than 90°, the effect on Fig. 8 would be to raise the zero-force axis, and a smaller per-unit net force would result. When θ is zero, the two waves are symmetrical about the zero-force axis, and no net force is produced. If we let θ be negative, which is to say that ϕ_2 is lagging ϕ_1, the zero-force axis is raised still higher and net force in the opposite direction is produced. However, for a given value of θ, the net force is the same at each instant.

In some induction-type relays one of the two fluxes does not react with rotor currents produced by the other flux. The force expression for such a relay has only one of the components inside the brackets of equation 2. The *average* force of such a relay may still be expressed by equation 3, but the instantaneous force is variable, as shown by omitting one of the waves of Fig. 8. Except when θ is 90° lead or lag, the instantaneous force will actually reverse during parts of the cycle; and, when $\theta = 0$, the average negative force equals the average positive force. Such a relay has a tendency to vibrate, particularly at values of θ close to zero.

Reference 2 of the bibliography at the end of this chapter gives more detailed treatment of induction-motor theory that applies also to induction relays.

TYPES OF ACTUATING STRUCTURE

The different types of structure that have been used are commonly called: (1) the "shaded-pole" structure; (2) the "watthour-meter" structure; (3) the "induction-cup" and the "double-induction-loop" structures; (4) the "single-induction-loop" structure.

Shaded-Pole Structure. The shaded-pole structure, illustrated in Fig. 9, is generally actuated by current flowing in a single coil on a magnetic structure containing an air gap. The air-gap flux produced by this current is split into two out-of-phase components by a

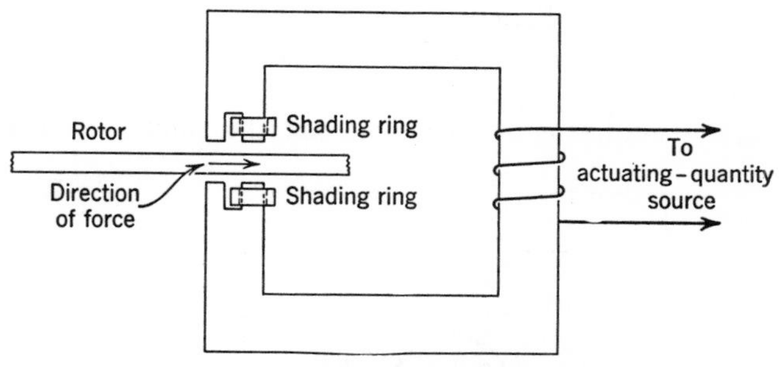

Fig. 9. Shaded-pole structure.

so-called "shading ring," generally of copper, that encircles part of the pole face of each pole at the air gap. The rotor, shown edgewise in Fig. 9, is a copper or aluminum disc, pivoted so as to rotate in the air gap between the poles. The phase angle between the fluxes piercing the disc is fixed by design, and consequently it does not enter into application considerations.

The shading rings may be replaced by coils if control of the operation of a shaded-pole relay is desired. If the shading coils are short-circuited by a contact of some other relay, torque will be produced; but, if the coils are open-circuited, no torque will be produced because there will be no phase splitting of the flux. Such torque control is employed where "directional control" is desired, which will be described later.

Watthour-Meter Structure. This structure gets its name from the fact that it is used for watthour meters. As shown in Fig. 10, this structure contains two separate coils on two different magnetic circuits, each of which produces one of the two necessary fluxes for driving the rotor, which is also a disc.

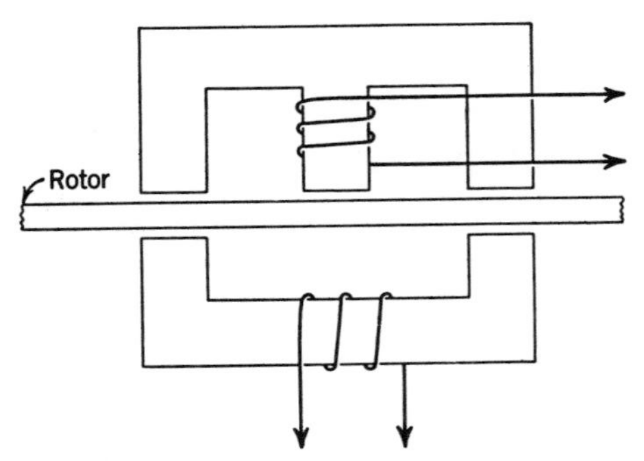

Fig. 10. Watthour-meter structure.

Induction-Cup and Double-Induction-Loop Structures. These two structures are shown in Figs. 11 and 12. They most closely resemble an induction motor, except that the rotor iron is stationary, only the rotor-conductor portion being free to rotate. The cup structure employs a hollow cylindrical rotor, whereas the double-loop structure

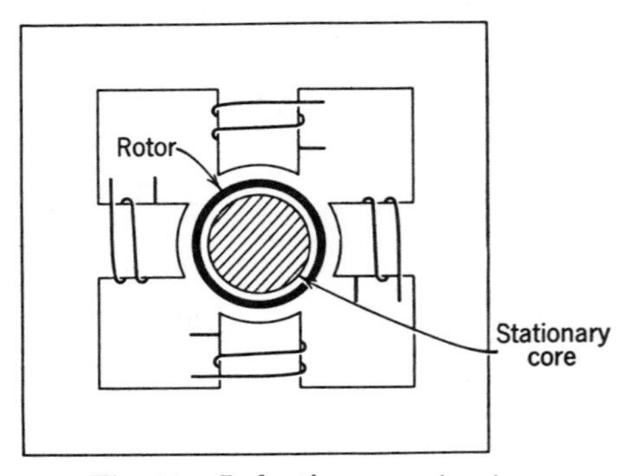

Fig. 11. Induction-cup structure.

employs two loops at right angles to one another. The cup structure may have additional poles between those shown in Fig. 11. Functionally, both structures are practically identical.

These structures are more efficient torque producers than either the shaded-pole or the watthour-meter structures, and they are the type used in high-speed relays.

Single-Induction-Loop Structure. This structure, shown in Fig. 13, is the most efficient torque-producing structure of all the induction types that have been described. However, it has the rather serious disadvantage that its rotor tends to vibrate as previously described for a relay in which the actuating force is expressed by only one component inside the brackets of equation 2. Also, the torque varies somewhat with the rotor position.

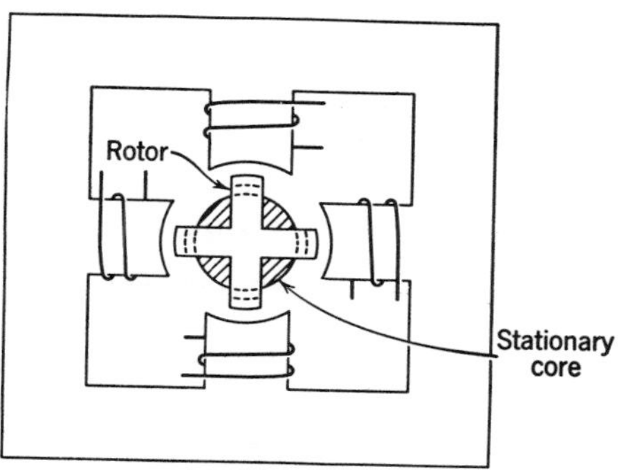

Fig. 12. Double-induction-loop structure.

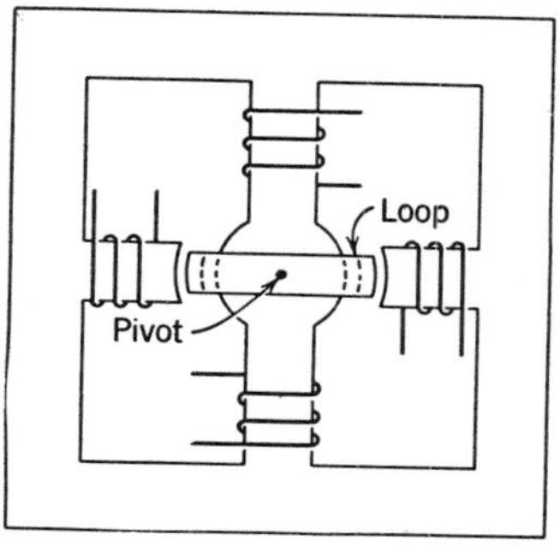

Fig. 13. Single-induction-loop structure.

ACCURACY

The accuracy of an induction relay recommends it for protective-relaying purposes. Such relays are comparable in accuracy to meters used for billing purposes. This accuracy is not a consequence of the induction principle, but because such relays invariably employ jewel bearings and precision parts that minimize friction.

Single-Quantity Induction Relays

A single-quantity relay is actuated from a single current or voltage source. Any of the induction-relay actuating structures may be used. The shaded-pole structure is used only for single-quantity relays. When any of the other structures is used, its two actuating circuits are connected in series or in parallel; and the required phase angle between the two fluxes is obtained by arranging the two circuits to have different X/R (reactance-to-resistance) ratios by the use of auxiliary resistance and/or capacitance in combination with one of the circuits. Neglecting the effect of saturation, the torque of all such relays may be expressed as:

$$T = K_1 I^2 - K_2$$

where I is the rms magnitude of the total current of the two circuits.

The phase angle between the individual currents is a design constant, and it does not enter into the application of these relays.

If the relay is actuated from a voltage source, its torque may be expressed as:

$$T = K_1V^2 - K_2$$

where V is the rms magnitude of the voltage applied to the relay.

TORQUE CONTROL

Torque control with the structures of Figs. 10, 11, 12, or 13 is obtained simply by a contact in series with one of the circuits if they are in parallel, or in series with a portion of a circuit if they are in series.

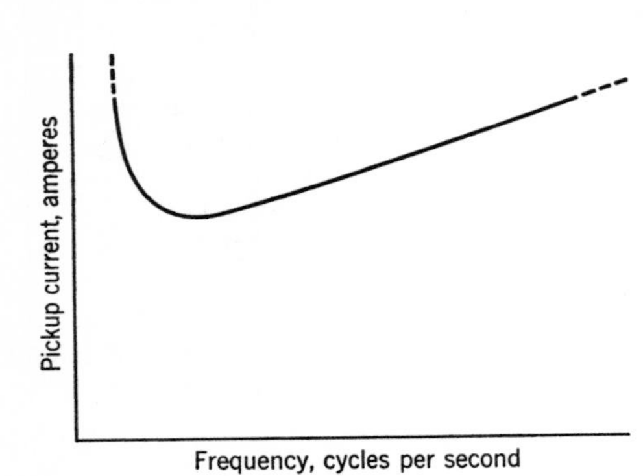

Fig. 14. Effect of frequency on the pickup of a single-quantity induction relay.

EFFECT OF FREQUENCY

The effect of frequency on the pickup of a single-quantity relay is shown qualitatively by Fig. 14. So far as possible, a relay is designed to have the lowest pickup at its rated frequency. The effect of slight changes in frequency normally encountered in power-system operation may be neglected. However, distorted wave form may produce significant changes in pickup and time characteristics. This fact is particularly important in testing relays at high currents; one should be sure that the wave form of the test currents is as good as that obtained in actual service, or else inconsistent results will be obtained.[3]

EFFECT OF D-C OFFSET

The effect of d-c offset may be neglected with inverse-time single-quantity relays. High-speed relays may or may not be affected, depending on the characteristics of their circuit elements. Generally, the pickup of high-speed relays is made high enough to compensate for any tendency to "overreach," as will be seen later, and no attempt is made to evaluate the effect of d-c offset.

RATIO OF RESET TO PICKUP

The ratio of reset to pickup is inherently high in induction relays, because their operation does not involve any change in the air gap of the magnetic circuit. This ratio is between 95% and 100%, friction and imperfect compensation of the control-spring torque being the only things that keep the ratio from being 100%. Moreover, this ratio is unaffected by the pickup adjustment where tapped current coils provide the pickup adjustment.

RESET TIME

Where fast automatic reclosing of circuit breakers is involved, the reset time of an inverse-time relay may be a critical characteristic in obtaining selectivity. If all relays involved do not have time to reset completely after a circuit breaker has been tripped and before the breaker recloses, and if the short circuit that caused tripping is re-established when the breaker recloses, certain relays may operate too quickly and trip unnecessarily. Sometimes the drop-out time may also be important with high-speed reclosing.

TIME CHARACTERISTICS

Inverse-time curves are obtained with relays whose rotor is a disc and whose actuating structure is either the shaded-pole type or the watthour-meter type. High-speed operation is obtained with the induction-cup or the induction-loop structures.

Directional Induction Relays

Contrasted with single-quantity relays, directional relays are actuated from two different independent sources, and hence the angle θ of equation 3 is subject to change and must be considered in the application of these relays. Such relays use the actuating structures of Figs. 10, 11, 12, or 13.

TORQUE RELATIONS IN TERMS OF ACTUATING QUANTITIES

Current-Current Relays. A current-current relay is actuated from two different current-transformer sources. Assuming no saturation, we may substitute the actuating currents for the fluxes of equation 3, and the expression for the torque becomes:

$$T = K_1 I_1 I_2 \sin\theta - K_2 \qquad (4)$$

where I_1 and I_2 = the rms values of the actuating currents.
θ = the phase angle between the rotor-piercing fluxes produced by I_1 and I_2.

An actuating current is not in phase with the rotor-piercing flux that it produces, for the same reason that the primary current of a transformer is not in phase with the mutual flux. (In fact, the equivalent circuit of a transformer may be used to represent each actuating circuit of an induction relay.) But in some relays, such as the induction-cylinder and double-induction-loop types, the rotor-piercing (or mutual) fluxes are at the same phase angle with respect to their actuating currents. For such so-called "symmetrical" structures, θ of equation 4 may be defined also as the phase angle between the actuating currents. For the wattmetric type of structure, the phase angle between the actuating currents may be significantly different

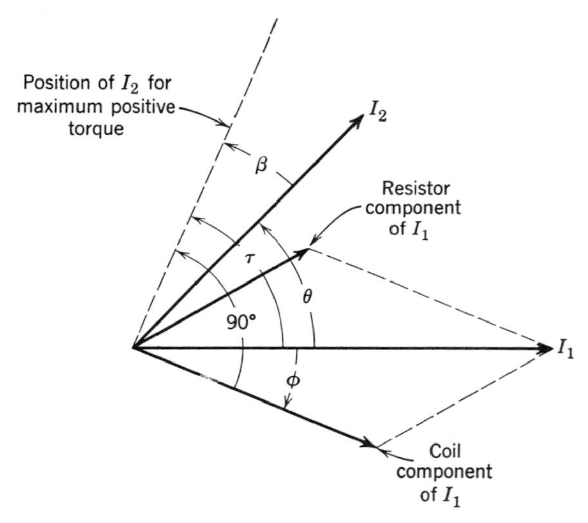

Fig. 15. Vector diagram for maximum torque in a current-current induction-type directional relay.

from the phase angle between the fluxes. For the moment, we shall assume that we are dealing with symmetrical structures, and that θ may be defined as the phase angle between I_1 and I_2 of equation 4.

However, it is usually desirable that maximum torque occur at some value of θ other than 90°. To this end, one of the actuating coils may be shunted by a resistor or a capacitor. Maximum torque will still occur when the coil currents are 90° out of phase; but, in terms of the currents supplied from the actuating sources, maximum torque will occur at some angle other than 90°.

Figure 15 shows the vector relations for a relay with a resistor shunting the I_1 coil. I_1 will now be defined as the total current supplied by the source to the coil and resistor in parallel. If the angle θ by which I_2 leads I_1 is defined as positive, the angle ϕ by which the coil component of I_1 lags I_1 will be negative, and the expression for the torque will be:

$$T = K_1 I_1 I_2 \sin (\theta - \phi) - K_2 \tag{5}$$

For example, if we let $\theta = 45°$ and $\phi = -30°$, the torque for the relations of Fig. 15 will be:

$$T = K_1 I_1 I_2 \sin 75° - K_2$$

The angle "τ" of Fig. 15 is called the "angle of maximum torque" since it is the value of θ at which maximum *positive* torque occurs. It is customary to specify this angle rather than ϕ when describing this characteristic of directional relays. The two angles are directly related by the fact that they add numerically to 90° in symmetrical structures such as we have assumed thus far. But, if we use τ as the design constant of a directional relay rather than ϕ, we can write the torque expression in such a way that it will apply to all relays whether symmetrical or not, as follows:

$$T = K_1 I_1 I_2 \cos (\theta - \tau) - K_2$$

where τ is positive when maximum positive torque occurs for I_2 leading I_1, as in Fig. 15. Or the torque may be expressed also as:

$$T = K_1 I_1 I_2 \cos \beta - K_2$$

where β is the angle between I_2 and the maximum-torque position of I_2, or $\beta = (\theta - \tau)$. These two equations will be used from now on because they are strictly true for any structure.

If a capacitor rather than a resistor is used to adjust the angle of maximum torque, it may be connected to the secondary of a transformer whose primary is connected across the coil and whose ratio is such that the secondary voltage is much higher than the primary voltage. The purpose of this is to permit the use of a small capacitor. Or, to accomplish the same purpose, another winding with many more turns than the current coil may be put on the same magnetic circuit with the current coil, and with a capacitor connected across this winding.

Current-Voltage Relays. A current-voltage relay receives one actuating quantity from a current-transformer source and the other actuating quantity from a voltage-transformer source. Equation 5 applies approximately for the currents in the two coils. However, in terms of the actuating quantities, the torque is strictly:

$$T = K_1 V I \cos (\theta - \tau) - K_2 \tag{6}$$

where V = the rms magnitude of the voltage applied to the voltage-coil *circuit*.
I = the rms magnitude of the current-coil current.
θ = the angle between I and V.
τ = the angle of maximum torque.

For whatever relation between I and V that we call θ positive, we should also call τ positive for that same relation. These quantities are shown in Fig. 16, together with the voltage-coil current I_V and the approximate angle ϕ by which I_V lags V.

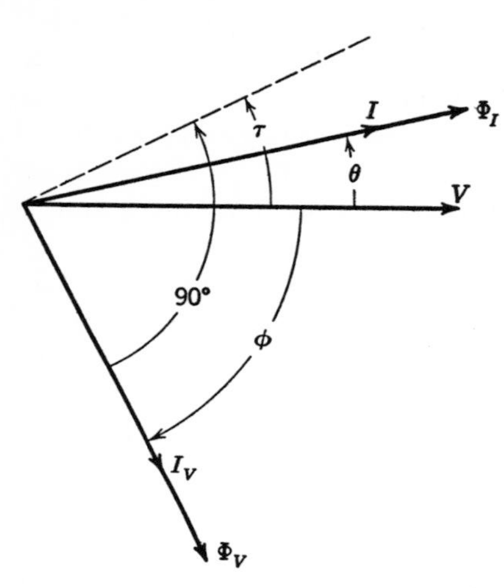

Fig. 16. Vector diagram for maximum torque in a current-voltage induction-type directional relay.

The value of ϕ is of the order of 60 to 70° lagging for most voltage coils, and therefore τ will be of the order of 30 to 20° leading if there is no impedance in series with the voltage coil. By inserting a combination of resistance and capacitance in series with the voltage coil, we can change the angle between the applied voltage and I_V to almost any value either lagging or leading V without changing the magnitude of I_V. A limited change in ϕ can be made with resistance alone, but the magnitude of I_V will be decreased, and hence the pickup will be increased. Hence, the angle of maximum torque can be made almost any desired value. By other supplementary means, which we shall not discuss here, the angle of maximum torque can be made any desired value. It is emphasized that V of equation 6 is the voltage applied to the voltage-coil *circuit;* it is the voltage-coil voltage only if no series impedance is inserted.

Voltage-Voltage Relays. It is not necessary to consider a relay actuated from two different voltage sources, since the principles already described will apply.

THE SIGNIFICANCE OF THE TERM "DIRECTIONAL"

A-c directional relays are used most extensively to recognize the difference between current being supplied in one direction or the other in an a-c circuit, and the term "directional" is derived from this usage. Basically, an a-c directional relay can recognize certain differences in phase angle between two quantities, just as a d-c directional relay recognizes differences in polarity. This recognition, as reflected in the contact action, is limited to differences in phase angle exceeding 90° from the phase angle at which maximum torque is developed, as already described.

THE POLARIZING QUANTITY OF A DIRECTIONAL RELAY

The quantity that produces one of the fluxes is called the "polarizing" quantity. It is the reference against which the phase angle of the other quantity is compared. Consequently, the phase angle of the polarizing quantity must remain more or less fixed when the other quantity suffers wide changes in phase angle. The choice of a suitable polarizing quantity will be discussed later, since it does not affect our present considerations.

THE OPERATING CHARACTERISTIC OF A DIRECTIONAL RELAY

Consider, for example, the torque relation expressed by equation 6 for a current-voltage directional relay. At the balance point when the relay is on the verge of operating, the net torque is zero, and we have:

$$VI\cos(\theta - \tau) = \frac{K_2}{K_1} = \text{constant}$$

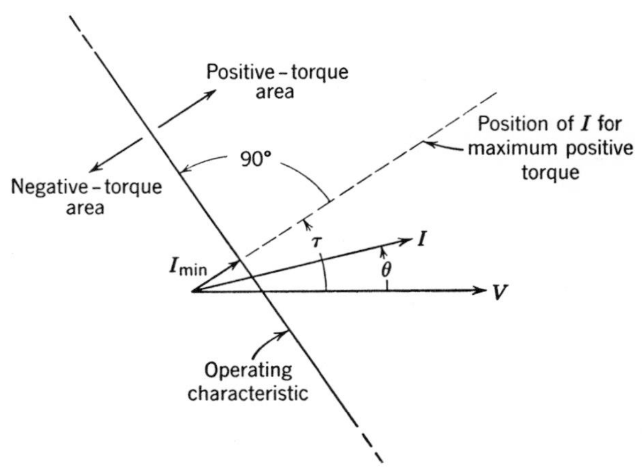

Fig. 17. Operating characteristic of a directional relay on polar coordinates.

This operating characteristic can be shown on a polar-coordinate diagram, as in Fig. 17. The polarizing quantity, which is the voltage for this type of relay, is the reference; and its magnitude is assumed to be constant. The operating characteristic is seen to be a straight line offset from the origin and perpendicular to the maximum positive-torque position of the current. This line is the plot of the relation:

$$I \cos (\theta - \tau) = \text{constant}$$

which is obtained when the magnitude of V is assumed to be constant, and it is the dividing line between the development of net positive and negative torque in the relay. Any current vector whose head lies in the positive-torque area will cause pickup; the relay will not pick up, or it will reset, for any current vector whose head lies in the negative-torque area.

For a different magnitude of the reference voltage, the operating characteristic will be another straight line parallel to the one shown and related to it by the expression:

$$VI_{\text{min}} = \text{constant}$$

where I_{min}, as shown in Fig. 17, is the smallest magnitude of all current vectors whose heads terminate on the operating characteristic. I_{min} is called "the minimum pickup current," although strictly speaking the current must be slightly larger to cause pickup. Thus, there is an infinite possible number of such operating characteristics, one for each possible magnitude of the reference voltage.

The operating characteristic will depart from a straight line as the phase angle of the current approaches 90° from the maximum-torque phase angle. For such large angular departures, the pickup current becomes very large, and magnetic saturation of the current element requires a different magnitude of current to cause pickup from the one that the straight-line relation would indicate.

The operating characteristic for current-current or voltage-voltage directional relays can be similarly shown.

THE "CONSTANT-PRODUCT" CHARACTERISTIC

The relation $VI_{\text{min}} = \text{constant}$ for the current-voltage relay (and similar expressions for the others) is called the "constant-product" characteristic. It corresponds closely to the pickup current or voltage of a single-quantity relay and is used as the basis for plotting the time characteristics. This relation holds only so long as saturation does not occur in either of the two magnetic circuits. When either of the two quantities begins to exceed a certain magnitude, the quantity producing saturation must be increased beyond the value indicated by the constant-product relation in order to produce net positive torque.

EFFECT OF D-C OFFSET AND OTHER TRANSIENTS

The effect of transients may be neglected with inverse-time relays, but, with high-speed relays, certain transients may have to be guarded against either in the design of the relay or in its application. Generally, an increase in pickup or the addition of one or two cycles (60-cycle-per-second basis) time delay will avoid undesired operation. This subject is much too complicated to do justice to here. Suffice it to say, trouble of this nature is extremely rare and is not generally a factor in the application of established relay equipments.

THE EFFECT OF FREQUENCY

Directional relays are affected like single-quantity relays by changes in frequency of both quantities. The angle of maximum torque is affected, owing to changes in the X/R ratio in circuits containing inductance or capacitance. The effect of slight changes in frequency such as are normally encountered, however, may be neglected. If the frequencies of the two quantities supplied to the relay are different, a sinusoidal torque alternating between positive and negative will be produced; the net torque for each torque cycle will be zero, but, if the frequencies are nearly equal and if a high-speed relay is involved, the relay may respond to the reversals in torque.

TIME CHARACTERISTICS

Disc-type relays are used where inverse-time characteristics are desired, and cup-type or loop-type relays are used for high-speed operation. When time delay is desired, it is often provided by another relay associated with the directional relay.

The Universal Relay-Torque Equation

As surprising as it may seem, we have now completed our examination of all the essential fundamentals of protective-relay operation. All relays yet to be considered are merely combinations of the types that have been described. At this point, we may write the universal torque equation as follows:

$$T = K_1 I^2 + K_2 V^2 + K_3 VI \cos(\theta - \tau) + K_4$$

By assigning plus or minus signs to certain of the constants and letting others be zero, and sometimes by adding other similar terms, the operating characteristics of all types of protective relays can be expressed.

Various factors that one must generally take into account in applying the different types have been presented. Little or no quantitative data have been given because it is of little consequence for a clear understanding of the subject. Such data are easily obtained for any particular type of relay; the important thing is to know how to use such data. In the following chapters, by applying the fundamental principles that have been given here, we shall learn how the various types of protective relays operate.

Problems

1. A 60-cycle single-phase directional relay of the current-voltage type has a voltage coil whose impedance is $230 + j560$ ohms. When connected as shown in Fig. 18, the relay develops maximum positive torque when load of a leading power factor is being supplied in a given direction.

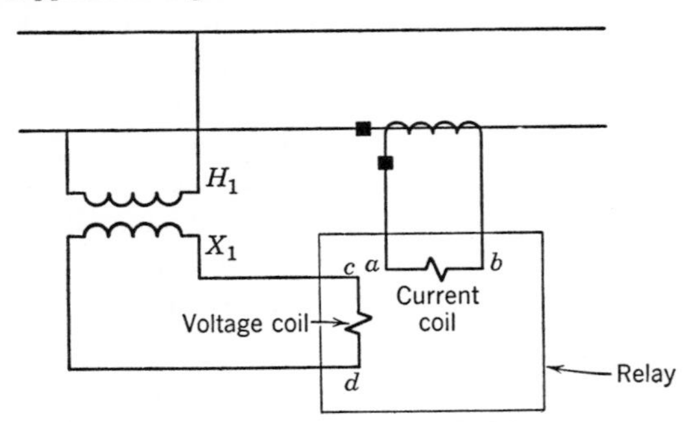

Fig. 18. Illustration for Problem 1.

It is desired to modify this relay so that it will develop maximum positive torque for load in the same direction as before, but at 45° lagging power factor. Moreover, it is desired to maintain the same minimum pickup current as before.

Draw a connection diagram similar to that given, showing the modifications you would make, and giving quantitative values. Assume that the relay has a symmetrical structure.

2. Given an induction-type directional relay in which the frequency of one flux is n times that of the other. Derive the equation for the torque of this relay at any instant if at zero time the relay is developing maximum torque.

3. What will the torque equation of an induction-type directional relay be if one flux is direct current and the other is alternating current?

Bibliography

1. "Relays Associated with Electric Power Apparatus," *Publ. C37.1–1950*, and "Graphical Symbols for Electric Power and Control," *Publ. Z–32.3–1946*, American Standards Assoc., Inc., 70 East 45th Street, New York 17, N. Y.

"Standards for Power Switchgear Assemblies," *Publ. SG–5–1950*, National Electrical Manufacturers Assoc., 155 East 44th Street, New York 17, N. Y.

2. "The Revolving Field Theory of the Capacitor Motor," by Wayne J. Morrill, *AIEE Trans., 48* (1929), pp. 614–629. Discussions, pp. 629–632.

Closing discussion by V. N. Stewart in "A New High-Speed Balanced-Current Relay," by V. N. Stewart, *AIEE Trans., 62* (1943), pp. 553–555. Discussions, pp. 972–974.

"Principles of Induction-Type Relay Design," by W. E. Glassburn and W. K. Sonnemann, *AIEE Trans., 72* (1953), pp. 23–27.

3. "Harmonics May Delay Relay Operation," by M. A. Fawcett and C. A. Keener, *Elec. World, 109* (April 9, 1935), p. 1226.

Relay Systems, by I. T. Monseth and P. H. Robinson, McGraw-Hill Book Co., New York, 1935.

3 CURRENT, VOLTAGE, DIRECTIONAL, CURRENT (OR VOLTAGE)- BALANCE, AND DIFFERENTIAL RELAYS

Chapter 2 described the operating principles and characteristics of the basic relay elements. All protective-relay types are derived either directly from these basic elements, by combining two or more of these elements in the same enclosing case or in the same circuit with certain electrical interconnections, or by directly adding the torques of two or more such elements to control a single set of contacts. Various external connections and auxiliary equipments are employed that may tend to obscure the true identity of the elements and make the equipment appear complicated. However, if one will examine the equipment, he will invariably recognize the basic elements.

General Protective-Relay Features

Certain features and capabilities apply generally to all types of protective relays. These will be discussed briefly before we consider the various types of relays.

CONTINUOUS AND SHORT-TIME RATINGS

All relays carry current- and/or voltage-coil ratings as a guide to their proper application. For relays complying with present standards, the continuous rating specifies what a relay will withstand under continuous operation in an ambient temperature of 40° C. Relays having current coils also carry a 1-second current rating, since such relays are usually subjected to momentary overcurrents. Such relays should not be subjected to currents in excess of the 1-second rating without the manufacturer's approval because either thermal or mechanical damage may result. Overcurrents lower than the 1-second-rating value are permissible for longer than 1 second, so long as the I^2t value of the 1-second rating is not exceeded. For example, if a relay will withstand 100 amperes for 1 second, it will withstand $100\sqrt{\frac{1}{2}}$ amperes for 2 seconds. It is not always safe to assume that a relay will withstand any current that it can get from current transformers for as long as it takes a circuit breaker to interrupt a short circuit after the relay has operated to trip the circuit breaker. Also, should a relay fail to succeed in tripping a circuit breaker, thermal damage should be expected unless back-up relays can stop the flow of short-circuit current soon enough to prevent such damage.

CONTACT RATINGS

Protective-relay contacts are rated on their ability to close and to open inductive or non-inductive circuits at specified magnitudes of circuit current and a-c or d-c circuit voltage. As stated in Chapter 2, protective relays that trip circuit breakers are not permitted to interrupt the flow of trip-coil current, and hence they require only a circuit-closing and momentary current-carrying rating. If a breaker fails to trip, the contacts of the relay will almost certainly be damaged. The circuit-opening rating is applicable only when a protective relay controls the operation of another relay, such as a timing relay or an auxiliary relay; in such a case, the protective relay should not have a holding coil or else it may not be able to open its contacts once they have closed. If a seal-in relay is used, the current taken by the controlled relay must be less than the pickup of the seal-in relay.

When a relay of the "over-and-under" type with *"a"* and *"b"* contacts is used to control the operation of some other device, the relay can be relieved of any circuit-breaking duty by the arrangement of Fig. 1. When the protective relay picks up, it causes an auxiliary relay to pick up and seal itself in around the protective-relay contacts. Other auxiliary-relay contacts may be used, as shown, for control purposes, thereby relieving the protective-relay contacts of this duty. When the protective relay resets, it shorts the auxiliary-relay coil, thereby causing the auxiliary relay to reset.

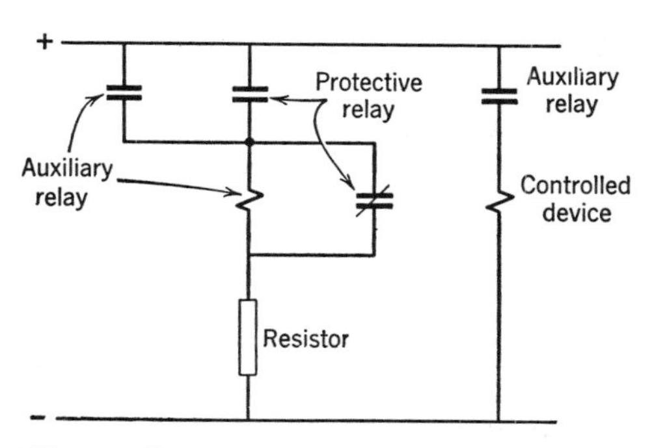

Fig. 1. Control circuit for relay of the "over-and-under" type.

HOLDING-COIL OR SEAL-IN-RELAY AND TARGET RATINGS

Two different current ratings are generally available either in the same relay or in different relays. The higher current rating is for use when the protective relay trips a circuit breaker directly, and the lower current rating is for use when the relay trips a circuit breaker indirectly through an auxiliary relay. In either event, one should be sure that the rating is low enough so that reliable seal-in and target operation will be obtained should two or more protective relays close contacts together, thereby dividing the total available trip-circuit current between the parallel protective-relay-contact circuits. Also, depending on the tripping speed of the breaker, the trip-circuit current may not have time to build up to its steady-state value. The resistances of the seal-in and target coils are given to permit one to calculate the trip-circuit currents.

BURDENS

The impedance of relay-actuating coils must be known to permit one to determine if the relay's voltage- or current-transformer sources will have sufficient capacity and suitable accuracy to supply the relay load together with any other loads that may be imposed on the transformers. These relay impedances are listed in relay publications. This subject will be treated further when we examine the characteristics of voltage and current transformers.

Overcurrent, Undercurrent, Overvoltage, and Undervoltage Relays

Overcurrent, undercurrent, overvoltage, and undervoltage relays are derived directly from the basic single-quantity electromagnetic-attraction or induction types described in Chapter 2. The prefix "over" means that the relay picks up to close a set of "*a*" contacts when the actuating quantity exceeds the magnitude for which the relay is adjusted to operate. Similarly, the prefix "under" means that the relay resets to close a set of "*b*" contacts when the actuating quantity decreases below the reset magnitude for which the relay is adjusted to operate. Some relays have both "*b*" and "*a*" contacts, and the prefix before the actuating quantity in their name is "over-and-under." In protective-relay terminology, a "current" relay is one whose actuating source is a current in a circuit supplied to the relay either directly or from a current transformer. A "voltage" relay is one whose actuating source is a voltage of the circuit obtained either directly or from a voltage transformer.

Because all these relays are derived directly from the single-quantity types described in Chapter 2, there is no need to consider further their principle of operation.

ADJUSTMENT

Pickup or Reset. Most overcurrent relays have a range of adjustment to make them adaptable to as wide a range of application circumstances as possible. The range of adjustment is limited, however, because of coil-space limitations and to simplify the relay construction. Hence, various relays are available, each having a different range of adjustment. The adjustment of plunger or attracted-armature relays may be by adjustment of the initial air gap, adjustment of restraining-spring tension, adjustable weights, or coil taps. The adjustment of current-actuated induction relays is generally by coil taps, and that of voltage-actuated relays by taps on series resistors or by auxiliary autotransformer taps.

Voltage relays and undercurrent relays do not generally have as wide a range of adjustment because they are expected to operate within a limited range from the normal magnitude of the actuating quantity. The normal magnitude does not vary widely because relay ratings are usually chosen with respect to the ratios of current and voltage transformers so that the relay current is normally slightly less than rated relay current and the relay voltage is approximately rated relay voltage, regardless of the application.

Time. Except for the "over-and-under" types, the operating time of inverse-time induction relays is usually adjustable by choosing the amount of travel of the rotor from its reset position to its pickup position. It is accomplished by adjustment of the position of the reset stop. A so-called "time lever" or "time dial" with an evenly divided scale provides this adjustment.

The slight increase in restraining torque of the control spring, as the reset stop is advanced toward the pickup position, is compensated for by the shape of the disc. A disc whose periphery is in the form of a spiral, or a disc having a fixed radius but with peripheral slots, the bottoms of which are on a spiral, provides this compensation by varying the active area of the disc between the poles. Similarly, holes of different diameter may be used. As the disc turns toward the pickup position when the reset stop is advanced, or whenever the relay operates to pick up, the increase in the amount of the disc area between the poles of the actuating structure causes an increase in the electrical torque that just balances the increase in the control-spring torque.

When a bellows is used for producing time delay, adjustment is made by varying the size of an orifice through which the air escapes from the bellows.

TIME CHARACTERISTICS

A typical time curve for a high-speed relay is shown in Fig. 2. It will be noted that this is an inverse curve, but that a 3-cycle (60-cycle-per-second basis) operating time is achieved only slightly above the pickup value, which permits the relay to be called "high speed."

Figure 3 shows a family of inverse-time curves of one widely used induction-type relay. A curve is shown for each major division of the adjustment scale. Any intermediate curves can be obtained by interpolation since the adjustment is continuous.

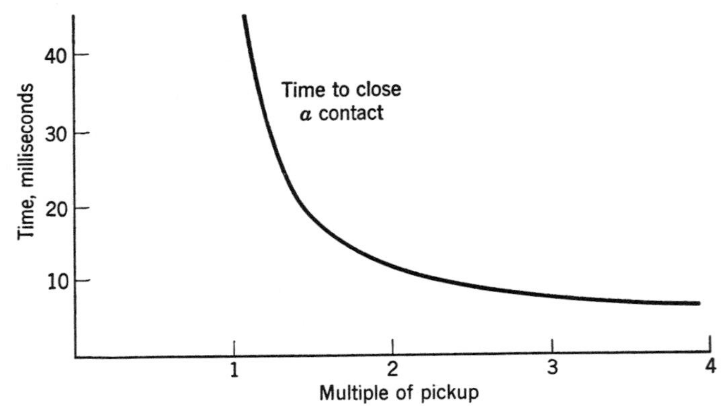

Fig. 2. Time curve of a high-speed relay.

It will be noted that both Fig. 2 and Fig. 3 are plotted in terms of multiples of the pickup value, so that the same curves can be used for any value of pickup. This is possible with induction-type relays where the pickup adjustment is by coil taps, because the ampere-turns at pickup are the same for each tap. Therefore, at a given multiple of pickup, the coil ampere-turns, and hence the torque, are the same regardless of the tap used. Where air-gap or restraining-spring pickup adjustment is used, the shape of the time curve varies with the pickup.

One should not rely on the operation of any relay when the magnitude of the actuating quantity is only slightly above pickup, because the net actuating force is so low that any additional friction may prevent operation, or may increase the operating time. Even though the relay closes its contacts, the contact pressure may be so low that contamination of the contact surface may prevent electrical contact. This is particularly true in inverse-time relays where there may not be much impact when the contacts close. It is the practice to apply relays in such a way that, when their operation must be reliable, their actuating quantity will be at least 1.5 times pickup. For this reason, some time curves are not shown for less than 1.5 times pickup.

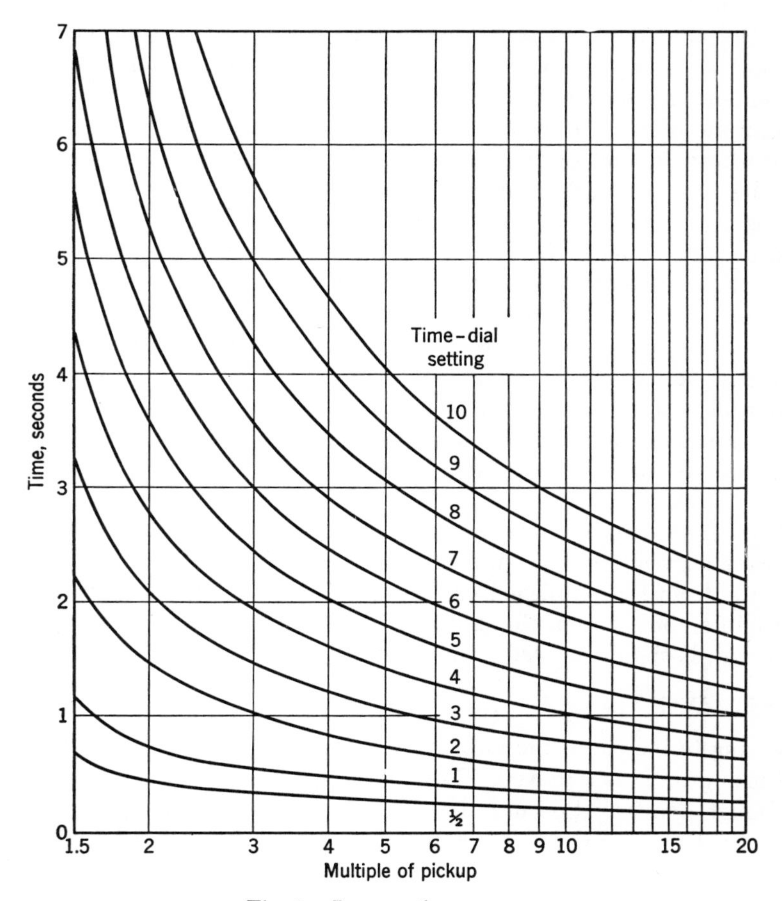

Fig. 3. Inverse-time curves.

The time curves of Fig. 3 can be used to estimate not only how long it will take the relay to close its contacts at a given multiple of pickup and for any time adjustment but also how far the relay disc will travel toward the contact-closed position within any time interval. For example, assume that the No. 5 time-dial adjustment is used, and that the multiple of pickup is 3. It will take the relay 2.45 seconds to close its contacts. We see that in 1.45 seconds, the relay would close its contacts if the No. 3 time-dial adjustment were used. In other words, in 1.45 seconds, the disc travels a distance corresponding to 3.0 time-dial divisions, or three-fifths of the total distance to close the contacts.

This method of analysis is useful to estimate whether a relay will pick up, and, if so, what its time delay will be when the magnitude of the actuating quantity is changing as, for example, during the current-inrush period when a motor is starting. The curve of the rms magnitude of current versus time can be studied for short successive time intervals, and the disc travel during each interval can be found for the average current magnitude during that interval. For each successive interval, the disc should be assumed to start from the position that it had reached at the end of the preceding interval.

For the most effective use of an inverse-time relay, its pickup should be chosen so that the relay will be operating on the most inverse part of its time curve over the range of magnitude of the actuating quantity for which the relay must operate. In other words, the minimum value of the actuating quantity for which the relay must operate should be at least 1.5 times pickup, but not too much more. This will become more evident when we consider the application of these relays.

The time curves illustrated in manufacturers' publications are average curves, and the time characteristics of individual relays vary slightly from the published curves. Ordinarily, this variation will be negligible, but, when the most accurate adjustment of a relay is required, it should be determined by test.

OVERTRAVEL

Owing to inertia of the moving parts, motion will continue when the actuating force is removed. This characteristic is called "overtravel." Although overtravel occurs in all relays, its effect is usually important only in time-delay relays, and particularly for inverse-time overcurrent relays, where selectivity is obtained on a time-delay basis. The basis for specifying overtravel is best described by an example, as follows. Suppose that, for a given adjustment and at a given multiple of pickup, a relay will pick up and close its contacts in 2.0 seconds. Now suppose that we make several tests by applying that same multiple of pickup for time intervals slightly less than 2.0 seconds, and we find that, if the time interval is any longer than 1.9 seconds, the relay will still close its contacts. We would say, then, that the overtravel is 0.1 second. The higher the multiple of pickup, the longer the overtravel time will be. However, a constant overtravel time of approximately 0.1 second is generally assumed in the application of inverse-time relays; the manner of its use will be described when we consider the application of these relays.

RESET TIME

For accurate data, the manufacturer should be consulted. The reset time will vary directly with the time-dial adjustment. The method of analysis described under "Time Characteristics" for estimating the amount of disc travel during short time intervals, combined with the knowledge of reset time, will enable one to estimate the operation of inverse-time relays during successive application and removal of the actuating quantity, as when a motor is "plugged," or when a circuit is tripped and then automatically reclosed on a fault several times, or during power surges accompanying loss of synchronism.

COMPENSATION FOR FREQUENCY OR TEMPERATURE CHANGES IN VOLTAGE RELAYS

A voltage relay may be provided with a resistor in series with its coil circuit to decrease changes in pickup by decreasing the effect of changes in coil resistance with heating. Such a resistor will also help to decrease the effect on the characteristics of change in frequency. A series capacitor may be used to obtain series resonance at normal frequency when operation on harmonics is to be avoided.

COMBINATION OF INSTANTANEOUS AND INVERSE-TIME RELAYS

Frequently, an instantaneous relay and an inverse-time relay are furnished in one enclosing case because the two functions are so often required together. The two relays are independently adjustable, but are actuated by the same quantity, and their "*a*" contacts may be connected in parallel.

D-C Directional Relays

Such relays are derived directly from the basic electromagnetic-attraction type described in Chapter 2. The various types and their capabilities are as follows.

CURRENT-DIRECTIONAL RELAYS

The current-directional relay is identical with the basic type described in Chapter 2. It is used for protection in d-c power circuits, its armature coil being connected either directly in series with the circuit or across a shunt in series with the circuit, so that the relay will respond to a certain direction of current flow. Such relays may be polarized either by a permanent magnet or by a field coil connected to be energized by the voltage of the circuit. A field coil would be used if the relay were calibrated to operate in terms of the magnitude of power (watts) in the circuit.

With adjustable calibration, the relay would also have overcurrent (or overpower) or undercurrent (or underpower) characteristics, or both, in addition to being directional.

VOLTAGE-DIRECTIONAL RELAYS

Voltage-directional relays are the same as current-directional relays except for the number of turns and the resistance of the armature coil, and possibly except for the polarizing source. Such relays are used in d-c power circuits to respond to a certain polarity of the voltage across the circuit or across some part of the circuit. If the relay is intended to respond to reversal of the circuit-voltage polarity, it is polarized by a permanent magnet, there being no other suitable polarizing source unless a storage battery is available for the purpose. Otherwise, either permanent-magnet polarization or a field coil energized from the circuit voltage would be used.

When such a relay is connected across a circuit breaker to permit closing the breaker only when the voltage across the open breaker has a certain polarity, the relay may be called a "differential" relay because it operates only in response to a predetermined difference between the magnitudes of the circuit voltages on either side of the breaker.

VOLTAGE-AND-CURRENT-DIRECTIONAL RELAYS

The voltage-and-current-directional relay has two armature coils. Such a relay, for example, controls the closing and opening of a circuit breaker in the circuit between a d-c generator and a bus to which another source of voltage may be connected, so as to avoid motoring of the generator. The voltage armature coil is connected across the breaker and picks up the relay to permit closing the breaker only if the generator voltage is a certain amount greater than the bus voltage. The current armature coil is connected in series with the circuit, or across a shunt, and resets the relay to trip the breaker whenever a predetermined amount of current starts to flow from the bus into the generator.

VOLTAGE-BALANCE-DIRECTIONAL RELAYS

A relay with two voltage coils encircling the armature may be used to protect a three-wire d-c circuit against unbalanced voltages. The two coils are connected in such a way that their magnetomotive forces are in opposition. Such a relay has double-throw contacts and two restraining springs to provide calibration for movement of the armature in either direction. When one voltage exceeds the other by a predetermined amount, the armature will move one way to close one set of contacts; if the other voltage is the higher, the armature will close the other set of contacts.

Such a relay may also be used to respond to a difference in circuit-voltage magnitudes on either side of a circuit breaker, instead of the single-coil type previously described.

CURRENT-BALANCE-DIRECTIONAL RELAYS

A relay like a voltage-balance type except with two current coils encircling the armature may be used for current-balance protection of a three-wire d-c circuit, or to compare the loads of two different circuits.

DIRECTIONAL RELAYS FOR VACUUM-TUBE OR RECTIFIED A-C CIRCUITS

Both the single-coil and the two-coil types of voltage relays previously described have been used to respond to the output of vacuum-tube or rectifying circuits. Such relays are generally called "polarized" relays, the term "directional" having no significance since the actuating quantity is always of the same polarity; the directional type of relay is used only for its high sensitivity. A relay used for this purpose may be polarized by a permanent magnet or from a suitable d-c source such as a station battery.

Another sometimes-useful characteristic of a d-c directional relay actuated from a rectified a-c source is that the torque of the relay is proportional to the first power of the actuating quantity. Thus, if two or more armature coils should be energized from different rectified a-c quantities, the relay torque would be proportional to the arithmetic sum (or difference, if desired) of the rms values of the a-c quantities, regardless of the phase angles between them. Such operation cannot be obtained with a-c relays.

POLARIZING MAGNET VERSUS FIELD COIL

Except where a permanent magnet is the only suitable polarizing source available, a field coil is generally preferred. It has already been said that a field coil is required if a relay is to respond to watts. A coil provides more flexibility of adjustment since series resistors can be used to vary the polarizing mmf. Also, it is necessary to remove polarization to permit some types of relays to reset after they have operated, and this requires a field coil.

SHUNTS

The rating of a shunt and the resistance of the leads from the shunt to a current-directional relay affect the calibration of the relay. It is customary to specify the rating of the shunt and the resistance of the leads necessary for a given range of calibration of the relay.

TIME DELAY

As mentioned in Chapter 2, d-c directional relays are inherently fast. For some applications, an auxiliary time-delay relay is necessary to prevent undesired operation on momentary reversals of the actuating quantity.

A-C Directional Relays

In Chapter 2, a-c directional relays were said to be able to distinguish between the flow of current in one direction or the other in an a-c circuit by recognizing differences in the phase angle between the current and the polarizing quantity. We shall see that the ability to distinguish between the flow of current in one direction or the other depends on the choice of the polarizing quantity and on the angle of maximum torque, and that all the variations in function provided by a-c directional relays depend on these two quantities. This will become evident on further examination of some typical types.

POWER RELAYS

Relays that must respond to power are generally used for protecting against conditions other than short circuits. Such relays are connected to be polarized by a voltage of a circuit, and the current connections and the relay characteristics are chosen so that maximum torque in the relay occurs when unity-power-factor load is carried by the circuit. The relay will then pick up for power flowing in one direction through the circuit and will reset for the opposite direction of power flow.

If a single-phase circuit is involved, a directional relay is used having maximum torque when the relay current is in phase with the relay voltage. The same relay can be used on a three-phase circuit if the load is sufficiently well balanced; in that event, the polarizing voltage must be in phase with the current in one of the three phases at unity-power-factor load. (For simplicity, the term "phase" will be used frequently where the term "phase conductor" would be more strictly correct.) Such an in-phase voltage will be available if phase-to-neutral voltage is available; otherwise, a connection like that in

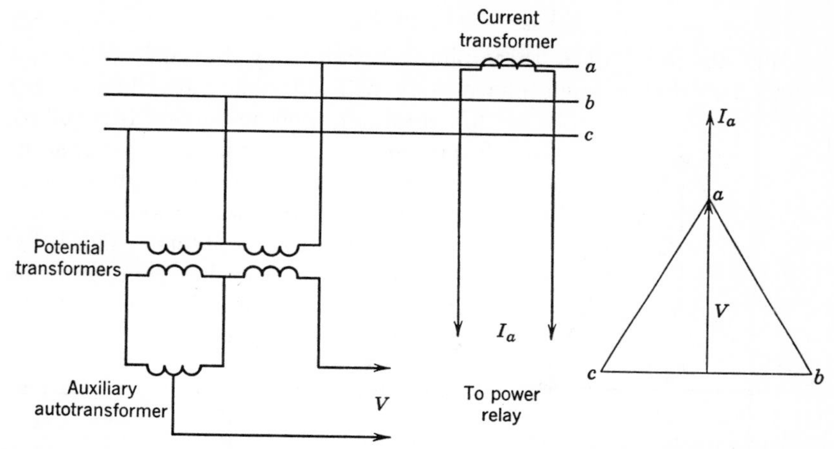

Fig. 4. Connections and vector diagram for a power relay where phase-to-neutral voltage is not available.

Fig. 4 will provide a suitable polarizing voltage. Or, a relay having maximum torque when its current leads its voltage by 30° can be connected to use V_{ac} and I_a, as in Fig. 5.

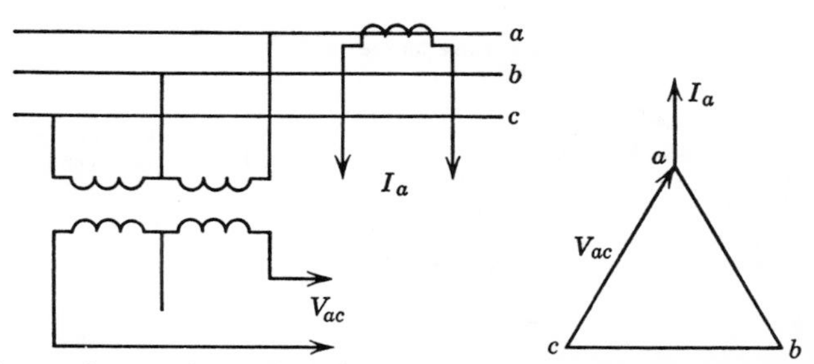

Fig. 5. Connections and vector diagram for a power relay using phase-to-phase voltage.

The conventions and nomenclature used throughout this book in dealing with three-phase voltage and current vector diagrams is shown in Fig. 6. The voltages of Fig. 6 are defined as follows:

$$V_{ab} = V_a - V_b$$

$$V_{bc} = V_b - V_c$$

$$V_{ca} = V_c - V_a$$

From these definitions, it follows that:

$$V_{ba} = -V_{ab} = V_b - V_a$$
$$V_{cb} = -V_{bc} = V_c - V_b$$
$$V_{ac} = -V_{ca} = V_a - V_c$$

When the load of a three-phase circuit may be sufficiently unbalanced so that a single-phase relay will not suffice, or when a very low minimum pickup current is required, a polyphase relay is used, having, actually or in effect, three single-phase relay elements whose torques are added to control a single set of contacts. The actuating quantities of such a relay may be any of several combinations, the following being frequently used:

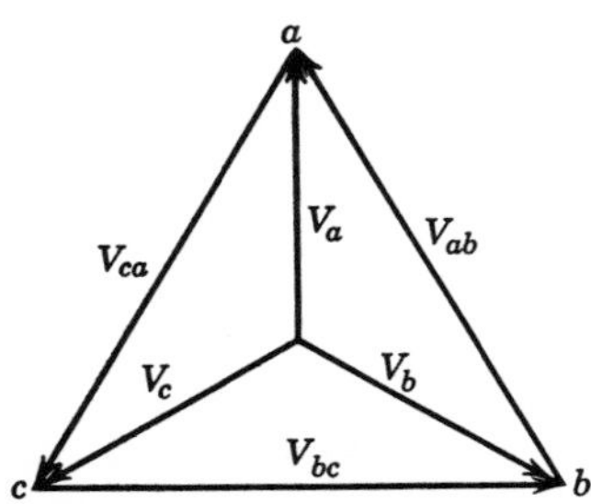

Fig. 6. Conventions and nomenclature for three-phase voltage vector diagrams.

Element No.	Voltage	Current
1	V_{ac}	I_a
2	V_{cb}	I_c
3	V_{ba}	I_b

However it may be accomplished, a power relay will distinguish between the flow of power in one direction or the other by developing positive (or pickup) torque for one direction, and negative (or reset) torque for the other. The unity-power-factor component of the current will reverse as the direction of power flow reverses, as illustrated in Fig. 7 for a relay connected as in Fig. 5.

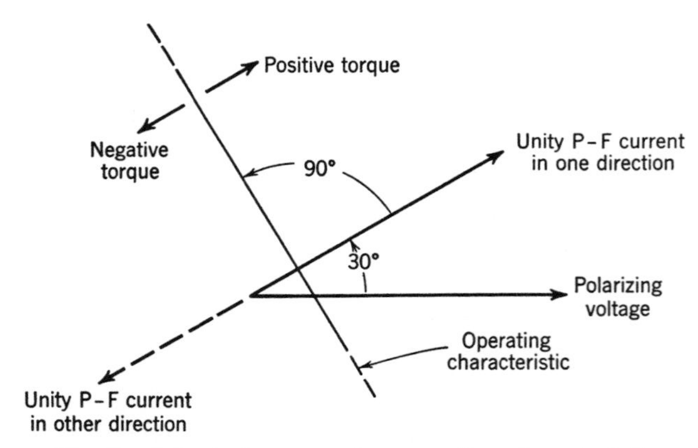

Fig. 7. A typical power-relay operating characteristic.

Power relays are used generally for responding to a certain direction of current flow under approximately balanced three-phase conditions and for approximately normal voltage magnitudes. Consequently, any combination of voltage and current may be used so long as the relay has the necessary angle of maximum torque so that maximum torque will be developed for unity-power-factor current in the three-phase system.

Power relays are available having adjustable minimum pickup currents. They may be calibrated either in terms of minimum pickup amperes at rated voltage or in terms of minimum pickup watts. Therefore, such relays may be adjusted to respond to any desired amount of power being supplied in a given direction. In effect, these relays are watthour meters with their dial mechanisms replaced by contacts, and having a control spring. Some power relays have actually been constructed directly from watthour-meter parts.

Power relays usually have time-delay characteristics to avoid undesired operation during momentary power reversals, such as generator synchronizing-power surges or power reversals when short circuits occur. This time delay may be an inherent inverse-time characteristic of the relay itself, or it may be provided by a separate time-delay relay.

DIRECTIONAL RELAYS FOR SHORT-CIRCUIT PROTECTION

Because short circuits involve currents that lag their unity-power-factor positions, usually by large angles, it is desirable that directional relays for short-circuit protection be arranged to develop maximum torque under such lagging-current conditions. The technique for obtaining any desired maximum-torque adjustment was described in Chapter 2. The problem is straightforward for a single-phase circuit. Exactly the same technique can be applied to three-phase circuits, but there are a number of possible solutions, and not all of them are good. The problem is somewhat different from that with power relays. With power relays, we are dealing with approximately balanced three-phase conditions, and where the polarizing voltage is maintained approximately at its normal value; any of the alternative ways of obtaining maximum torque at unity-power-factor-load current is equally acceptable from a functional standpoint. If three-phase short circuits were the only kind with which we had to contend, any of the many possible arrangements for obtaining maximum torque at a given angle would also be equally acceptable. But the choice of connections for obtaining correct directional discrimination for unbalanced short circuits (i.e., phase-to-phase, phase-to-ground, and two-phase-to-ground) is severely restricted.

Three conventional current-and-voltage combinations that are used for phase relays are illustrated by the vector diagrams of Fig. 8, in which the quantities shown are for one of three single-phase relays, or for one of the three elements of a polyphase relay. The other

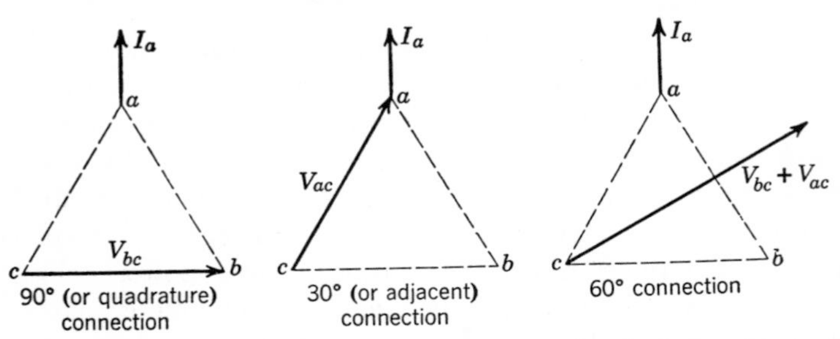

Fig. 8. Conventional connections of directional phase relays.

two relays or elements would use the other two corresponding voltage-and-current combinations. The names of these three combinations, as given in Fig. 8, will be recognized as describing the phase relation of the current-coil current to the polarizing voltage under balanced three-phase unity-power-factor conditions.

The relations shown in Fig. 8 are for the relay or element that provides directional discrimination when short circuits occur involving phases a and b. Note that the voltage V_{ab} is not used by the relay or element on which dependence for protection is placed. For such a short circuit, one or both of the other two relays will also develop torque. It would be highly undesirable if one of these others should develop contact-closing torque when the conditions were such that it would cause unnecessary tripping of a circuit breaker. It is to avoid this possibility, and yet to assure operation when it is required, that the many possible alternative connections are narrowed down to the three shown. Even with these, there are circumstances when incorrect operation is sometimes possible unless additional steps are taken to avoid it; this whole subject will be treated further when we consider the application of such relays to the protection of lines. It will probably be evident, however, that, since in a polyphase relay the torques of the three elements are added, it is only necessary that the net torque be in the right direction to avoid undesired operation. The bibliography[1] gives reference material for further study of polyphase directional-relay connections and their effect on relay behavior, but it is a bit advanced, in view of our present status.

With directional relays for protection against short circuits involving ground, there is no problem similar to that just described for phase relays. As will be seen later, only a single-phase relay is necessary, and the connections are such that, no matter which phase is involved, the quantities affecting relay operation have the same phase relation. Moreover, a ground relay is unaffected by other than ground faults because, for such other faults, the actuating quantities are not present unless the CT's (current transformers) fail to transform their currents accurately.

Except for circuit arrangements for providing the desired maximum-torque relations, a directional relay for ground protection is essentially the same as a single-phase directional relay for phase-fault protection. Such relays are available with or without time delay, and for current or voltage polarization, or for polarization by both current and voltage simultaneously.

Directional relays for short-circuit protection are generally used to supplement other relays. The directional relays permit tripping only for a certain direction of current flow, and the other relays determine (1) if it is a short circuit that is causing the current to flow, and (2) if the short circuit is near enough so that the relays should trip their circuit breaker. Such directional relays have no intentional time delay, and their pickup is non-adjustable but low enough so that the directional relays will always operate when their associated relays must operate. Some directional relays combine the directional with the fault-detecting and locating function; then, the directional relay will have adjustable pickup and either instantaneous or inverse-time characteristics.

Some directional relays have adjustment of their maximum-torque angle to permit their use with various connections of voltage-transformer sources, or to match their maximum-torque angle more accurately to the actual fault-current angle.

DIRECTIONAL-OVERCURRENT RELAYS

Directional-overcurrent relays are combinations of directional and overcurrent relay units in the same enclosing case. Any combination of directional relay, inverse-time overcurrent relay, and instantaneous overcurrent relay is available for phase- or ground-fault protection.

"Directional control" is a design feature that is highly desirable for this type of relay. With this feature, an overcurrent unit is inoperative, no matter how large the current may be, unless the contacts of the directional unit are closed. This is accomplished by connecting the directional-unit contacts in series with the shading-coil circuit or with one of the two flux-producing circuits of the overcurrent unit. When this circuit is open, no operating torque is developed in the overcurrent unit. The contacts of the overcurrent unit alone are in the trip circuit.

Without directional control, the contacts of the directional and overcurrent units would merely be connected in series, and there would be a possibility of incorrect tripping under certain circumstances. For example, consider the situation when a very large current, flowing to a short circuit in the non-tripping direction, causes the overcurrent unit to pick up. Then, suppose that the tripping of some circuit breaker causes the direction of current flow to reverse. The directional unit would immediately pick up and undesired tripping would result; even if the overcurrent unit should have a tendency to reset, there would be a race between the closing of the directional-unit contacts and the opening of the overcurrent-unit contacts.

Separate directional and overcurrent units are generally preferred because they are easier to apply than directional relays with inherent time characteristics and adjustable pickup. The operating time with separate units is simply a function of the current in the overcurrent unit; the pickup and time delay of the directional unit are so small that they can be neglected. But the operating time of the directional relay is a function of the product of its actuating and polarizing quantities and of the phase angle between them. However, the relay composed of separate directional and overcurrent units is somewhat larger, and it imposes somewhat more burden on its current-transformer source.

Current (or Voltage) - Balance Relays

Two basically different types of current-balance relay are used. Based on the production of actuating torque, one may be called the "overcurrent" type and the other the "directional" type.

OVERCURRENT TYPE

The overcurrent type of current-balance relay has one overcurrent element arranged to produce torque in opposition to another overcurrent element, both elements acting on the same moving structure. Figure 9 shows schematically an electromagnetic-attraction "balanced-beam" type of structure. Another commonly used structure is an induction-type relay having two overcurrent elements acting in opposition on a rotor.

If we neglect the negative-torque effect of the control spring, the torque equation of either type is:

$$T = K_1I_1^2 - K_2I_2^2$$

When the relay is on the verge of operating, the net torque is zero, and:

$$K_1I_1^2 = K_2I_2^2$$

Therefore, the operating characteristic is·

$$\frac{I_1}{I_2} = \sqrt{\frac{K_2}{K_1}} = \text{constant}$$

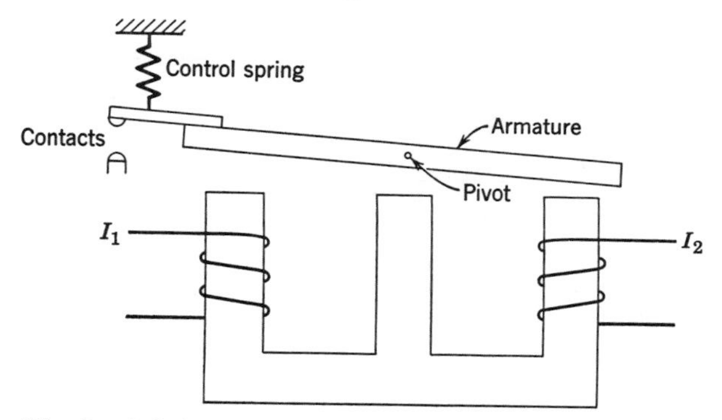

Fig. 9. A balanced-beam type of current-balance relay.

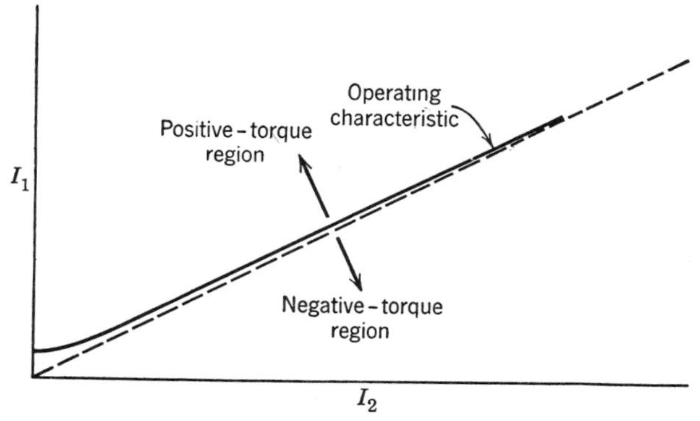

Fig. 10. Operating characteristic of a current-balance relay.

The operating characteristic of such a relay, including the effect of the control spring, is shown in Fig. 10.

The effect of the control spring is to require a certain minimum value of I_1 for pickup when I_2 is zero, but the spring effect becomes less and less noticeable at the higher values of current. The relay will pick up for ratios of I_1 to I_2 represented by points above the operating characteristic.

Such an operating characteristic is specified by expressing in percent the ratio of I_1 to I_2 required for pickup when the relay is operating on the straight part of the characteristic, and by giving the minimum pickup value of I_1 when I_2 is zero. I_1 is called the "operating"

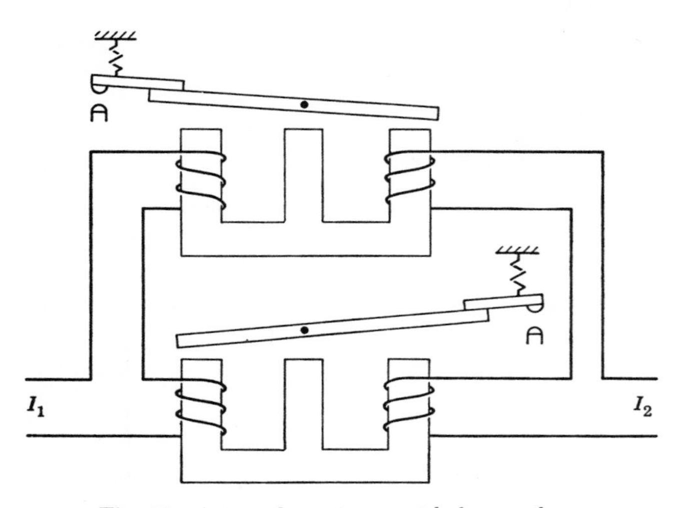

Fig. 11. A two-element current-balance relay.

current since it produces positive, or pickup, torque; I_2 is called the "restraining" current. By proportioning the number of turns on the operating and the restraining coils, one can obtain any desired "percent slope," as it is sometimes called.

Should it be desired to close an "*a*" contact circuit when either of two currents exceeds the other by a given percentage, two elements are used, as illustrated schematically in Fig. 11. For some applications, the contacts of the two elements may be arranged to trip different circuit breakers, depending on which element operates.

By these means, the currents in the different phases of a circuit, in different circuit branches of the same phase, or between corresponding phases of different circuits, can be compared. When applied between circuits where the ratio of one of the currents to the other never exceeds a certain amount except when a short circuit occurs in one of the circuits, a current-balance relay provides inherently selective protection.

Although the torque equations were written on the assumption that the phase angle between the two balanced quantities had no effect, the characteristics of such relays may be somewhat affected by the phase angle. In other words, the actual torque relation may be:

$$T = K_1 I_1{}^2 - K_2 I_2{}^2 + K_3 I_1 I_2 \cos(\theta - \tau)$$

where the effect of the control spring is neglected, and where θ and τ are defined as for directional relays. The constant K_3 is small, the production of directional torque by the interaction between the induced currents and stray fluxes of the two elements being incidental

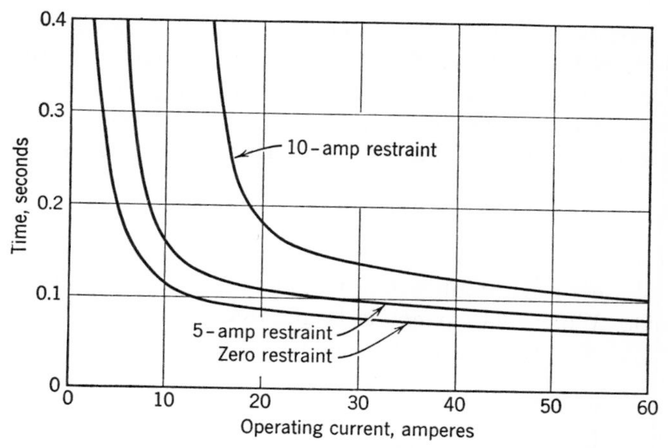

Fig. 12. Time-current curves of a current-balance relay.

and often purposely minimized by design. With only rare exceptions, the directional effect can be neglected. It is mentioned here in passing merely for completeness of the theoretical considerations. It will not be mentioned again when other relays that balance one quantity against another are considered, but the effect is sometimes there nevertheless.

It will be evident that the characteristics of a voltage-balance relay may be expressed as for the current-balance relay if we substitute V_1 and V_2 for I_1 and I_2. Also, whereas the current-balance relay operates when one current exceeds a normal value in comparison with the other current, the voltage-balance relay is generally arranged to operate when one voltage drops below a normal value.

Relays are available having high-speed characteristics or inverse-time characteristics with or without an adjustable time dial. A typical set of time curves is shown in Fig. 12, where the effect of different values of restraining currents on the shape of the time curve is shown for one time adjustment. Such curves cannot be plotted on a multiple-of-pickup basis because the pickup is different for each value of restraining current. It will be noted that each curve is asymptotic to the pickup current for the given value of restraining current.

High-speed relays may operate undesirably on transient unbalances if the percent slope is too nearly 100%, and for this reason such relays may require higher percent-slope characteristics than inverse-time relays.

DIRECTIONAL TYPE

The directional type of current-balance relay uses a current-current directional element in which the polarizing quantity is the vector difference of two currents, and the actuating quantity is the vector sum of the two currents. If we assume that the currents are in phase, and neglect the effect of the control spring, the torque is:

$$T = K_1 (I_1 + I_2)(I_1 - I_2)$$

where I_1 and I_2 are rms values. Therefore, when the two currents are in phase and are of equal magnitude, no operating torque is developed. When one current is larger than the other, torque is developed, its direction depending on which current is the larger. If

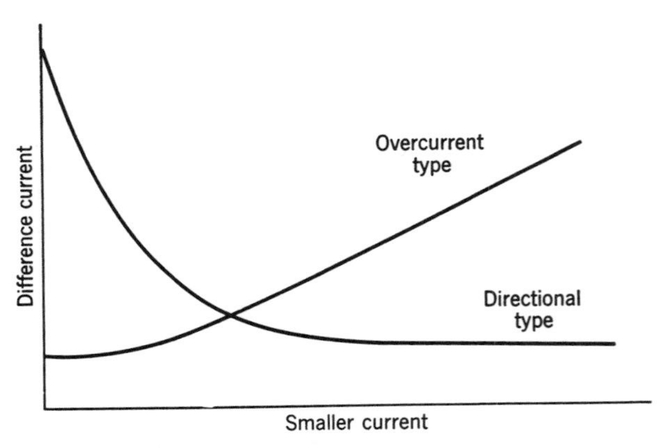

Fig. 13. Comparison of current-balance-relay characteristics

the two currents are 180° out of phase, the direction of torque for a given unbalance will be the same as when the currents are in phase, as can be seen by changing the sign of either current in the torque equation. This type of relay may have double-throw contacts both of which are normally open, the control spring being arranged to produce restraint against movement in either direction from the midposition.

This relay is not a current-balance relay in the same sense as the overcurrent type, as shown by a comparison of their operating characteristics in Fig. 13. The directional type is more sensitive to unbalance when the two currents are large, and is less sensitive when they are small. This is advantageous under one circumstance and disadvantageous under another. For parallel-line protection, which is the principal use of the directional type, auxiliary means are not required to prevent undesirable operation on load currents during switching; this is because the pickup is inherently higher when one line is out of service, at which time one of the two currents is zero. On the other hand, the directional type is more apt to operate undesirably on transient current-transformer unbalances when short circuits occur beyond the ends of the parallel lines; this is because the relay is more sensitive to current unbalance under high-current conditions when the errors of current transformers are apt to be greatest.

Differential Relays

Differential relays take a variety of forms, depending on the equipment they protect. The definition of such a relay is "one that operates when the vector difference of two or more similar electrical quantities exceeds a predetermined amount."[2] It will be seen later that almost any type of relay, when connected in a certain way, can be made to operate as a differential relay. In other words, it is not so much the relay construction as the way the relay is connected in a circuit that makes it a differential relay.

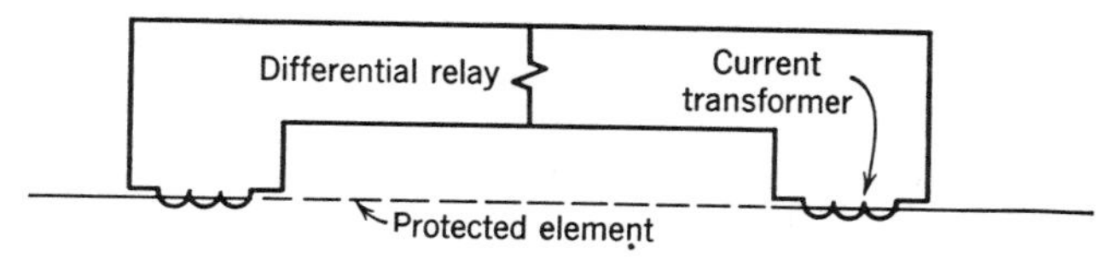

Fig. 14. A simple differential-relay application.

Most differential-relay applications are of the "current-differential" type. The simplest example of such an arrangement is shown in Fig. 14. The dashed portion of the circuit of Fig. 14 represents the system element that is protected by the differential relay. This system element might be a length of circuit, a winding of a generator, a portion of a bus, etc. A current transformer (CT) is shown in each connection to the system element. The secondaries of the CT's are interconnected, and the coil of an overcurrent relay is connected across the CT secondary circuit. This relay could be any of the a-c types that we have considered.

Now, suppose that current flows through the primary circuit either to a load or to a short circuit located at X. The conditions will be as in Fig. 15. If the two current transformers have the same ratio,

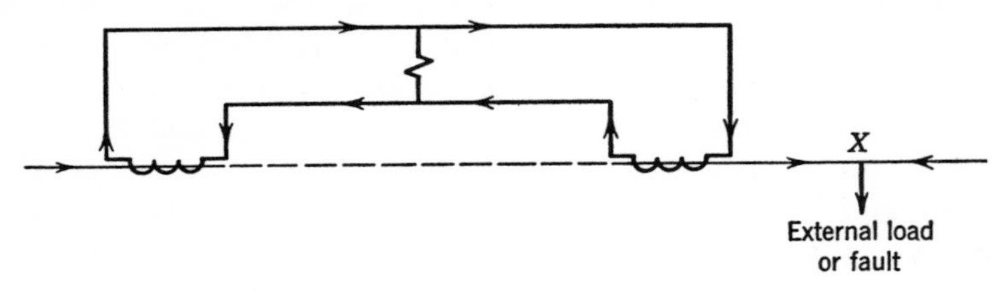

Fig. 15. Conditions for an external load or fault.

and are properly connected, their secondary currents will merely circulate between the two CT's as shown by the arrows, and no current will flow through the differential relay.

But, should a short circuit develop anywhere between the two CT's, the conditions of Fig. 16 will then exist. If current flows to the short circuit from both sides as shown, the sum of the CT secondary currents will flow through the differential relay. It is not necessary that short-circuit current flow to the fault from both sides to cause secondary current to flow through the differential relay. A flow on

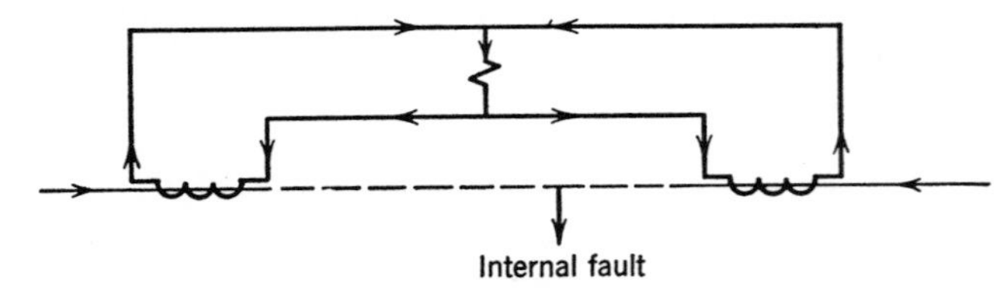

Fig. 16. Conditions for an internal fault.

one side only, or even some current flowing out of one side while a larger current enters the other side, will cause a differential current. In other words, the differential-relay current will be proportional to the vector difference between the currents entering and leaving the protected circuit; and, if the differential current exceeds the relay's pickup value, the relay will operate.

It is a simple step to extend the principle to a system element having several connections. Consider Fig. 17, for example, in which three connections are involved. It is only necessary, as before, that all the CT's have the same ratio, and that they be connected so that the relay receives no current when the total current leaving the circuit element is equal vectorially to the total current entering the circuit element.

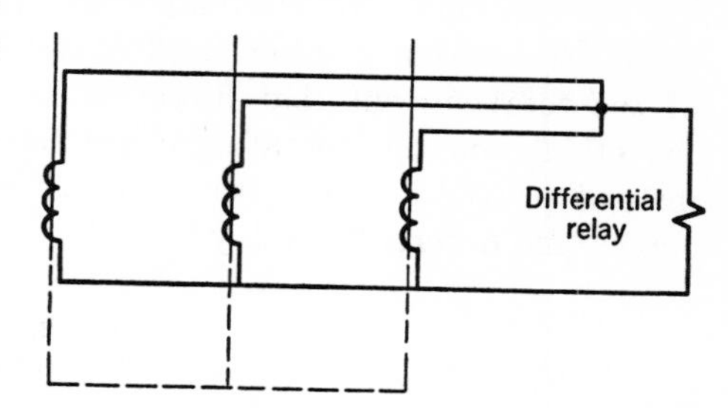

Fig. 17. A three-terminal current-differential application.

The principle can still be applied where a power transformer is involved, but, in this case, the ratios and connections of the CT's on opposite sides of the power transformer must be such as to compensate for the magnitude and phase-angle change between the power-transformer currents on either side. This subject will be treated in detail when we consider the subject of power-transformer protection.

A most extensively used form of differential relay is the "percentage-differential" type. This is essentially the same as the overcurrent type of current-balance relay that was described earlier, but it is connected in a differential circuit, as shown in Fig. 18. The

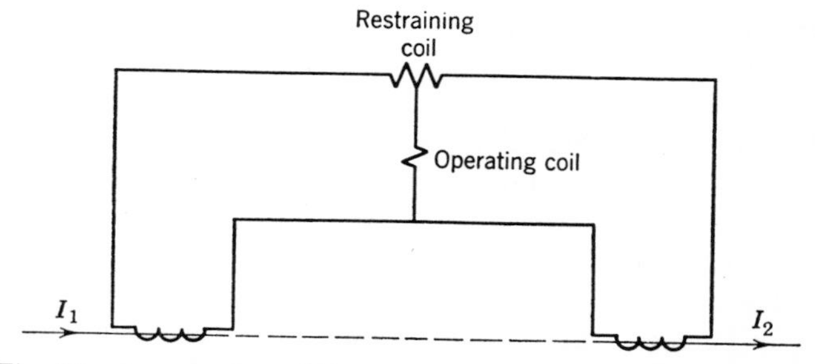

Fig. 18. A percentage-differential relay in a two-terminal circuit.

differential current required to operate this relay is a variable quantity, owing to the effect of the restraining coil. The differential current in the operating coil is proportional to $I_1 - I_2$, and the equivalent current in the restraining coil is proportional to $(I_1 + I_2)/2$, since the operating coil is connected to the midpoint of the restraining coil; in other words, if we let N be the number of turns on the restraining coil, the total ampere-turns are $I_1N/2 + I_2N/2$, which is the same as if $(I_1 + I_2)/2$ were to flow through the whole coil. The operating characteristic of such a relay is shown in Fig. 19. Thus, except for the slight effect of the control spring at low currents, the ratio of the differential operating current to the average restraining current is a fixed percentage, which explains the name of this relay. The term "through" current is often used to designate I_2, which is the portion of the total current that flows through the circuit from one end to the other, and the operating characteristics may be plotted using I_2 instead of $(I_1 + I_2)/2$, to conform with the ASA definition for a percentage differential relay.[2]

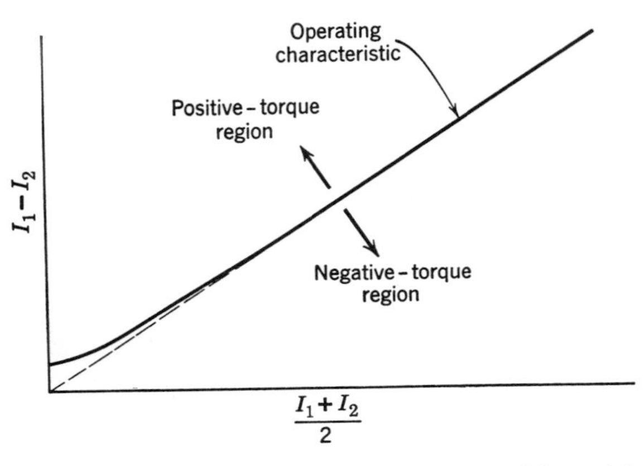

Fig. 19. Operating characteristic of a percentage-differential relay.

The advantage of this relay is that it is less likely to operate incorrectly than a differentially connected overcurrent relay when a short circuit occurs external to the protected zone. Current transformers of the types normally used do not transform their primary currents so accurately under transient conditions as for a short time after a short circuit occurs. This is particularly true when the short-circuit current is offset. Under such conditions, supposedly identical current transformers may not have identical secondary currents, owing to slight differences in magnetic properties or to their having different amounts of residual magnetism, and the difference current may be greater, the larger the magnitude of short-circuit current. Even if the short-circuit current to an external fault is not offset, the CT secondary currents may differ owing to differences in the CT types or loadings, particularly in power-transformer protection. Since the percentage-differential relay has a rising pickup characteristic as the magnitude of through current increases, the relay is restrained against operating improperly.

Figure 20 shows the comparison of a simple overcurrent relay with a percentage-differential relay under such conditions. An overcurrent relay having the same minimum pickup as a percentage-differential

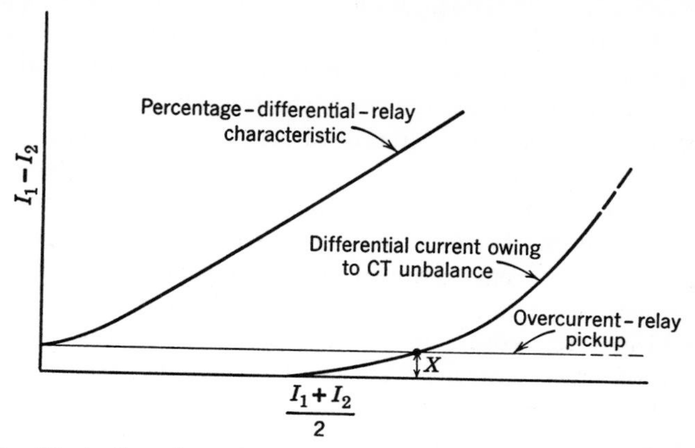

Fig. 20. Illustrating the value of the percentage-differential characteristic.

relay would operate undesirably when the differential current barely exceeded the value X, whereas there would be no tendency for the percentage-differential relay to operate.

Percentage-differential relays can be applied to system elements having more than two terminals, as in the three-terminal application of Fig. 21. Each of the three restraining coils of Fig. 21 has the

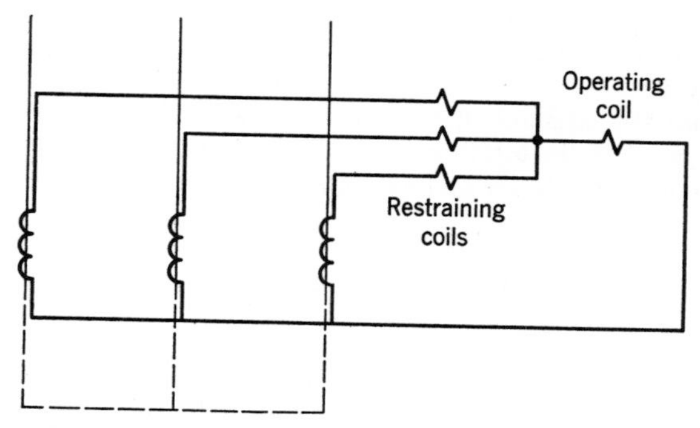

Fig. 21. Three-terminal application of a percentage-differential relay.

same number of turns, and each coil produces restraining torque independently of the others, and their torques are added arithmetically. The percent-slope characteristic for such a relay will vary with the distribution of currents between the three restraining coils.

Percentage-differential relays are usually instantaneous or high speed. Time delay is not required for selectivity because the percentage-differential characteristic and other supplementary features to be described later make these relays virtually immune to the effects of transients when the relays are properly applied.

The adjustments provided with some percentage-differential relays will be described in connection with their application.

Several other types of differential-relay arrangements could be mentioned. One of these uses a directional relay. Another has additional restraint obtained from harmonics and the d-c component of the differential current. Another type uses an overvoltage relay instead of an overcurrent relay in the differential circuit. Special current transformers may be used having little or no iron in their magnetic circuit to avoid errors in transformation caused by the d-c component of offset current waves. All these types are extensions of the fundamental principles that have been described, and they will be treated later in connection with their specific applications.

There has been great activity in the development of the differential relay because this form of relay is inherently the most selective of all the conventional types. However, each kind of system element presents special problems that have thus far made it impossible to devise a differential-relaying equipment having universal application.

Problems

1. Given a polyphase directional relay, each element of which develops maximum torque when the current in its current coil leads the voltage across its voltage coil by 20°.

Assuming that the constant in the torque equation is 1.0, and neglecting the spring torque, calculate the total torque for each of the three conventional connections for the following voltages and currents:

$$V_{bc} = 90 + j0$$

$$V_{ab} = -30 + j50$$

$$V_{ca} = -60 - j50$$

$$I_a = 25 + j7$$

$$I_b = -15 - j18$$

$$I_c = -2 + j10$$

2. Figure 22 shows a percentage-differential relay applied for the protection of a generator winding. The relay has a 0.1-ampere minimum pickup and a 10% slope (defined as in Fig. 19). A high-resistance ground fault has occurred as shown near the grounded-neutral end of the generator winding while the generator is

carrying load. As a consequence, the currents in amperes flowing at each end of the generator winding have the magnitudes and directions as shown on Fig. 22.

Assuming that the CT's have a 400/5-ampere ratio and no inaccuracies, will the relay operate to trip the generator breaker under this condition? Would the relay operate at the given value of fault current if the generator were carrying no load with its breaker open? On the same diagram, show the relay operating characteristic and the points that represent the operating and restraining currents in the relay for the two conditions.

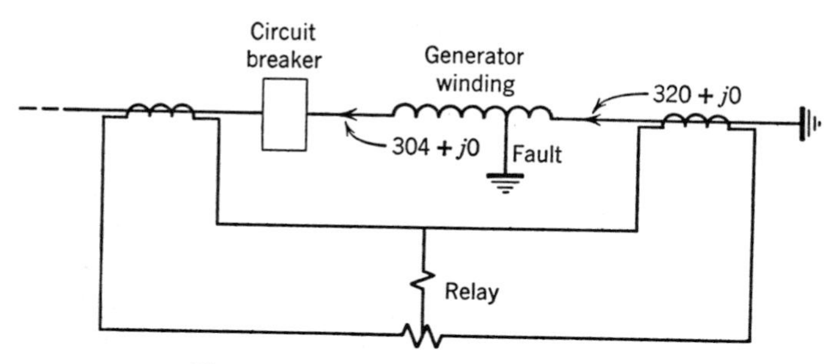

Fig. 22. Illustration for Problem 2.

3. Given two circuits carrying a-c currents having rms values of I_1 and I_2, respectively. Show a relay arrangement that will pick up on a constant magnitude of the arithmetic (not vector) sum of I_1 and I_2.

Bibliography

1. "A Single-Element Polyphase Directional Relay," by A. J. McConnell, *AIEE Trans., 56* (1937), pp. 77–80. Discussions, pp. 1025–1028.

"Factors Which Influence the Behavior of Directional Relays," by T. D. Graybeal, *AIEE Trans., 61* (1942), pp. 942–952.

"An Analysis of Polyphase Directional Relay Torques," by C. J. Baldwin, Jr., and B. N. Gafford, *AIEE Trans., 72,* Part III (1953), pp. 752–757. Discussions, pp. 757–759.

2. "Relays Associated with Electric Power Apparatus," *Publ. C37.1–1950,* American Standards Assoc., Inc., 70 East 45th St., New York 17, N. Y.

4 DISTANCE RELAYS

Perhaps the most interesting and versatile family of relays is the distance-relay group. In the preceding chapter, we examined relays in which one current was balanced against another current, and we saw that the operating characteristic could be expressed as a ratio of the two currents. In distance relays, there is a balance between voltage and current, the ratio of which can be expressed in terms of impedance. Impedance is an electrical measure of distance along a transmission line, which explains the name applied to this group of relays.

The Impedance-Type Distance Relay

Since this type of relay involves impedance-type units, let us first became acquainted with them. Generally speaking, the term "impedance" can be applied to resistance alone, reactance alone, or a combination of the two. In protective-relaying terminology, however, an impedance relay has a characteristic that is different from that of a relay responding to any component of impedance. And hence, the term "impedance relay" is very specific.

In an impedance relay, the torque produced by a current element is balanced against the torque of a voltage element. The current element produces positive (pickup) torque, whereas the voltage element produces negative (reset) torque. In other words, an impedance relay is a voltage-restrained overcurrent relay. If we let the control-spring effect be $-K_3$, the torque equation is:

$$T = K_1 I^2 - K_2 V^2 - K_3$$

where I and V are rms magnitudes of the current and voltage, respectively. At the balance point, when the relay is on the verge of operating, the net torque is zero, and

$$K_2 V^2 = K_1 I^2 - K_3$$

Dividing by K_2I^2, we get:

$$\frac{V^2}{I^2} = \frac{K_1}{K_2} - \frac{K_3}{K_2I^2}$$

$$\frac{V}{I} = Z = \sqrt{\frac{K_1}{K_2} - \frac{K_3}{K_2I^2}}$$

It is customary to neglect the effect of the control spring, since its effect is noticeable only at current magnitudes well below those normally encountered. Consequently, if we let K_3 be zero, the preceding equation becomes:

$$Z = \sqrt{\frac{K_1}{K_2}} = \text{constant}$$

In other words, an impedance relay is on the verge of operating at a given constant value of the ratio of V to I, which may be expressed as an impedance.

The operating characteristic in terms of voltage and current is shown in Fig. 1, where the effect of the control spring is shown as causing a noticeable bend in the characteristic only at the low-current end. For all practical purposes, the dashed line, which represents a

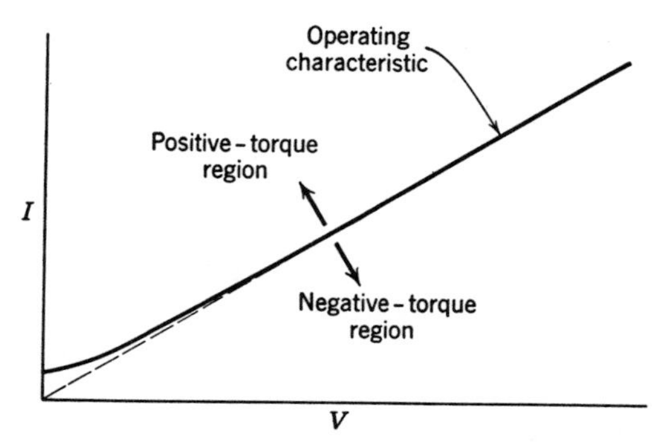

Fig. 1. Operating characteristic of an impedance relay.

constant value of Z, may be considered the operating characteristic. The relay will pick up for any combination of V and I represented by a point above the characteristic in the positive-torque region, or, in other words, for any value of Z less than the constant value represented by the operating characteristic. By adjustment, the slope of the operating characteristic can be changed so that the relay will respond to all values of impedance less than any desired upper limit.

A much more useful way of showing the operating characteristic of distance relays is by means of the so-called "impedance diagram" or "*R-X* diagram." Reference 1 provides a comprehensive treatment of this method of showing relay characteristics. The operating characteristic of the impedance relay, neglecting the control-spring effect, is shown in Fig. 2 on this type of diagram. The numerical value of

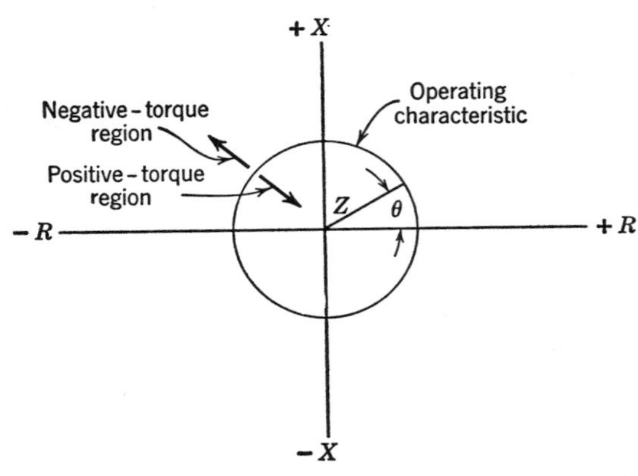

Fig. 2. Operating characteristic of an impedance relay on an *R-X* diagram.

the ratio of V to I is shown as the length of a radius vector, such as Z, and the phase angle θ between V and I determines the position of the vector, as shown. If I is in phase with V, the vector lies along the $+R$ axis; but, if I is 180 degrees out of phase with V, the vector lies along the $-R$ axis. If I lags V, the vector has a $+X$ component; and, if I leads V, the vector has a $-X$ component. Since the operation of the impedance relay is practically or actually independent of the phase angle between V and I, the operating characteristic is a circle with its center at the origin. Any value of Z less than the radius of the circle will result in the production of positive torque, and any value of Z greater than this radius will result in negative torque, regardless of the phase angle between V and I.

At very low currents where the operating characteristic of Fig. 1 departs from a straight line because of the control spring, the effect on Fig. 2 is to make the radius of the circle smaller. This does not have any practical significance, however, since the proper application of such relays rarely if ever depends on operation at such low currents.

Although impedance relays with inherent time delay are encountered occasionally, we shall consider only the high-speed type. The operating-time characteristic of a high-speed impedance relay is shown qualitatively in Fig. 3. The curve shown is for a particular value of current magnitude. Curves for higher currents will lie under this curve, and curves for lower currents will lie above it. In general, however, the operating times for the currents usually encountered in

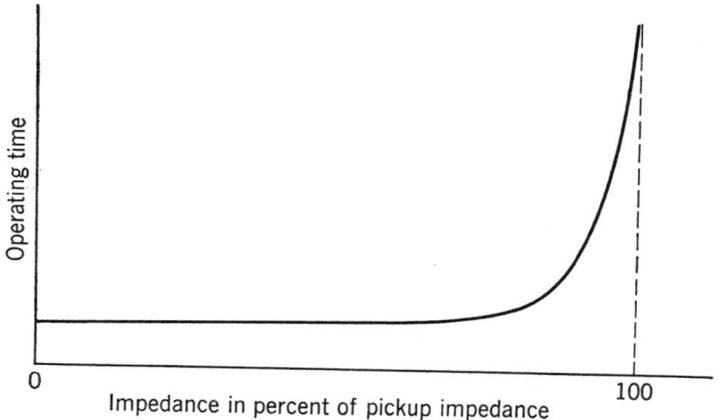

Fig. 3. Operating-time-versus-impedance characteristic of a high-speed impedance relay for one value of current.

normal applications of distance relays are so short as to be within the definition of high speed, and the variations with current are neglected. In fact, even the increase in time as the impedance approaches the pickup value is often neglected, and the time curve is shown simply as in Fig. 4.

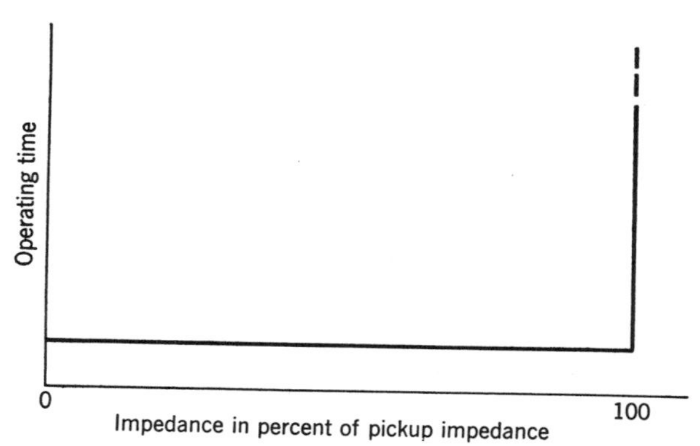

Fig. 4. Simplified representation of Fig. 3.

Various types of actuating structure are used in the construction of impedance relays. Inverse-time relays use the shaded-pole or the watt-metric structures. High-speed relays may use a balance-beam magnetic-attraction structure or an induction-cup or double-loop structure.

For transmission-line protection, a single-phase distance relay of the impedance type consists of a single-phase directional unit, three high-speed impedance-relay units, and a timing unit, together with the usual targets, seal-in unit, and other auxiliaries. Figure 5 shows very schematically the contact circuits of the principal units. The three impedance units are labeled Z_1, Z_2, and Z_3. The operating characteristics of these three units are independently adjustable. On the R-X diagram of Fig. 6, the circle for Z_1 is the smallest, the circle for Z_3 is the largest, and the circle for Z_2 is intermediate. It will be

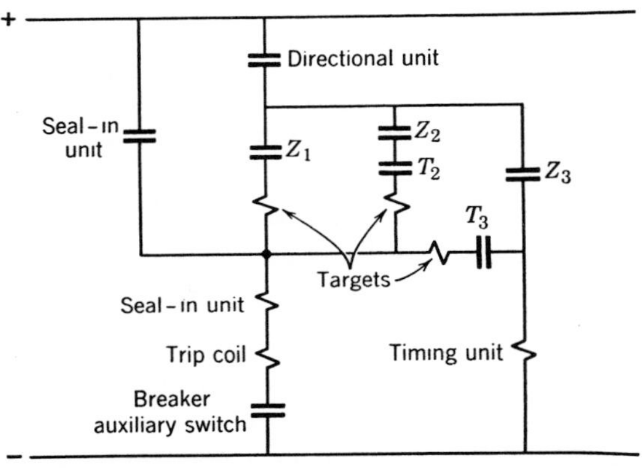

Fig. 5. Schematic contact-circuit connections of an impedance-type distance relay.

evident, then, that any value of impedance that is within the Z_1 circle will cause all three impedance units to operate. The operation of Z_1 and the directional unit will trip a breaker directly in a very short time, which we shall call T_1. Whenever Z_3 and the directional unit operate, the timing unit is energized. After a definite delay, the timing unit will first close its T_2 contact, and later its T_3 contact, both time delays being independently adjustable. Therefore, it can be seen that a value of impedance within the Z_2 circle, but outside the Z_1 circle, will result in tripping in T_2 time. And finally, a value of Z outside the Z_1 and Z_2 circles, but within the Z_3 circle, will result in tripping in T_3 time.

It will be noted that, if tripping is somehow blocked, the relay will make as many attempts to trip as there are characteristic circles around a given impedance point. However, use may not be made of this possible feature.

Figure 6 shows also the relation of the directional-unit operating characteristic to the impedance-unit characteristics on the same R-X

diagram. Since the directional unit permits tripping only in its positive-torque region, the inactive portions of the impedance-unit characteristics are shown dashed. The net result is that tripping will occur only for points that are both within the circles and above the directional-unit characteristic.

Because this is the first time that a simple directional-unit characteristic has been shown on the R-X diagram, it needs some explanation. Strictly speaking, the directional unit has a straight-line

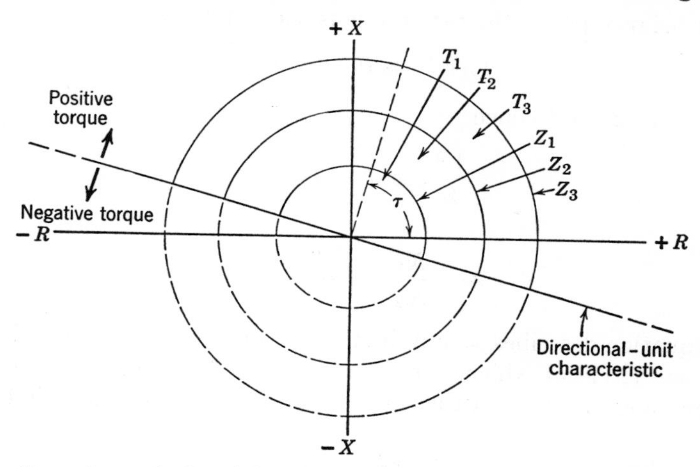

Fig. 6. Operating and time-delay characteristics of an impedance-type distance relay.

operating characteristic, as shown, only if the effect of the control spring is neglected, which is to assume that there is no restraining torque. It will be recalled that, if we neglect the control-spring effect, the torque of the directional unit is:

$$T = K_1 VI \cos(\theta - \tau)$$

When the net torque is zero,

$$K_1 VI \cos(\theta - \tau) = 0$$

Since K_1, V, or I are not necessarily zero, then, in order to satisfy this equation,

$$\cos(\theta - \tau) = 0$$

or

$$(\theta - \tau) = \pm 90°$$

Hence, $\theta = \tau \pm 90°$ describes the characteristic of the relay. In other words, the head of any radius vector Z at 90° from the angle of maximum torque lies on the operating characteristic, and this describes the straight line shown on Fig. 6, the particular value of τ having been chosen for reasons that will become evident later.

We should also develop the operating characteristic of a directional relay when the control-spring effect *is* taken into account. The torque equation as previously given is:

$$T = K_1 VI \cos(\theta - \tau) - K_2$$

At the balance point, the net torque is zero, and hence:

$$K_1 VI \cos(\theta - \tau) = K_2$$

But $I = V/Z$, and hence:

$$\frac{V^2}{Z} \cos(\theta - \tau) = \frac{K_2}{K_1}$$

or

$$Z = \frac{K_1}{K_2} V^2 \cos(\theta - \tau)$$

This equation describes an infinite number of circles, one for each value of V, one circle of which is shown in Fig. 7 for the same relay connections and the same value of τ as in Fig. 6. The fact that some

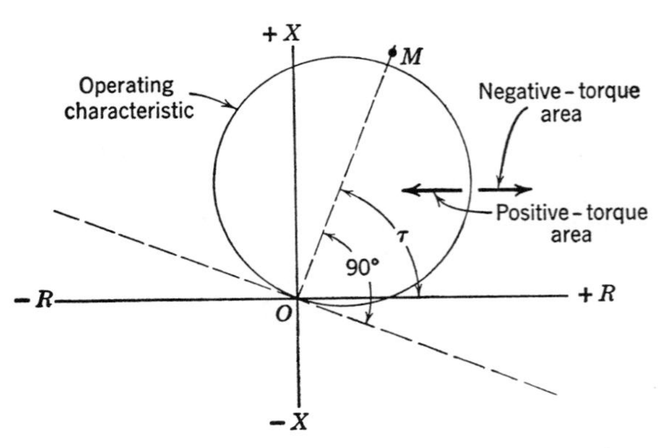

Fig. 7. The characteristics of a directional relay for one value of voltage.

values of θ will give negative values of Z should be ignored. Negative Z has no significance and cannot be shown on the R-X diagram.

The centers of all the circles will lie on the dashed line directed from O through M, which is at the angle of maximum torque. The diameter of each circle will be proportional to the square of the voltage. At normal voltage, and even at considerably reduced voltages, the diameter will be so large that for all practical purposes we may assume the straight-line characteristic of Fig. 6.

Looking somewhat ahead to the application of distance relays for transmission-line protection, we can show the operating-time-versus-impedance characteristic as in Fig. 8. This characteristic is generally

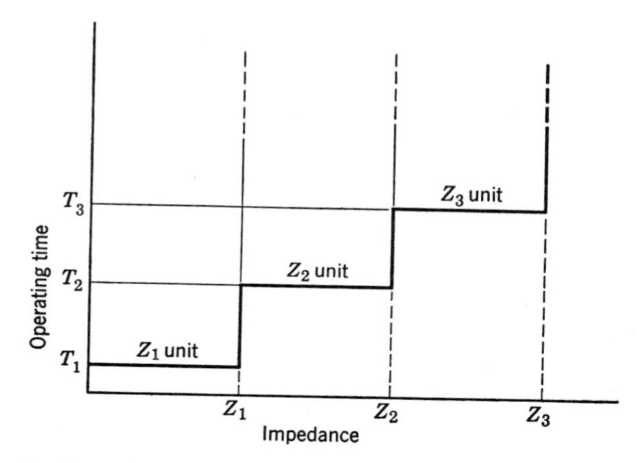

Fig. 8. Operating time versus impedance for an impedance-type distance relay.

called a "stepped" time-impedance characteristic. It will be shown later that the Z_1 and Z_2 units provide the primary protection for a given transmission-line section, whereas Z_2 and Z_3 provide back-up protection for adjoining busses and line sections.

The Modified Impedance-Type Distance Relay

The modified impedance-type distance relay is like the impedance type except that the impedance-unit operating characteristics are shifted, as in Fig. 9. This shift is accomplished by what is called a "current bias," which merely consists of introducing into the voltage supply an additional voltage proportional to the current,[2] making the torque equation as follows:

$$T = K_1 I^2 - K_2 (V + CI)^2$$

The term $(V + CI)$ is the rms magnitude of the vector addition of V and CI, involving the angle θ between V and I as well as a constant angle in the constant C term. This is the equation of a circle whose center is offset from the origin, as shown in Fig. 9. By such biasing, a characteristic circle can be shifted in any direction from the origin, and by any desired amount, even to the extent that the origin is outside the circle. Slight variations may occur in the biasing, owing to saturation of the circuit elements. For this reason, it is not the prac-

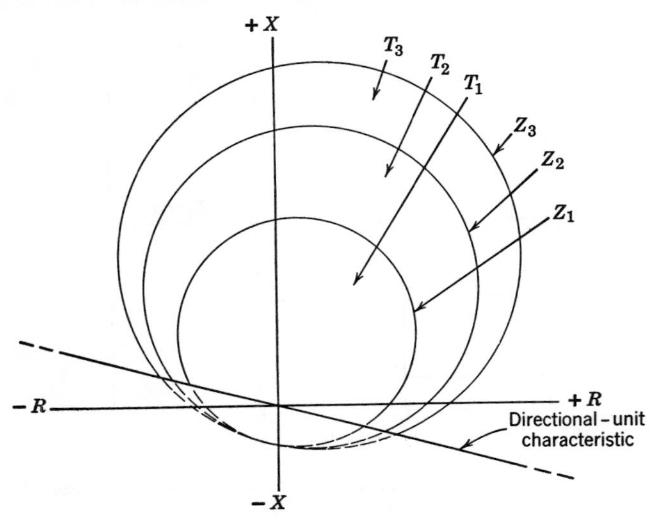

Fig. 9. Operating characteristic of a modified impedance-type distance relay.

tice to try to make the circles go through the origin, and therefore a separate directional unit is required as indicated in Fig. 9.

Since this relay is otherwise like the impedance-type relay already described, no further description will be given here.

The Reactance-Type Distance Relay

The reactance-relay unit of a reactance-type distance relay has, in effect, an overcurrent element developing positive torque, and a current-voltage directional element that either opposes or aids the overcurrent element, depending on the phase angle between the current and the voltage. In other words, a reactance relay is an overcurrent relay with directional restraint. The directional element is arranged to develop maximum negative torque when its current lags its voltage by 90°. The induction-cup or double-induction-loop structures are best suited for actuating high-speed relays of this type.

If we let the control-spring effect be $-K_3$, the torque equation is:

$$T = K_1 I^2 - K_2 VI \sin \theta - K_3$$

where θ is defined as positive when I lags V. At the balance point, the net torque is zero, and hence:

$$K_1 I^2 = K_2 VI \sin \theta + K_3$$

Dividing both sides of the equation by I^2, we get:

$$K_1 = K_2 \frac{V}{I} \sin \theta + \frac{K_3}{I^2}$$

or

$$\frac{V}{I} \sin \theta = Z \sin \theta = X = \frac{K_1}{K_2} - \frac{K_3}{K_2 I^2}$$

If we neglect the effect of the control spring,

$$X = \frac{K_1}{K_2} = \text{constant}$$

In other words, this relay has an operating characteristic such that all impedance radius vectors whose heads lie on this characteristic have a constant X component. This describes the straight line of Fig. 10. The significant thing about this characteristic is that the resistance

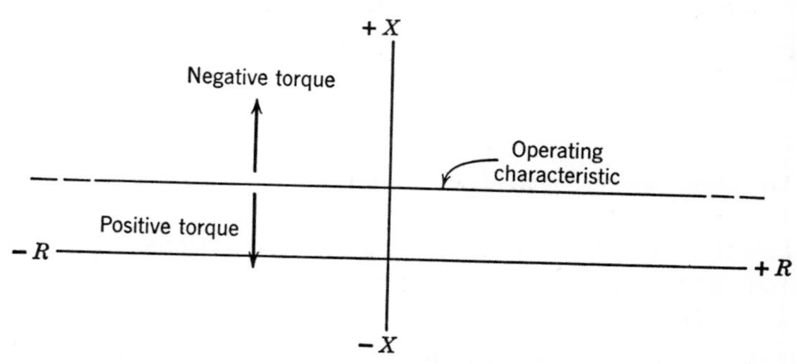

Fig. 10. Operating characteristic of a reactance relay.

component of the impedance has no effect on the operation of the relay; the relay responds solely to the reactance component. Any point below the operating characteristic—whether above or below the R axis—will lie in the positive-torque region.

Taking into account the effect of the control spring would lower the operating characteristic toward the R axis and beyond at very low values of current. This effect can be neglected in the normal application of reactance relays.

It should be noted in passing that, if the torque equation is of the general form $T = K_1 I^2 - K_2 VI \cos (\theta - \tau) - K_3$, and if τ is made some value other than 90°, a straight-line operating characteristic will still be obtained, but it will not be parallel to the R axis. This general form of relay has been called an "angle-impedance" relay.

A reactance-type distance relay for transmission-line protection could not use a simple directional unit as in the impedance-type relay, because the reactance relay would trip under normal load conditions at or near unity power factor, as will be seen later when we consider what different system-operating conditions "look" like on the R-X diagram. The reactance-type distance relay requires a directional

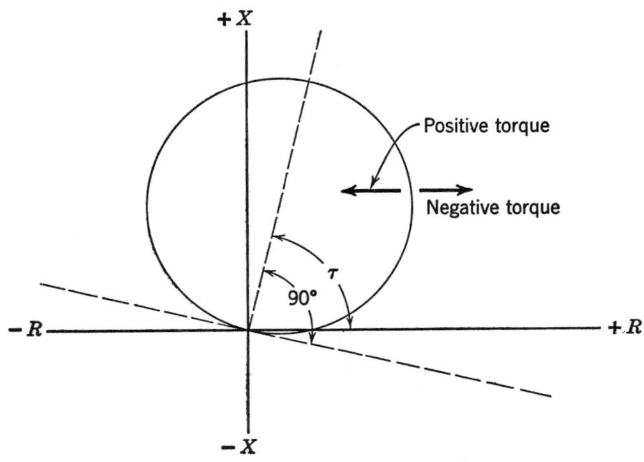

Fig. 11. Operating characteristic of a directional relay with voltage restraint.

unit that is inoperative under normal load conditions. The type of unit used for this purpose has a voltage-restraining element that opposes a directional element, and it is called an "admittance" or "mho" unit or relay. In other words, this is a voltage-restrained directional relay. When used with a reactance-type distance relay, this unit has also been called a "starting unit." If we let the control-spring effect be $-K_3$, the torque of such a unit is:

$$T = K_1VI\cos(\theta - \tau) - K_2V^2 - K_3$$

where θ and τ are defined as positive when I lags V. At the balance point, the net torque is zero, and hence:

$$K_2V^2 = K_1VI\cos(\theta - \tau) - K_3$$

Dividing both sides by K_2VI, we get:

$$\frac{V}{I} = Z = \frac{K_1}{K_2}\cos(\theta - \tau) - \frac{K_3}{K_2VI}$$

If we neglect the control-spring effect,

$$Z = \frac{K_1}{K_2} \cos (\theta - \tau)$$

It will be noted that this equation is like that of the directional relay when the control-spring effect is included, but that here there is no voltage term, and hence the relay has but one circular characteristic.

The operating characteristic described by this equation is shown in Fig. 11. The diameter of this circle is practically independent of voltage or current, except at very low magnitudes of current or voltage when the control-spring effect is taken into account, which causes the diameter to decrease.

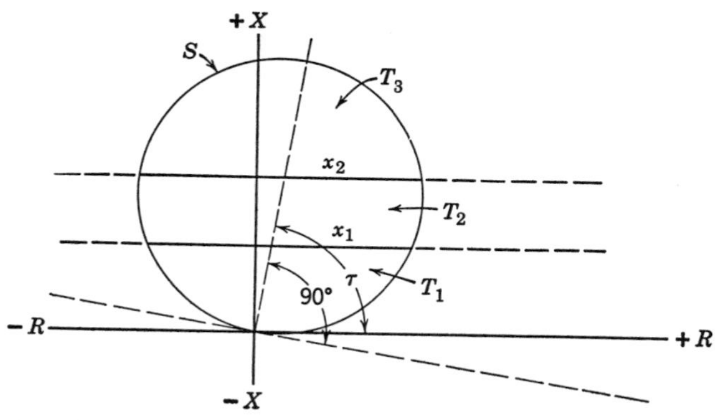

Fig. 12. Operating characteristics of a reactance-type distance relay.

The complete reactance-type distance relay has operating characteristics as shown in Fig. 12. These characteristics are obtained by arranging the various units as described in Fig. 5 for the impedance-type distance relay. It will be observed here, however, that the directional or starting unit (S) serves double duty, since it not only provides the directional function but also provides the third step of distance measurement with inherent directional discrimination.

The time-versus-impedance characteristic is the same as that of Fig. 8.

The Mho-Type Distance Relay

The mho unit has already been described, and its operating characteristic was derived in connection with the description of the starting unit of the reactance-type distance relay. The induction-cylinder or double-induction-loop structures are used in this type of relay.

The complete distance relay for transmission-line protection is composed of three high-speed mho units (M_1, M_2, and M_3) and a timing unit, connected in a manner similar to that shown for an impedance-type distance relay, except that no separate directional unit is required, since the mho units are inherently directional.[3] The operating characteristic of the entire relay is shown on Fig. 13.

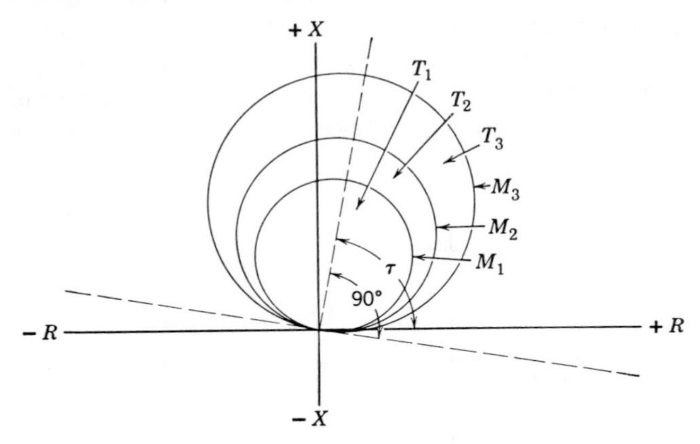

Fig. 13. Operating characteristics of a mho-type distance relay.

The operating-time-versus-impedance characteristic of the mho-type distance relay is the same as that of the impedance-type distance relay, Fig. 8.

By means of current biasing similar to that described for the offset impedance relay, a mho-relay characteristic circle can be offset so that either it encircles the origin of the R-X diagram or the origin is outside the circle.

General Considerations Applicable to All Distance Relays

OVERREACH

When a short circuit occurs, the current wave is apt to be offset initially. Under such conditions, distance relays tend to "overreach," i.e., to operate for a larger value of impedance than that for which they are adjusted to operate under steady-state conditions. This tendency is greater, the more inductive the impedance is. Also, the tendency is greater in electromagnetic-attraction-type relays than in induction-type relays. The tendency to overreach is minimized in the design of the relay-circuit elements, but it is still necessary to compensate for some tendency to overreach in the adjustment of the relays. Compensation for overreach as well as for inaccuracies in the current and voltage sources is obtained by adjusting the relays to operate at 10% to 20% lower impedance than that for which they would otherwise be adjusted. This will be further discussed when we consider the application of these relays.

MEMORY ACTION

Relays in which voltage is required to develop pickup torque, such as mho-type relays or directional units of other relays, may be provided with "memory action." Memory action is a feature that can be obtained by design in which the current flow in a voltage-polarizing coil does not cease immediately when the voltage on the high-voltage side of the supply-voltage transformer is instantly reduced to zero. Instead, the stored energy in the voltage circuit causes sinusoidal current to flow in the voltage coil for a short time. The frequency of this current and its phase angle are for all practical purposes the same as before the high-tension voltage dropped to zero, and therefore the relay is properly polarized since, in effect, it "remembers" the voltage that had been impressed on it. It will be evident that memory action is usable only with high-speed relays that are capable of operating within the short time that the transient polarizing current flows. It will also be evident that a relay must have voltage applied to it initially for memory action to be effective; in other words, memory action is ineffective if a distance relay's voltage is obtained from the line side of a line circuit breaker and the breaker is closed when there is a short circuit on the line.

Actually, it is a most rare circumstance when a short circuit reduces the relay supply voltage to zero. The short circuit must be exactly at the high-voltage terminals of the voltage transformer, and there must be no arcing in the short circuit. About the only time that this can happen in practice is when maintenance men have forgotten to remove protective grounding devices before the line breaker is closed. The voltage across an arcing short circuit is seldom less than about 4% of normal voltage, and this is sufficient to assure correct distance-relay operation even without the help of memory action.

Memory action does not adversely affect the distance-measuring ability of a distance relay. Such ability is important only for impedance values near the point for which the operating time steps from T_1 to T_2 or from T_2 to T_3. For such impedances, the primary voltage at the relay location does not go to zero, and the effect of the transient is "swamped."

THE VERSATILITY OF DISTANCE RELAYS

It is probably evident from the foregoing that on the R-X diagram we can construct any desired distance-relay operating characteristic composed of straight lines or circles. The characteristics shown here have been those of distance relays for transmission-line protection. But, by using these same characteristics or modifications of them, we can encompass any desired area on the R-X diagram, or we can divide the diagram into various areas, such that relay operation can be obtained only for certain relations between V, I, and θ. That this is a most powerful tool will be seen later when we learn what various types of abnormal system conditions "look" like on the R-X diagram.

THE SIGNIFICANCE OF Z

Since we are accustomed to associating impedance with some element such as a coil or a circuit of some sort, one might well ask what the significance is of the impedance expressed by the ratio of the voltage to the current supplied to a distance relay. To answer this question completely at this time would involve getting too far ahead of the story. It depends, among other things, on how the voltage and current supplied to the relay are obtained. For the protection of transmission lines against short circuits, which is the largest field of application of distance relays, this impedance is proportional, within certain limits, to the physical distance from the relay to the short circuit. However, the relay will still be energized by voltage and current under other than short-circuit conditions, such as when a system is carrying normal load, or when one part of a system loses synchronism with another, etc. Under any such condition, the impedance has a different significance from that during a short circuit. This is a most fascinating part of the story, but it must wait until we consider the application of distance relays.

At this point, one may wonder why there are different types of distance relays for transmission-line protection such as those described. The answer to this question is largely that each type has its particular field of application wherein it is generally more suitable than any other type. This will be discussed when we examine the application of these relays. These fields of application overlap more or less, and, in the overlap areas, which relay is chosen is a matter of personal preference for certain features of one particular type over another.

Problems

1. On an R-X diagram, show the impedance radius vector of a line section having an impedance of $2.8 + j5.0$ ohms. On the same diagram, show the operating characteristics of an impedance relay, a reactance relay, and a mho relay, each of which is adjusted to just operate for a zero-impedance short circuit at the end of the line section. Assume that the center of the mho relay's operating characteristic lies on the line-impedance vector.

Assuming that an arcing short circuit having an impedance of $1.5 + j0$ ohms can occur anywhere along the line section, show and state numerically for each type of relay the maximum portion of the line section that can be protected.

2. Derive and show the operating characteristic of an overcurrent relay on an R-X diagram.

3. A current-voltage directional relay has maximum torque when the current leads the voltage by 90°. The voltage coil is energized through a voltage regulator that maintains at the relay terminals a voltage that is always in phase with —and of the same frequency as—the system voltage, and that is constant in magnitude regardless of changes in the system voltage.

Derive the equation for the relay's operating characteristic in terms of the system voltage and current, and show this characteristic on an R-X diagram.

4. Write the torque equation, and derive the operating characteristic of a resistance relay.

Bibliography

1. "A Comprehensive Method of Determining the Performance of Distance Relays," by J. H. Neher, *AIEE Trans., 56* (1937), pp. 833–844. Discussions, p. 1515.

2. "A Distance Relay with Adjustable Phase-Angle Discrimination," by S. L. Goldsborough, *AIEE Trans., 63* (1944), pp. 835–838. Discussions, pp. 1471–1472.

"Application of the Ohm and Mho Principles to Protective Relays," by A. R. van C. Warrington, *AIEE Trans., 65* (1945), pp. 378–386. Discussions, p. 491.

3. "The Mho Distance Relay," by R. M. Hutchinson, *AIEE Trans., 65* (1945), pp. 353–360.

5 WIRE-PILOT RELAYS

Pilot relaying is an adaptation of the principles of differential relaying for the protection of transmission-line sections. Differential relaying of the type described in Chapter 3 is not used for transmission-line protection because the terminals of a line are separated by too great a distance to interconnect the CT secondaries in the manner described. Pilot relaying provides primary protection only; back-up protection must be provided by supplementary relaying.

The term "pilot" means that between the ends of the transmission line there is an interconnecting channel of some sort over which information can be conveyed. Three different types of such a channel are presently in use, and they are called "wire pilot," "carrier-current pilot," and "microwave pilot." A wire pilot consists generally of a two-wire circuit of the telephone-line type, either open wire or cable; frequently, such circuits are rented from the local telephone company. A carrier-current pilot for protective-relaying purposes is one in which low-voltage, high-frequency (30 kc to 200 kc) currents are transmitted along a conductor of a power line to a receiver at the other end, the earth and ground wire generally acting as the return conductor. A microwave pilot is an ultra-high-frequency radio system operating above 900 megacycles. A wire pilot is generally economical for distances up to 5 or 10 miles, beyond which a carrier-current pilot usually becomes more economical. Microwave pilots are used when the number of services requiring pilot channels exceeds the technical or economic capabilities of carrier current.

In the following, we shall first examine the fundamental principles of pilot relaying, and then see how these apply to some actual wire-pilot relaying equipments.

Why Current-Differential Relaying Is Not Used

Because the current-differential relays described in Chapter 3 for the protection of generators, transformers, busses, etc., are so selective, one might wonder why they are not used also for transmission-line relaying. The principal reason is that there would have to be too many interconnections between current transformers (CT's) to make current-differential relaying economically feasible over the usual distances involved in transmission-line relaying. For a three-phase line, six pilot conductors would be required, one for each phase CT and one for the neutral connection, and two for the trip circuit. Because even a two-wire pilot much more than 5 to 10 miles long becomes more costly than a carrier-current pilot, we could conclude that, on this basis alone, current-differential relaying with six pilot wires would be limited to very short lines.

Other reasons for not using current-differential relaying like that described in Chapter 3 are: (1) the likelihood of improper operation owing to CT inaccuracies under the heavy loadings that would be involved, (2) the effect of charging current between the pilot wires, (3) the large voltage drops in the pilot wires requiring better insulation, and (4) the pilot currents and voltages would be excessive for pilot circuits rented from a telephone company. Consequently, although the fundamental principles of current-differential relaying will still apply, we must take a different approach to the problem.

Purpose of a Pilot

Figure 1 is a one-line diagram of a transmission-line section connecting stations *A* and *B*, and showing a portion of an adjoining line section beyond *B*. Assume that you were at station *A*, where very accurate meters were available for reading voltage, current, and the

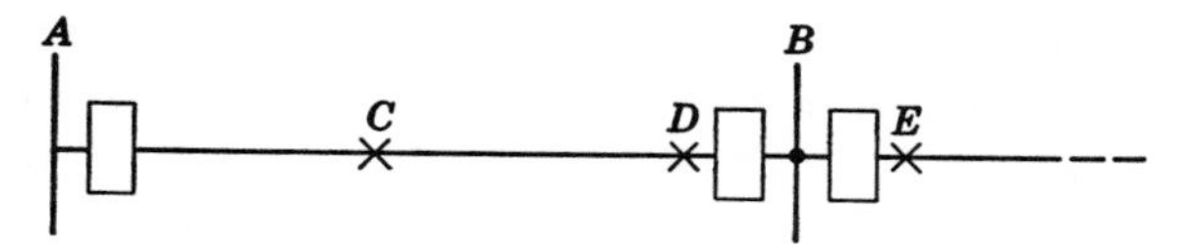

Fig. 1. Transmission-line sections for illustrating the purpose of a pilot.

phase angle between them for the line section *AB*. Knowing the impedance characteristics per unit length of the line, and the distance from *A* to *B*, you could, like a distance relay, tell the difference between a short circuit at *C* in the middle of the line and at *D*, the far end of the line. But you could not possibly distinguish between a fault at *D* and a fault at *E* just beyond the breaker of the adjoining line section, because the impedance between *D* and *E* would be so small as to produce a negligible difference in the quantities that you were measuring. Even though you might detect a slight difference, you could not be sure how much of it was owing to inaccuracies, though slight, in your meters or in the current and voltage transformers supplying your meters. And certainly, you would have difficulties if offset current waves were involved. Under such circumstances, you would hardly wish to accept the responsibility of tripping your circuit breaker for the fault at *D* and not tripping it for the fault at *E*.

But, if you were at station *B*, in spite of errors in your meters or source of supply, or whether there were offset waves, you could determine positively whether the fault was at *D* or *E*. There would be practically a complete reversal in the currents, or, in other words, approximately a 180° phase-angle difference.

What is needed at station *A*, therefore, is some sort of indication when the phase angle of the current at station *B* (with respect to the current at *A*) is different by approximately 180° from its value for faults in the line section *AB*. The same need exists at station *B* for faults on either side of station *A*. This information can be provided either by providing each station with an appropriate sample of the actual currents at the other station, or by a signal from the other station when its current phase angle is approximately 180° different from that for a fault in the line section being protected.

Tripping and Blocking Pilots

Having established that the purpose of a pilot is to convey certain information from one end of a line section to another in order to make selective tripping possible, the next consideration is the use to be made of the information. If the relaying equipment at one end of the line must receive a certain signal or current sample from the other end in order to prevent tripping at the one end, the pilot is said to be a "blocking" pilot. However, if one end cannot trip without receiving a certain signal or current sample from the other end, the pilot is said to be a "tripping" pilot. In general, if a pilot-relaying equipment at one end of a line can trip for a fault in the line with the breaker at the other end closed, but with no current flowing at that other end, it is a blocking pilot—otherwise it is like a tripping pilot.

It is probably evident from the foregoing that a blocking pilot is the preferred—if not the required—type. Other advantages of the blocking pilot will be given later.

D-C Wire-Pilot Relaying

Scores of different wire-pilot-relaying equipments have been devised, and many are in use today, where d-c signals in one form or another have been transmitted over pilot wires, or where pilot wires have constituted an extended contact-circuit interconnection between relaying equipments at terminal stations. For certain applications, some such arrangement has advantages—particularly where the distances are short and where a line may be tapped to other stations

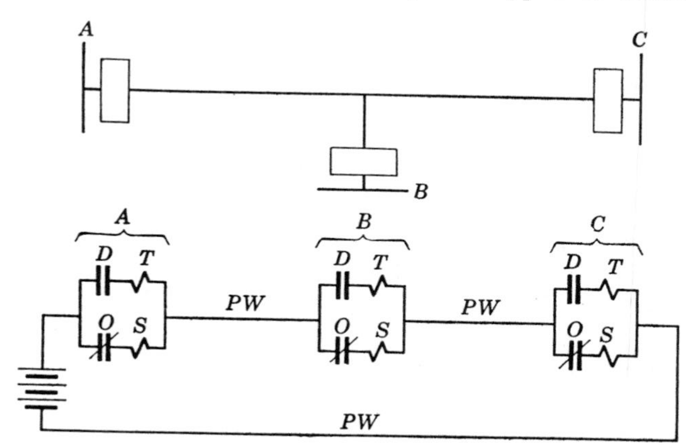

Fig. 2. Schematic illustration of a d-c wire-pilot relaying equipment. D = voltage-restrained directional (mho) relay; O = overcurrent relay; T = auxiliary tripping relay; S = auxiliary supervising relay; PW = pilot wire.

at one or more points. However, d-c wire-pilot relaying is nearly obsolete for other than very special applications. Nevertheless, a study of this type will reveal certain fundamental requirements that apply to modern pilot-relaying equipments, and will serve to prepare us better for understanding still other fundamentals.

An example of d-c wire-pilot relaying is shown very schematically in Fig. 2. The relaying equipments at the three stations are connected in a series circuit, including the pilot wires and a battery at station *A*. Normally, the battery causes current to flow through the "*b*" contacts of the overcurrent relay and the coil of the supervising relay at each station. Should a short circuit occur in the transmission-line section, the overcurrent relay will open its "*b*" contact at any station where there is a flow of short-circuit current. If the short-circuit-current flow at a given station is into the line, the directional relay at that station will close its "*a*" contact. The circuit at this station is thereby shifted to include the auxiliary tripping relay instead of the supervising relay. If this occurs at the other stations, current will flow through the tripping auxiliaries at all stations, and the breakers at all the line terminals will trip. But should a fault occur external to the protected-line section, the overcurrent relay at the station nearest the fault will pick up, but the directional relay will not close its contact because of the direction of current flow, and the circuit will be open

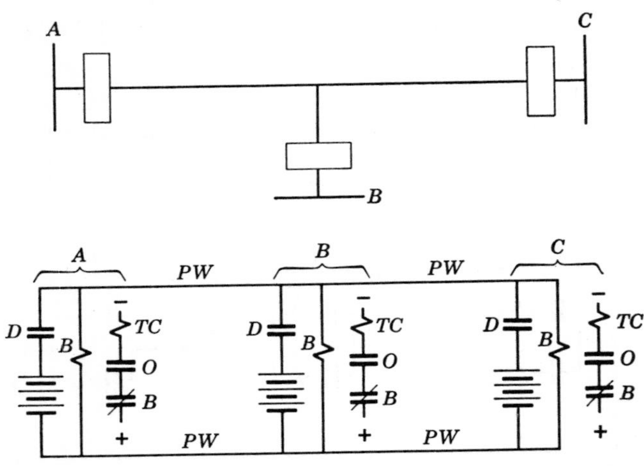

Fig. 3. Schematic illustration of a d-c wire-pilot scheme where information is transmitted over the pilot. D = voltage-restrained directional (mho) relay; B = auxiliary blocking relay; O = overcurrent relay; TC = trip coil; PW = pilot wire.

at that point, thereby preventing tripping at the other stations. If an internal fault occurs for which there may be no short-circuit-current flow at one of the stations, the overcurrent relay at that station will not pick up; but pilot-wire current will flow through the supervising auxiliary relay (whose resistance is equal to that of the tripping auxiliary relay), and tripping will still occur at the other two stations. (The supervising relays not only provide a path for current to flow so that tripping will occur as just described but also can be used to actuate an alarm should the pilot wires become open circuited or short circuited.) Therefore, this arrangement has the characteristics of a blocking pilot where the blocking signal is an interruption of current flow in the pilot. However, if the overcurrent and the supervising relays were removed from the circuit, it would be a tripping pilot, because tripping could not occur at any station unless all the directional relays operated to close their contacts, and tripping would be impossible if there was no flow of short-circuit current into one end.

An example of a blocking pilot, where positive blocking information is transmitted by the pilot, is shown in Fig. 3. Here, the directional relay at each station is arranged to close its contact when short-circuit current flows out of the line as to an external fault. It can be seen that, for an external fault beyond any station, the closing of the directional-relay contact at that station will cause a d-c voltage to be impressed on the pilot that will pick up the blocking relay at each station. The opening of the blocking relay "*b*" contact in series with the trip circuit will prevent tripping at each station. For an internal fault, no directional relay will operate, and hence no blocking relay will pick up, and tripping will occur at all stations where there is sufficient short-circuit current flowing to pick up the overcurrent relay.

Additional Fundamental Considerations

Now that we are a little better acquainted with pilot relaying, we are prepared to consider some other fundamentals that apply to certain modern types.

Whenever tripping by a relay at one station has to be blocked by the operation of a relay at another station, the blocking relay should be more sensitive than the tripping relay. The reason for this is to be certain that any time the tripping relay can pick up for an external fault the blocking relay will be sure to pick up also, or else undesired tripping will occur.

The matter of contact "races" must also be considered. For example, refer to Fig. 3 where the "*b*" contact of the blocking relay must open before the overcurrent contact closes, when tripping must be blocked. With the scheme as shown, the overcurrent relay must be given sufficient time delay to make this a safe race. An ingenious scheme can be used to avoid the necessity for adding time delay, but this will be described later in connection with carrier-current-pilot relaying.

A further complication arises because of the necessity for using separate phase and ground relays in order to obtain sufficient sensitivity under all short-circuit conditions. This makes it necessary to be sure that any tendency of a phase relay to operate improperly for a ground fault will not interfere with the proper operation of the equipment. To overcome this possibility, the principle of "ground preference" is employed where necessary. "Ground preference" means that operation of a ground relay takes blocking and tripping control away from the phase relays. This principle will be illustrated in connection with carrier-current-pilot relaying.

Some pilot-relaying equipments utilizing the blocking-and-tripping-relay principle must have additional provision against improper tripping during severe power swings or loss of synchronism. Such provision will be described later.

A-C Wire-Pilot Relaying

A-c wire-pilot relaying is the most closely akin to current-differential relaying. However, in modern a-c wire-pilot relaying, the magnitude of the current that flows in the pilot circuit is limited, and only a two-wire pilot is required. These two features make a-c wire-pilot relaying economically feasible over greater distances than current-differential relaying. They also introduce certain limitations in application that will be discussed later.

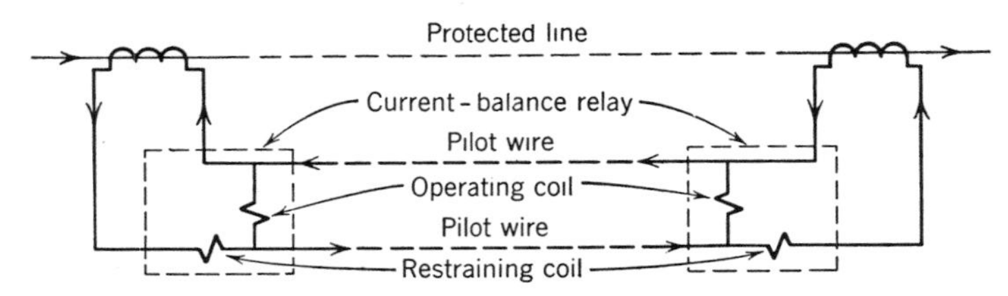

Fig. 4. Schematic illustration of the circulating-current principle of a-c wire-pilot relaying.

First, we should become acquainted with two new terms to describe the principle of operation: "circulating current" and "opposed voltage." Briefly, "circulating current" means that current circulates normally through the terminal CT's and the pilot, and "opposed voltage" means that current does not normally circulate through the pilot.

An adaptation of the current-differential type of relaying described in Chapter 3, employing the circulating-current principle, is shown schematically in Fig. 4. Except that a current-balance relay is used at each end of the pilot, this is essentially the same as the percentage-differential type described in Chapter 3. The only reason for having a relay at each end is to avoid having to run a tripping circuit the full length of the pilot.

A schematic illustration of the opposed-voltage principle is shown in Fig. 5. A current-balance type of relay is employed at each end, and the CT's are connected in such a way that the voltages across the restraining coils at the two ends of the pilot are in opposition for current flowing through the line section as to a load or an external fault. Consequently, no current flows in the pilot except charging current, if we assume that there is no unbalance between the CT outputs. The restraining coils serve to prevent relay operation owing to such unbalance currents. But should a short circuit occur on the protected line section, current will circulate in the pilot and operate the relays at both ends. Current will also flow through the restraining coils, but, in a proper application, this current will not be sufficient to prevent relay operation; the impedance of the pilot circuit will be the governing factor in this respect.

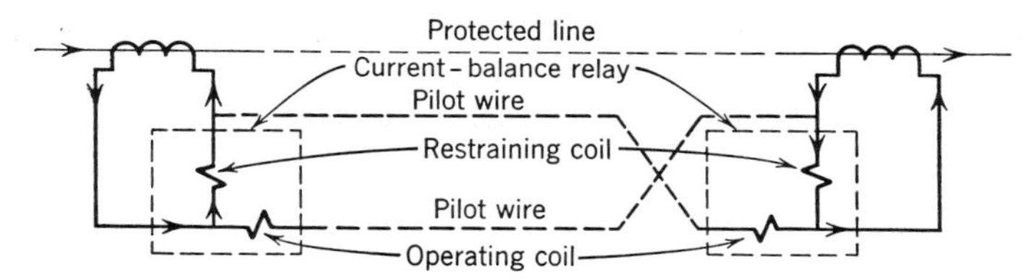

Fig. 5. Schematic illustration of the opposed-voltage principle of a-c wire-pilot relaying.

Short circuits or open circuits in the pilot wires have opposite effects on the two types of relaying equipment, as the accompanying table shows. Where it is indicated that tripping will be caused, tripping is contingent, of course, on the magnitude of the power-line current being high enough to pick up the relays.

	Effect of Shorts	Effect of Open Circuits
Opposed voltage	Cause tripping	Block tripping
Circulating current	Block tripping	Cause tripping

Both the opposed-voltage and the circulating-current principles permit tripping at both ends of a line for short-circuit current flow into one end only. However, the application of either principle may involve certain features that provide tripping only at the end having short-circuit-current flow, as will be seen when actual equipments are considered.

As has been said before, the feature that makes a-c wire-pilot relaying economically feasible, for the distances over which it is applied, is that only two pilot wires are used. In order to use only two wires, some means are required to derive a representative single-phase sample from the three phase and ground currents at the ends of a transmission line, so that these samples can be compared over the pilot. It would be a relatively simple matter to derive samples such that tripping would not occur for external faults for which the same currents that enter one end of a line go out the other end substantially unchanged. The real problem is to derive such samples that tripping will be assured for internal faults when the currents entering the line at the ends may be widely different. What must be avoided is a so-called "blind spot," as described in Reference 1 of the Bibliography. However, we are not yet ready to analyze such a possibility.

CIRCULATING-CURRENT TYPE

Figure 6 shows schematically a practical example of a circulating-current type of equipment.[2] The relay at each end of the pilot is a d-c permanent-magnet-polarized directional type. The coil marked

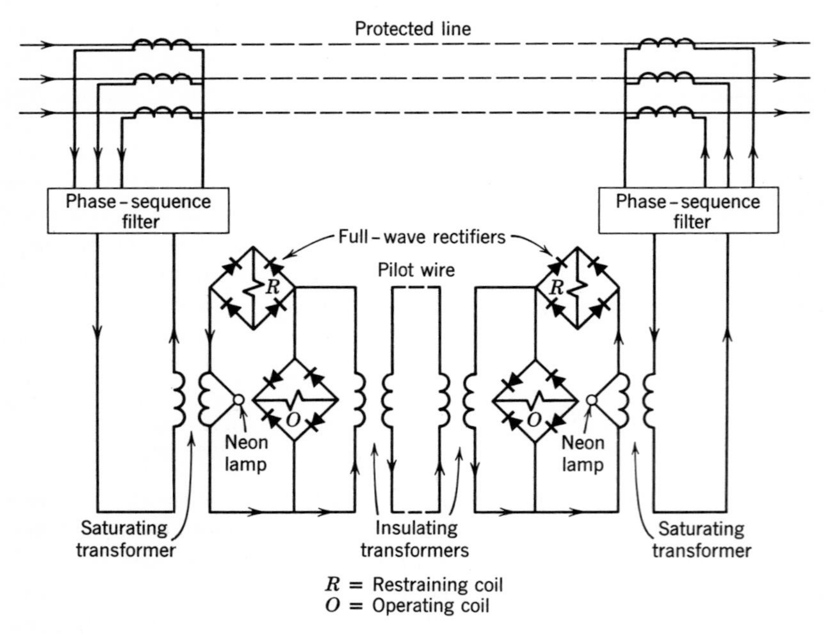

Fig. 6. Schematic connections of a circulating-current a-c wire-pilot relaying equipment.

"*O*" is an operating coil, and "*R*" is a restraining coil, the two coils acting in opposition on the armature of the polarized relay. These coils are energized from full-wave rectifiers. Here, a d-c directional relay is being used with rectified a-c quantities to get high sensitivity.

Although this relay is fundamentally a directional type, it is in effect a very sensitive current-balance relay. Phase-sequence filters convert the three phase and ground currents to a single-phase quantity. Saturating transformers limit the magnitude of the rms voltage impressed on the pilot circuit, and the neon lamps limit the peak voltages. Insulating transformers at the ends of the pilot insulate the terminal equipment from the pilot circuit for reasons that will be given later.

This equipment is capable of tripping the breakers at both ends of a line for an internal fault with current flowing at only one end. Whether tripping at both ends will actually occur will depend on the magnitude of the short-circuit current and on the impedance of the pilot circuit. This will be evident from an examination of Fig. 6 where, at the end where no short-circuit current flows, the operating coil and the pilot are in series, and this series circuit is in parallel with the operating coil at the other end. In other words, at the end where fault current flows, the current from the phase-sequence filter divides between the two operating coils, the larger portion going through the local coil. If the pilot impedance is too high, insufficient current will flow through the coil at the other end to cause tripping there.

Charging current between the pilot wires will tend to make the equipment less sensitive to internal faults, acting somewhat like a short circuit between the pilot wires, but with impedance in the short circuit.

OPPOSED-VOLTAGE TYPE

An example of an opposed-voltage type of equipment is shown schematically in Fig. 7.[1] The relay at each end of the pilot is an a-c directional-type relay having in effect two directional elements with a common polarizing source, the two directional elements acting in opposition. Except for the effect of phase angle, this is equivalent to a very sensitive balance-type relay. The "mixing" transformer at each end provides a single-phase quantity for all types of faults. Saturation in the mixing transformer limits the rms magnitude of the voltage that is impressed on the pilot circuit. The impedance of the circuit connected across the mixing transformer is low enough to limit the magnitude of peak voltages to acceptable values.

The equipment illustrated in Fig. 7 requires enough restraint to overcome a tendency to trip for charging current between the pilot wires, although the angle of maximum torque of the operating directional element is such that it minimizes this tripping tendency.

The equipment will not trip the breakers at both ends of a line for an internal fault if current flows into the line at only one end; it will trip only the end where there is fault current flowing. Current will circulate through the operating and restraining coils at the other

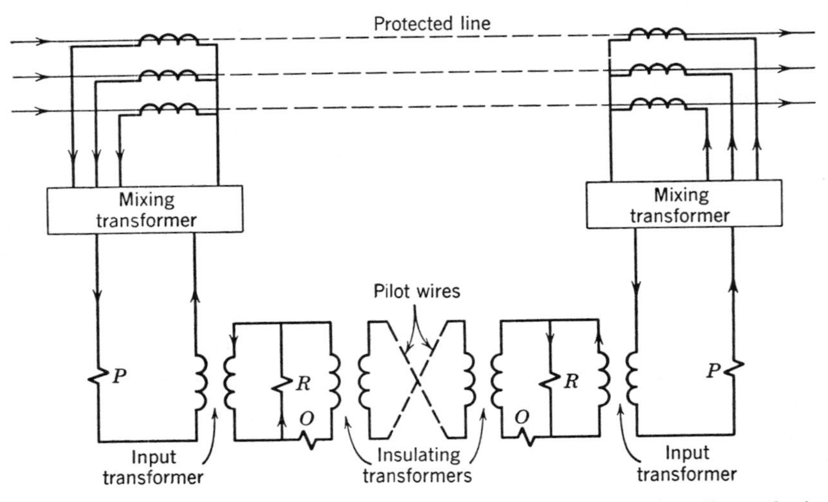

Fig. 7. Schematic connections of an opposed-voltage a-c wire-pilot relaying equipment. P = current polarizing coil; R = voltage restraining coil; O = current operating coil.

end, but there will be insufficient current in the polarizing coil at that end to cause operation there. This characteristic is seldom objectionable, and it has the compensating advantage of preventing undesired tripping because of induced pilot currents.

ADVANTAGES OF A-C OVER D-C WIRE-PILOT EQUIPMENTS

Certain problems described in connection with d-c wire-pilot relaying are not associated with the a-c type. Since separate blocking and tripping relays are not used, the problem of different levels of blocking and tripping sensitivity are avoided. Also, the problems associated with contact racing and ground preference do not exist. Moreover, a-c wire-pilot relaying is inherently immune to power swings or loss of synchronism. In view of the simplifications permitted by the elimination of these problems, one can understand why a-c wire-pilot relaying has largely superseded the d-c type.

LIMITATIONS OF A-C WIRE-PILOT EQUIPMENTS

Both the circulating-current and the opposed-voltage types that have been described are not always applicable to tapped or multiterminal lines, because both types use saturating transformers to limit the magnitudes of the pilot-wire current and voltage. The non-linear relation between the magnitudes of the power-system current and the output of the saturating transformer prevents connecting more than two equipments in series in a pilot-wire circuit except under certain restricted conditions. Since this subject involves so many details of different possible system conditions and ranges of adjustment of specific relaying equipments, it is impractical to discuss it further here. In general, the manufacturer's advice should be obtained before attempting to apply such a-c wire-pilot-relaying equipments to tapped or multiterminal lines.

SUPERVISION OF PILOT-WIRE CIRCUITS

Manual equipment is available for periodically testing the pilot circuit, and automatic equipment is available for continuously supervising the pilot circuit. The manual equipment provides means for measuring the pilot-wire quantities and the contribution from the ends. The automatic equipment superimposes direct current on the pilot circuit; trouble in the pilot circuit causes either an increase or a decrease in the d-c supervising current, which is detected by sensitive auxiliary relays.[6] The automatic equipment can be arranged not only to sound an alarm when the pilot wires become open circuited or short circuited but also to open the trip circuit so as to avoid undesired tripping; in such cases, it may be necessary to delay tripping slightly.

REMOTE TRIPPING OVER THE PILOT WIRES

Should it be desired to trip the remote breaker under any circumstance, it can be done by superimposing direct current on the pilot circuit. If automatic supervising equipment is in use, the magnitude of the d-c voltage imposed momentarily on the circuit for remote tripping is higher than that of the continuous voltage used for supervising purposes.[7] Parts of the automatic supervising equipment may be used in common for both purposes. A disadvantage of this method of remote tripping is the possibility of undesired tripping if, during testing, one inadvertently applies a d-c test voltage to the pilot wires. To avoid this, "tones" have been used over a separate pilot.

PILOT-WIRE REQUIREMENTS

Because pilot-wire circuits are often rented from the local telephone company, and because the telephone company imposes certain restrictions on the current and voltage applied to their circuits, these restrictions effectively govern wire-pilot-relaying-equipment design. The a-c equipments that have been described are suitable for telephone circuits since they impose no more than the permissible current and voltage on the pilot, and the wave forms are acceptable to the telephone companies.[8]

The equipments that have been described operate without special adjustment over pilot wires having as much as approximately 2000 ohms d-c loop resistance and 1.5 microfarads distributed shunt capacitance. However, one should determine these limitations in any application.

PILOT WIRES AND THEIR PROTECTION AGAINST OVERVOLTAGES

The satisfactory operation of wire-pilot relaying equipment depends primarily on the reliability of the pilot-wire circuit.[3] Protective-relaying requirements are generally more exacting than the requirements of any other service using pilot circuits. The ideal pilot circuit is one that is owned by the user and is constructed so as not to be exposed to lightning, mutual induction with other pilot or power circuits, differences in station ground potential, or direct contact with any power conductor. However, satisfactory operation can generally be obtained where these ideals are not entirely realized, if proper countermeasures are used.

The conventional a-c wire-pilot relaying equipments that have been described tolerate only about 5 to 15 volts induced between the two wires in the pilot loop. For this reason, the pilot wires should be a twisted pair if the mutual induction is high. For moderate induction, wires in spiraled quads will often suffice if the other pair in the quad will not carry high currents. In addition to other useful information, Reference 4 of the Bibliography contains a method for calculating voltages caused by mutual induction.

If supervising or remote-tripping equipment is not used, or, in other words, if there are no terminal-equipment connections to the pilot wires on the pilot-wire side of the insulating transformer, it is only a question of whether the insulating transformer and the pilot wires can withstand the voltage to ground that they will get from mutual induction and from differences in station ground potentials. The insulating transformers can generally be expected to have sufficient insulation, and only the pilot wires need to be critically examined. But if supervising equipment is involved, or if the pilot wires may otherwise be grounded at one end and do not have sufficient insulation, additional means, including neutralizing transformers, may be required to protect personnel or equipment.[3,4,5,8]

Pilot wires exposed to lightning overvoltages must be protected with lightning arresters. Similarly, pilot wires exposed to contact with a power circuit must be protected.

The subject of pilot-wire protection has too many ramifications to do justice to it here. The Bibliography gives references to much useful information on the subject. In general, the manufacturer of the relaying equipment should be consulted, and also the local telephone company, if a telephone circuit is to be used. The subject is complicated by the fact that it is necessary not only to protect the equipment or personnel from harm but also, in so doing, to do nothing that will interfere with the proper functioning of the relaying equipment. Such things as mutual induction, difference in station ground potentials, and lightning overvoltages generally occur when there is a fault on the protected line or in the immediate vicinity, at just the time when the proper operation of the relaying equipment is required.

Bibliography

1. "An Improved A-c Pilot-Wire Relay," by J. H. Neher and A. J. McConnell, *AIEE Trans., 60* (1941), pp. 12–17. Discussions, pp. 638–643.

2. "A Single-Element Differential Pilot-Wire Relay System," by E. L. Harder and M. A. Bostwick, *Elec. J., 35,* No. 11 (Nov., 1938), pp. 443–448.

3. "Pilot-Wire Circuits for Protective Relaying—Experience and Practice 1942–1950," by AIEE Committee, *AIEE Trans., 72,* Part III (1953), pp. 331–336. Discussions, p. 336.

4. "Protection of Pilot-Wire Circuits," by E. L. Harder and M. A. Bostwick, *AIEE Trans., 61* (1942), pp. 645–652. Discussions, pp. 996–998.

5. "Protection of Pilot Wires from Induced Potentials," by R. B. Killen and G. G. Law, *AIEE Trans., 65* (1946), pp. 267–270.

"Neutralizing Transformers," *EEI Engineering Report No. 44, Publ. H-12.*

"Pilot-Wire Relay Protection," by E. E. George and W. R. Brownlee, *AIEE Trans., 54* (1935), pp. 1262–1269. Discussions, *55* (1936), pp. 907–909.

"Neutralizing Transformer to Protect Power Station Communication," by E. E. George, R. K. Honaman, L. L. Lockrow, and E. L. Schwartz, *AIEE Trans., 55* (1936), pp. 524–529.

"Dependable Pilot-Wire Relay Operation," by M. A. Bostwick, *AIEE Trans., 72,* Part III (1953), pp. 1073–1076. Discussions, pp. 1076–1077.

6. "Supervisory Circuit Checks Relay System," by R. M. Smith, *Elec. World, 115* (May 3, 1941), p. 1507.

7. "Supervisory Circuit Performs Double Duty," by M. A. Bostwick, *Elec. World, 115* (June 28, 1941), p. 223*b*.

8. "Protection of Wire Communication Facilities Serving Power Stations and Substations," by T. W. Alexander, Jr., *AIEE Trans., 72,* Part I (1953), pp. 587–591.

6 CARRIER-CURRENT-PILOT AND MICROWAVE-PILOT RELAYS

Chapter 5 introduced the subject of pilot relaying, gave the fundamental principles involved, and described some typical wire-pilot relaying equipments. In this chapter, we shall deal with carrier-current-pilot and microwave-pilot relaying; for either type of pilot, the relay equipment is the same. Two types of relay equipment will be described, the "phase-comparison" type, which is much like the a-c wire-pilot types, and the "directional-comparison" type, which is similar to the d-c wire-pilot types.

The Carrier-Current Pilot

It is not necessary for one to understand the details of carrier-current transmitters or receivers in order to understand the fundamental relaying principles. All one needs to know is that when a voltage of positive polarity is impressed on the control circuit of the transmitter, it generates a high-frequency output voltage. In the United States, the frequency range allotted for this purpose is 30 to 200 kc. This output voltage is impressed between one phase conductor of the transmission line and the earth, as shown schematically in Fig. 1.

Each carrier-current receiver receives carrier current from its local transmitter as well as from the transmitter at the other end of the line. In effect, the receiver converts the received carrier current into a d-c voltage that can be used in a relay or other circuit to perform any desired function. This voltage is zero when carrier current is not being received.

"Line traps" shown in Fig. 1 are parallel resonant circuits having negligible impedance to power-frequency currents, but having very high impedance to carrier-frequency currents. Traps are used to keep the carrier currents in the desired channel so as to avoid interference with or from other adjacent carrier-current channels, and also to avoid loss of the carrier-current signal in adjoining power circuits for any reason whatsoever, external short circuits being a principal reason. Consequently, carrier current can flow only along the line section between the traps.

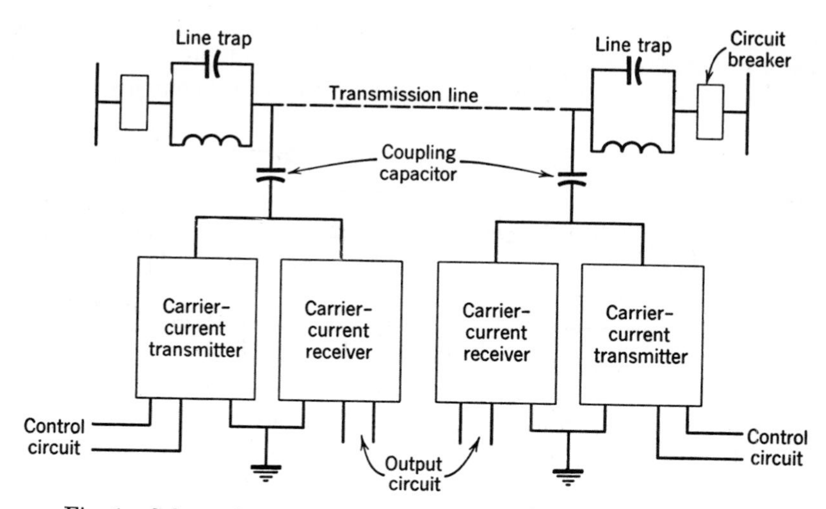

Fig. 1. Schematic illustration of the carrier-current-pilot channel.

The Microwave Pilot

The microwave pilot is an ultra-high-frequency radio system operating in allotted bands above 900 megacycles in the United States. The transmitters are controlled in the same manner as carrier-current transmitters, and the receivers convert the received signals into d-c voltage as carrier-current receivers do. With the microwave pilot, line coupling and trapping are eliminated, and, instead, line-of-sight antenna equipment is required.

The following descriptions of the relaying equipments assume a carrier-current pilot, but the relay equipment and its operation would be the same if a microwave pilot were used.

Phase-Comparison Relaying

Phase-comparison relaying equipment uses its pilot to compare the phase relation between current entering one terminal of a transmission-line section and leaving another. The current magnitudes are not compared. Phase-comparison relaying provides only primary protection; back-up protection must be provided by supplementary relaying equipment.

Figure 2 shows schematically the principal elements of the equipment at both ends of a two-terminal transmission line, using a carrier-current pilot. As in a-c wire-pilot relaying, the transmission-line current transformers feed a network that transforms the CT output currents into a single-phase sinusoidal output voltage. This voltage is applied to a carrier-current transmitter and to a "comparer." The

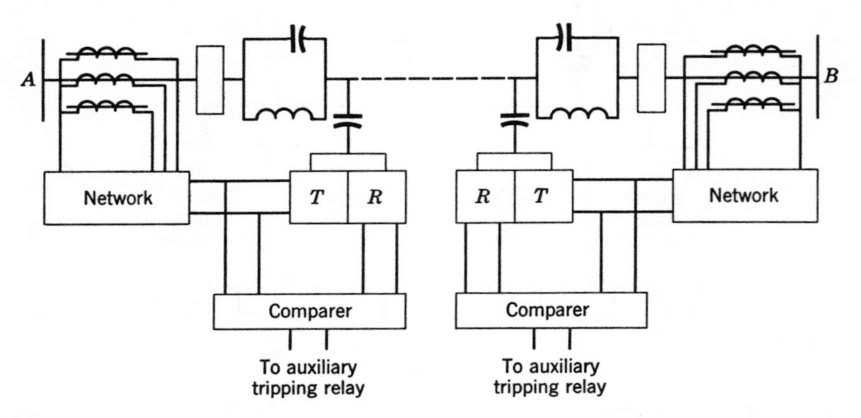

Fig. 2. Schematic representation of phase-comparison carrier-current-pilot-relaying equipment. T = carrier-current transmitter; R = carrier-current receiver.

output of a carrier-current receiver is also applied to the comparer. The comparer controls the operation of an auxiliary relay for tripping the transmission-line circuit breaker. These elements provide means for transmitting and receiving carrier-current signals for comparing at each end the relative phase relations of the transmission-line currents at both ends of the line.

Let us examine the relations between the network output voltages at both ends of the line and also the carrier-current signals that are transmitted during external and internal fault conditions. These relations are shown in Fig. 3. It will be observed that for an external fault at D, the network output voltages at stations A and B (waves a and c) are 180° out of phase; this is because the current-transformer connections at the two stations are reversed. Since an a-c voltage is used to control the transmitter, carrier current is transmitted only during the half cycles of the voltage wave when the polarity is positive. The carrier-current signals transmitted from A and B (waves b and d) are displaced in time, so that there is always a carrier-current signal being sent from one end or the other. However, for the internal fault at C, owing to the reversal of the network output voltage at station B caused by the reversal of the power-line currents there, the carrier-current signals (waves b and f) are concurrent, and there is no signal from either station every other half cycle.

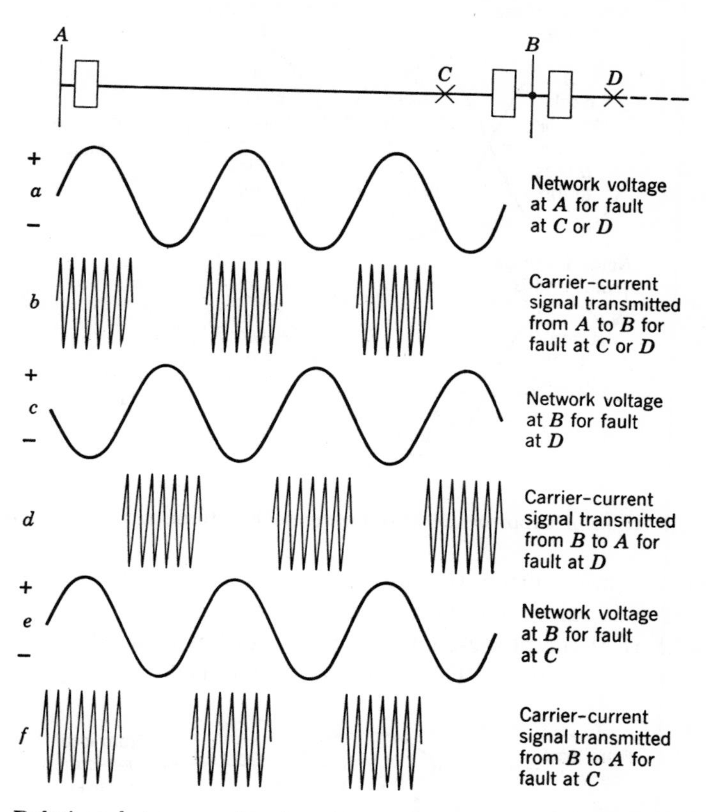

Fig. 3. Relations between network output voltages and carrier-current signals.

Phase-comparison relaying acts to block tripping at both terminals whenever the carrier-current signals are displaced in time so that there is little or no time interval when a signal is not being transmitted from one end or the other. When the carrier-current signals are approximately concurrent, tripping will occur wherever there is sufficient short-circuit current flowing. This is illustrated in Fig. 4 where the network output voltages are superimposed, and the related tripping and blocking tendencies are shown. As indicated in Figs. 3 and 4, the equipment at one station transmits a blocking carrier-current signal during one half cycle, and then stops transmitting and tries to trip during the next half cycle; if carrier current is not received from the other end of the line during this half cycle, the equipment operates to trip its breaker. But, if carrier current is received from the other

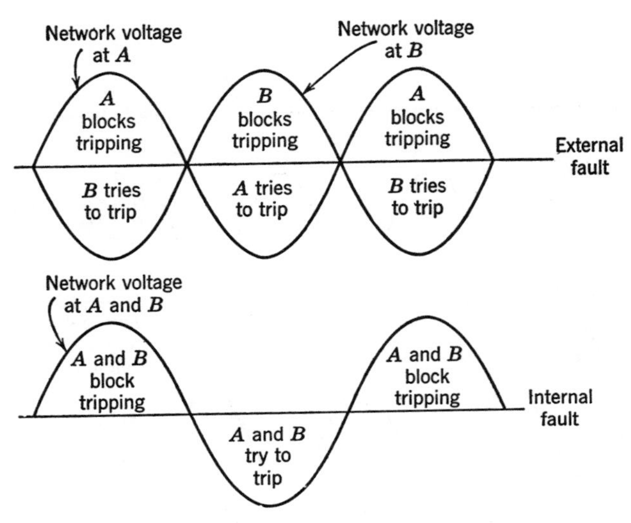

Fig. 4. Relation of tripping and blocking tendencies to network output voltages.

end of the line during the interval when the local carrier-current transmitter is idle, tripping does not occur.

The heart of the phase-comparison system lies in what is sometimes called the "comparer." The comparer of one type of relaying equip-

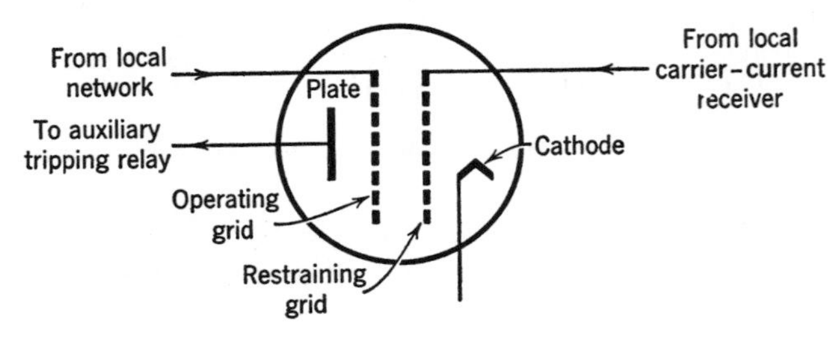

Fig. 5. Schematic representation of the comparer.

ment, shown schematically in Fig. 5, is a vacuum-tube equipment at each end of the line, which is here represented as a single tube. When voltage of positive polarity is impressed on the "operating" grid by the local network, the tube conducts if voltage of negative polarity is not concurrently impressed on the "restraining" grid by the local carrier-current receiver by virtue of carrier current received from the other end of the line. When the tube conducts, an auxiliary tripping relay picks up and trips the local breaker. Positive polarity is impressed on the operating grid of the local comparer during the negative half cycle of the network output of Fig. 3 when the local transmitter is idle. Therefore, the local transmitter cannot block local tripping. The voltage from the carrier-current receiver impressed on the restraining grid makes the tube non-conducting, whether the operating grid is energized or not, whenever carrier current is being received.

It is not necessary that the carrier-current signals be exactly interspersed to block tripping, nor must they be exactly concurrent to permit tripping. For blocking purposes, a phase shift of the order of 35° either way from the exactly interspersed relation can be tolerated. Considerably more phase shift can be tolerated for tripping purposes. It is necessary that more phase shift be permissible for tripping purposes because more phase shift is possible under tripping conditions than under blocking conditions. Phase shift under blocking conditions (i.e., when an external fault occurs) is caused by the small angular difference between the currents at the ends of the line, owing to the line-charging component of current, and also by the length of time it takes for the carrier-current signal to travel from one end of the line to the other, which is approximately at the speed of light. In a 60-cycle system, this travel time accounts for about 12° phase shift per 100 miles of line; it can be compensated for by shifting the phase of the voltage supplied by the network to the comparer by the same amount. No compensation can be provided for the charging-current effect, but this phase shift is negligible except with very long lines. The major part of the phase shift under tripping conditions (i.e., when an internal fault occurs) is caused by the generated voltages beyond the ends of the line being out of phase, and also by a different distribution of ground-fault currents between the two ends as compared with the distribution of phase-fault currents (as, for example, if the main source of generation is at one end of the line and the main ground-current source is at the other end); in addition, the travel time of the carrier-current signal is also a factor.

The principle of different levels of blocking and tripping sensitivity, described in connection with d-c wire-pilot relaying, applies also to phase-comparison pilot relaying. So-called "fault detectors," which may be overcurrent or distance relays, are employed to establish these two sensitivity levels. It is desirable that carrier current not be transmitted under normal conditions, to conserve the life of the vacuum tubes, and also to make the pilot available for other uses when not required by the relaying equipment. Consequently, one set of fault detectors is adjusted to pick up somewhat above maximum load current, to permit the transmission of carrier current. The other set of fault detectors picks up at still higher current, to permit tripping if called for by the comparer. The required pickup adjustment of these tripping fault detectors might be considerably higher for tapped-line applications; this will be treated in more detail when we consider the application of relays for transmission-line protection. Tripping for an internal fault will occur only at the ends of a line where sufficient short-circuit current flows to pick up the tripping fault detectors.

It will be evident from the foregoing that the phase-comparison pilot is a blocking pilot, since a pilot signal is not required to permit tripping. Without the agency of the pilot, phase-comparison relaying reverts to high-speed non-directional overcurrent relaying. Failure of the pilot will not prevent tripping, but tripping will not be selective under such circumstances; that is, undesired tripping may occur. A short circuit on the protected line between ground and the conductor to which the carrier-current equipment is coupled will not interfere with desired tripping, because carrier-current transmission is not required to permit tripping; external faults, being on the other side of a line trap, will not affect the proper transmission of carrier current when it is required.

Phase-comparison relaying is inherently immune to the effects of power surges or loss of synchronism between sources of generation beyond the ends of a protected line. Similarly, currents flowing in a line because of mutual induction from another nearby circuit will not affect the operation of the equipment. In both of these situations, the currents merely flow through the line as to an external load or to an external short circuit.

Directional-Comparison Relaying

Modern relaying equipment of the directional-comparison type operates in conjunction with distance relays because the distance relays will provide back-up protection, and because certain elements of the distance relays can be used in common with the directional-comparison equipment. However, for our immediate purposes, we shall consider only those elements that are essential to directional-comparison relaying.

With directional-comparison relaying, the pilot informs the equipment at one end of the line how a directional relay at the other end responds to a short circuit. Normally, no pilot signal is transmitted from any terminal. Should a short circuit occur in an immediately adjacent line section, a pilot signal is transmitted from any terminal where short-circuit current flows out of the line (i.e., in the non-tripping direction). While any station is transmitting a pilot signal, tripping is blocked at all other stations. But should a short circuit occur on the protected line, no pilot signal is transmitted and tripping occurs at any terminal where short-circuit current flows. Therefore, the pilot is a blocking pilot, since the reception of a pilot signal is not required to permit tripping.

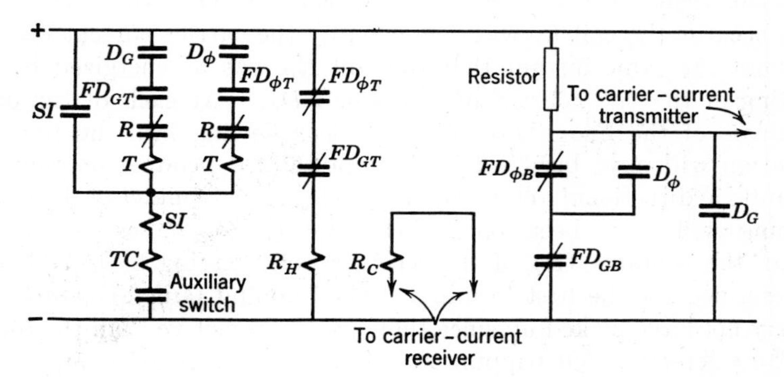

Fig. 6. Schematic diagram of essential contact circuits of directional-comparison relaying equipment. SI = seal-in relay; D_G = directional ground relay; $D\phi$ = directional phase relay; FD_{GT} = ground tripping fault-detector relay; $FD\phi_T$ = phase tripping fault-detector relay; R = receiver relay; R_H = d-c holding coil; R_C = carrier-current coil; T = target; TC = trip coil; FD_{GB} = ground blocking fault-detector relay; $FD\phi_B$ = phase blocking fault-detector relay.

The pilot signal is steady once it is started, and not every other half cycle as in phase-comparison relaying.

The essential relay elements at each end of a line are shown schematically in Fig. 6 for one type of equipment. With two exceptions, all the contacts are shown in the position that they take under normal conditions; the exceptions are that the receiver-relay contacts (R) are open because the receiver-relay holding coil (R_H) is energized normally, and the circuit-breaker auxiliary switch is closed when the breaker is closed. The phase directional-relay contacts (D_ϕ) may be closed or not, depending on the direction in which load current is flowing.

Now, let us assume that a short circuit occurs in an adjoining line back of the end where the equipment of Fig. 6 is located. If the magnitude of the short-circuit current is high enough to operate a blocking fault detector ($FD_{\phi B}$ for a phase fault or FD_{GB} for a ground fault), the operation of this fault detector opens the connection from the negative side of the d-c bus to the control circuit of the carrier-current transmitter. The polarity of this connection then becomes positive, owing to the connection through the resistor to the positive side of the d-c bus, and the carrier-current transmitter transmits a signal to block tripping at the other terminals of the line. There is no tendency to trip at this terminal because the current is flowing in the direction to open the directional-relay contacts (D_G or D_ϕ) in the tripping circuit, even though a tripping fault detector (FD_{GT} or $FD_{\phi T}$) may have operated. Moreover, the receiver relay contacts (R) will have stayed open because the coil R_C was energized by the carrier-current receiver at about the same instant that the coil R_H was de-energized by the opening of the "*b*" contact of FD_{GT} or $FD_{\phi T}$. At each of the other terminals of the line where the current is flowing into the line, the operation will have been similar, except that, depending on the type of fault, a directional relay will have closed its contacts. However, tripping will have been blocked by receipt of the carrier-current signal, the contacts (R) of the receiver relay having been held open as described for the first terminal. The tripping fault detectors may or may not have picked up since they are less sensitive than the blocking fault detectors, but tripping would have been blocked in any event. The operation of a blocking fault detector at one of these other terminals may have started carrier transmission from that terminal, but it would have been immediately stopped by the operation of a directional relay.

For a short circuit on the protected line, the directional relays at all terminals where short-circuit current flows will close their contacts, thereby stopping carrier transmission as soon as it is started by the blocking fault detectors. With no carrier signal to block tripping, all terminals will trip where there is sufficient fault current to pick up a tripping fault detector.

The directional-ground relay can stop carrier-current transmission whether it was started by either the phase blocking fault detector or the ground blocking fault detector, but the directional phase relay can stop transmission only if it was started by the phase blocking fault detector. This illustrates how "ground preference" is obtained if desired. The principle of ground preference is used when a directional phase relay is apt to operate incorrectly for a ground fault. Ground preference is not required if distance-type phase-fault detectors are used.

Figure 6 shows only the contacts of the phase relays of one phase. In the tripping and carrier-stopping circuits, the contact circuits for the other two phases would be in parallel with those shown. In the receiver-relay d-c holding-coil circuit and in the carrier-starting circuit, the contacts would be in series.

A feature that contributes to high-speed operation is the "normally blocked trip circuit." As shown in Fig. 6, this feature consists of providing the carrier-current receiver relay with a second coil (R_H), which, when energized, holds the receiver-relay contact open as when carrier current is being received. This auxiliary coil is normally energized through a series circuit consisting of a "*b*" contact on each tripping fault-detector relay. In earlier equipments without the normally blocked trip circuit, the reception of carrier current had to open the receiver-relay "*b*" contact before a tripping fault detector could close its "*a*" contact, and this race required a certain time delay in the tripping-fault-detector operation to avoid undesired tripping. With the normally blocked trip circuit, the receiver-relay contact is held open normally by the R_H coil; and, when a fault occurs, carrier-current transmission is started and the R_C coil is energized at approximately the same time that the R_H coil is de-energized. Thus, the flux keeping the relay picked up does not have time to change. Therefore, the tripping fault detector can be as fast as possible, and there is no objectionable contact race.

The term "intermittent," as contrasted with "continuous," identifies a type of pilot in which the transmission of a pilot signal occurs only when short circuits occur. A continuous-type pilot would not require the normally blocked trip circuit, but it would have the same disadvantage as a tripping pilot because there would be no way to stop the transmission of the pilot signal at a station where the breaker was closed and where there was no flow of short-circuit current for an internal fault. Therefore, it is evident that the directional-comparison pilot is of the intermittent type. As such, it has the same desirable features, described for the phase-comparison pilot, of conserving the life of vacuum tubes and of permitting other uses to be made of the pilot when not required by the relaying equipment.

The blocking-fault-detector function may be directional or not, but the tripping-fault-detector function must be directional. In other words, a carrier signal may be started at a given station whenever a short circuit occurs either in the protected line or beyond its ends, and may then be stopped immediately if the current at that station is in the tripping direction; or the carrier signal may be started only if the current is in the non-tripping direction. The phase-fault detectors are distance-type relays. When mho-type distance relays are used, the directional function is inherently provided, and the separate directional relays of Fig. 6 are not required. Overcurrent and directional relays are used for ground-fault detectors.

Directional-comparison relaying requires supplementary equipment to prevent tripping during severe power surges or when loss of synchronism occurs. In a later chapter we shall see what loss of synchronism "looks" like to protective relays, and how it is possible to differentiate between such a condition and a short circuit.

The ground-relaying portion of directional-comparison equipment is apt to cause undesired tripping because of mutual induction during ground faults on certain arrangements of closely paralleled power lines. The remedy for this tendency is described in a later chapter where the effects of mutual induction are described.

Looking Ahead

We have now completed our examination of the operating principles and characteristics of several types of commonly used protective-relaying equipments. Much more could have been said of present-day relays that might be helpful to one who intends to pursue this subject further. However, an attempt has been made to present the essential information as briefly as possible so as not to interfere with the continuity of the material. There are many more types of protective relays, some of which will be described later in connection with specific applications. However, these are merely those basic types that we have considered, but arranged in a slightly different way.

We are not yet ready to study the application of the various relays. We have learned how various relay types react to the quantities that actuate them. We must still know how to derive these actuating quantities and how they vary under different system-operating conditions. If one is able to ascertain the difference in these quantities between a condition for which relay operation is required and all other possible conditions for which a relay must not operate, he can then employ a particular relay, or a combination of relays with certain connections, that also can recognize the difference and operate accordingly.

Because protective relays receive their actuating quantities through the medium of current and voltage transformers, and because the connections and characteristics of these transformers have an important bearing on the response of protective relays, these transformers will be our next consideration.

Bibliography

"A New Carrier Relaying System," by T. R. Halman, S. L. Goldsborough, H. W. Lensner, and A. F. Drompp, *AIEE Trans., 63* (1944), pp. 568–572. Discussions, pp. 1423–1426.

"Phase-Comparison Carrier-Current Relaying," by A. J. McConnell, T. A. Cramer, and H. T. Seeley, *AIEE Trans., 64* (1945), pp. 825–832. Discussions, pp. 973–975.

"A Phase-Comparison Carrier-Current-Relaying System for Broader Application," by N. O. Rice and J. S. Smith, *AIEE Trans., 71,* Part III (1952), pp. 246–249. Discussions, p. 250.

7 CURRENT TRANSFORMERS

Protective relays of the a-c type are actuated by current and voltage supplied by current and voltage transformers. These transformers provide insulation against the high voltage of the power circuit, and also supply the relays with quantities proportional to those of the power circuit, but sufficiently reduced in magnitude so that the relays can be made relatively small and inexpensive.

The proper application of current and voltage transformers involves the consideration of several requirements, as follows: mechanical construction, type of insulation (dry or liquid), ratio in terms of primary and secondary currents or voltages, continuous thermal rating, short-time thermal and mechanical ratings, insulation class, impulse level, service conditions, accuracy, and connections. Application standards for most of these items are available.[1] Most of them are self-evident and do not require further explanation. Our purpose here and in Chapter 8 will be to concentrate on *accuracy* and *connections* because these directly affect the performance of protective relaying, and we shall assume that the other general requirements are fulfilled.

The accuracy requirements of different types of relaying equipment differ. Also, one application of a certain relaying equipment may have more rigid requirements than another application involving the same type of relaying equipment. Therefore, no general rules can be given for all applications. Technically, an entirely safe rule would be to use the most accurate transformers available, but few would follow the rule because it would not always be economically justifiable.

Therefore, it is necessary to be able to predict, with sufficient accuracy, how any particular relaying equipment will operate from any given type of current or voltage source. This requires that one know how to determine the inaccuracies of current and voltage transformers under different conditions, in order to determine what effect these inaccuracies will have on the performance of the relaying equipment.

Methods of calculation will be described using the data that are published by the manufacturers; these data are generally sufficient. A problem that cannot be solved by calculation using these data should be solved by actual test or should be referred to the manufacturer. This chapter is not intended as a text for a CT designer, but as a generally helpful reference for usual relay-application purposes.

The methods of connecting current and voltage transformers also are of interest in view of the different quantities that can be obtained from different combinations. Knowledge of the polarity of a current or voltage transformer and how to make use of this knowledge for making connections and predicting the results are required.

Types of Current Transformers

All types of current transformers[1] are used for protective-relaying purposes. The bushing CT is almost invariably chosen for relaying in the higher-voltage circuits because it is less expensive than other types. It is not used in circuits below about 5 kv or in metal-clad equipment. The bushing type consists only of an annular-shaped core with a secondary winding; this transformer is built into equipment such as circuit breakers, power transformers, generators, or switchgear, the core being arranged to encircle an insulating bushing through which a power conductor passes.

Because the internal diameter of a bushing-CT core has to be large to accommodate the bushing, the mean length of the magnetic path is greater than in other CT's. To compensate for this, and also for the fact that there is only one primary turn, the cross section of the core is made larger. Because there is less saturation in a core of greater cross section, a bushing CT tends to be more accurate than other CT's at high multiples of the primary-current rating. At low currents, a bushing CT is generally less accurate because of its larger exciting current.

Calculation of CT Accuracy

Rarely, if ever, is it necessary to determine the phase-angle error of a CT used for relaying purposes. One reason for this is that the load on the secondary of a CT is generally of such highly lagging power factor that the secondary current is practically in phase with the exciting current, and hence the effect of the exciting current on the phase-angle accuracy is negligible. Furthermore, most relaying applications can tolerate what for metering purposes would be an intolerable phase-angle error. If the ratio error can be tolerated, the phase-angle error can be neglected. Consequently, phase-angle errors will not be discussed further. The technique for calculating the phase-angle error will be evident, once one learns how to calculate the ratio error.

Accuracy calculations need to be made only for three-phase- and single-phase-to-ground-fault currents. If satisfactory results are thereby obtained, the accuracy will be satisfactory for phase-to-phase and two-phase-to-ground faults.

CURRENT-TRANSFORMER BURDEN

All CT accuracy considerations require knowledge of the CT burden. The *external* load applied to the secondary of a current transformer is called the "burden." The burden is expressed preferably in terms of the impedance of the load and its resistance and reactance components. Formerly, the practice was to express the burden in terms of volt-amperes and power factor, the volt-amperes being what would be consumed in the burden impedance at rated secondary current (in other words, rated secondary current squared times the burden impedance). Thus, a burden of 0.5-ohm impedance may be expressed also as "12.5 volt-amperes at 5 amperes," if we assume the usual 5-ampere secondary rating. The volt-ampere terminology is no longer standard, but it needs defining because it will be found in the literature and in old data.

The term "burden" is applied not only to the total external load connected to the terminals of a current transformer but also to elements of that load. Manufacturers' publications give the burdens of individual relays, meters, etc., from which, together with the resistance of interconnecting leads, the total CT burden can be calculated.

The CT burden impedance decreases as the secondary current increases, because of saturation in the magnetic circuits of relays and other devices. Hence, *a given burden may apply only for a particular value of secondary current.* The old terminology of "volt-amperes at 5 amperes" is most confusing in this respect since it is not necessarily the actual volt-amperes with 5 amperes flowing, but is what the volt-amperes would be at 5 amperes if there were no saturation. Manufacturers' publications give impedance data for several values of overcurrent for some relays for which such data are sometimes required. Otherwise, data are provided only for one value of CT secondary current. If a publication does not clearly state for what value of

current the burden applies, this information should be requested. Lacking such saturation data, one can obtain it easily by test. At high saturation, the impedance approaches the d-c resistance. Neglecting the reduction in impedance with saturation makes it appear that a CT will have more inaccuracy than it actually will have. Of course, if such apparently greater inaccuracy can be tolerated, further refinements in calculation are unnecessary. However, in some applications neglecting the effect of saturation will provide overly optimistic results; consequently, it is safer always to take this effect into account.

It is usually sufficiently accurate to add series burden impedances arithmetically. The results will be slightly pessimistic, indicating slightly greater than actual CT ratio inaccuracy. But, if a given application is so borderline that vector addition of impedances is necessary to prove that the CT's will be suitable, such an application should be avoided.

If the impedance at pickup of a tapped overcurrent-relay coil is known for a given pickup tap, it can be estimated for pickup current for any other tap. The reactance of a tapped coil varies as the square of the coil turns, and the resistance varies approximately as the turns. At pickup, there is negligible saturation, and the resistance is small compared with the reactance. Therefore, it is usually sufficiently accurate to assume that the impedance varies as the square of the turns. The number of coil turns is inversely proportional to the pickup current, and therefore the impedance varies inversely approximately as the square of the pickup current.

Whether CT's are connected in wye or in delta, the burden impedances are always connected in wye. With wye-connected CT's the neutrals of the CT's and of the burdens are connected together, either directly or through a relay coil, except when a so-called "zero-phase-sequence-current shunt" (to be described later) is used.

It is seldom correct simply to add the impedances of series burdens to get the total, whenever two or more CT's are connected in such a way that their currents may add or subtract in some common portion of the secondary circuit. Instead, one must calculate the sum of the voltage drops and rises in the external circuit from one CT secondary terminal to the other for assumed values of secondary currents flowing in the various branches of the external circuit. The effective CT burden impedance for each combination of assumed currents is the calculated CT terminal voltage divided by the assumed CT secondary current. This effective impedance is the one to use, and it may be larger or smaller than the actual impedance which would apply if no other CT's were supplying current to the circuit.

If the primary of an auxiliary CT is to be connected into the secondary of a CT whose accuracy is being studied, one must know the impedance of the auxiliary CT viewed from its primary with its secondary short-circuited. To this value of impedance must be added the impedance of the auxiliary CT burden as viewed from the primary side of the auxiliary CT; to obtain this impedance, multiply the actual burden impedance by the square of the ratio of primary to secondary turns of the auxiliary CT. It will become evident that, with an auxiliary CT that steps up the magnitude of its current from primary to secondary, very high burden impedances, when viewed from the primary, may result.

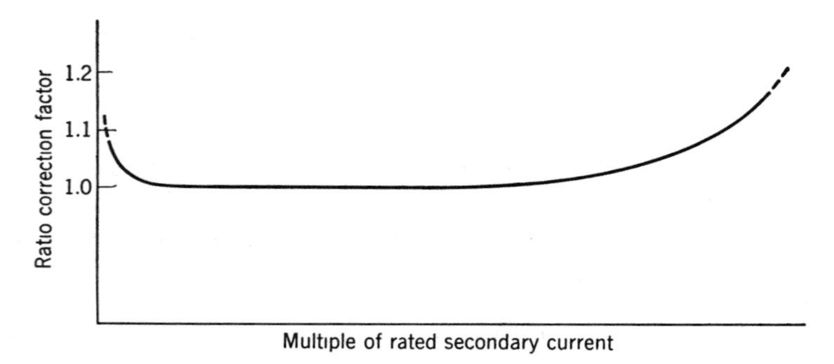

Fig. 1. Ratio-correction-factor curve of a current transformer.

RATIO-CORRECTION-FACTOR CURVES

The term "ratio-correction factor" is defined as "that factor by which the marked (or nameplate) ratio of a current transformer must be multiplied to obtain the true ratio."[1] The ratio errors of current transformers used for relaying are such that, for a given magnitude of primary current, the secondary current is less than the marked ratio would indicate; hence, the ratio-correction factor is greater than 1.0. A ratio-correction-factor curve is a curve of the ratio-correction factor plotted against multiples of rated primary or secondary current for a given constant burden, as in Fig. 1. Such curves give the most accurate results because the only errors involved in their use are the slight differences in accuracy between CT's having the same nameplate ratings, owing to manufacturers' tolerances. Usually, a family of such curves is provided for different typical values of burden.

To use ratio-correction-factor curves, one must calculate the CT

burden for each value of secondary current for which he wants to know the CT accuracy. Owing to variation in burden with secondary current because of saturation, no single RCF curve will apply for all currents because these curves are plotted for constant burdens; instead, one must use the applicable curve, or interpolate between curves, for each different value of secondary current. In this way, one can calculate the primary currents for various assumed values of secondary current; or, for a given primary current, he can determine, by trial and error, what the secondary current will be.

The difference between the actual burden power factor and the power factor for which the RCF curves are drawn may be neglected because the difference in CT error will be negligible. Ratio-correction-factor curves are drawn for burden power factors approximately like those usually encountered in relay applications, and hence there is usually not much discrepancy. Any application should be avoided where successful relay operation depends on such small margins in CT accuracy that differences in burden power factor would be of any consequence.

Extrapolations should not be made beyond the secondary current or burden values for which the RCF curves are drawn, or else unreliable results will be obtained.

Ratio-correction-factor curves are considered standard application data and are furnished by the manufacturers for all types of current transformers.

CALCULATION OF CT ACCURACY USING A SECONDARY-EXCITATION CURVE[2]

Figure 2 shows the equivalent circuit of a CT. The primary current is assumed to be transformed perfectly, with no ratio or phase-angle error, to a current I_P/N, which is often called "the primary current referred to the secondary." Part of the current may be considered consumed in exciting the core, and this current (I_e) is called "the secondary-excitation current." The remainder (I_s) is the true secondary current. It will be evident that the secondary-excitation current is a function of the secondary-excitation voltage (E_s) and the secondary-excitation impedance (Z_e). The curve that relates E_s and I_e is called "the secondary-excitation curve," an example of which is shown in Fig. 3. It will also be evident that the secondary current is a function of E_s and the total impedance in the secondary circuit. This total impedance is composed of the effective resistance and the leakage reactance of the secondary winding and the impedance of the burden.

Figure 2 shows also the primary-winding impedance, but this impedance does not affect the ratio error. It affects only the magnitude of current that the power system can pass through the CT primary,

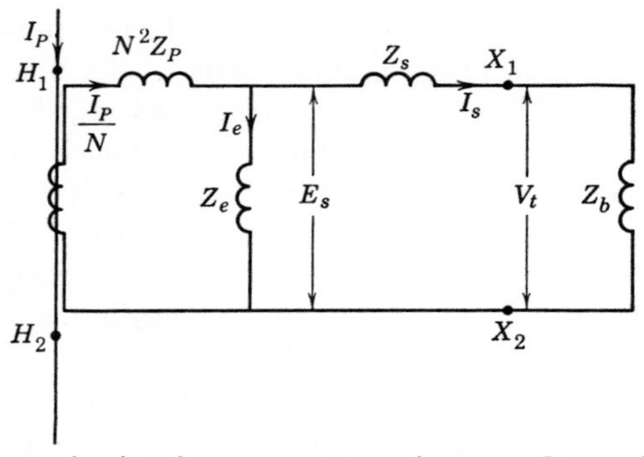

Fig. 2. Equivalent circuit of a current transformer. I_P = primary current in rms amperes; N = ratio of secondary to primary turns; Z_p = primary-winding impedance in ohms; I_e = secondary-excitation current in rms amperes; Z_e = secondary-excitation impedance in ohms; E_s = secondary-excitation voltage in rms volts; Z_s = secondary-winding impedance in ohms; I_s = secondary current in rms amperes; V_t = secondary terminal voltage in rms volts; Z_b = burden impedance in ohms.

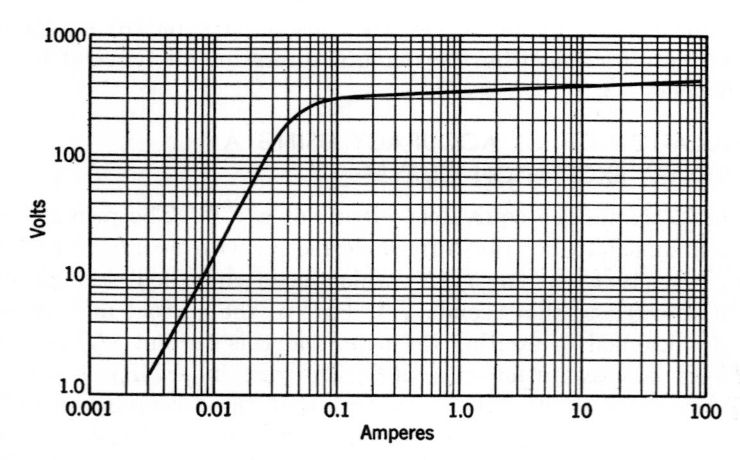

Fig. 3. Secondary-excitation characteristic. Frequency, 60; internal resistance, 1.08 ohms; secondary turns, 240.

and is of importance only in low-voltage circuits or when a CT is connected in the secondary of another CT.

If the secondary-excitation curve and the impedance of the secondary winding are known, the ratio accuracy can be determined for any

burden. It is only necessary to assume a magnitude of secondary current and to calculate the total voltage drop in the secondary winding and burden for this magnitude of current. This total voltage drop is equal numerically to E_s. For this value of E_s, the secondary-excitation curve will give I_e. Adding I_e to I_s gives I_P/N, and multiplying I_P/N by N gives the value of primary current that will produce the assumed value of I_s. The ratio-correction factor will be I_P/NI_s. By assuming several values of I_s and obtaining the ratio-correction factor for each, one can plot a ratio-correction-factor curve. It will be noted that adding I_s arithmetically to I_e may give a ratio-correction factor that is slightly higher than the actual value, but the refinement of vector addition is considered to be unnecessary.

The secondary resistance of a CT may be assumed to be the d-c resistance if the effective value is not known. The secondary leakage reactance is not generally known except to CT designers; it is a variable quantity depending on the construction of the CT and on the degree of saturation of the CT core. Therefore, the secondary-excitation-curve method of accuracy determination does not lend itself to general use except for bushing-type, or other, CT's with completely distributed secondary windings, for which the secondary leakage reactance is so small that it may be assumed to be zero. In this respect, one should realize that, even though the total secondary winding is completely distributed, tapped portions of this winding may not be completely distributed; to ignore the secondary leakage reactance may introduce significant errors if an undistributed tapped portion is used.

The secondary-excitation-curve method is intended only for current magnitudes or burdens for which the calculated ratio error is approximately 10% or less. When the ratio error appreciably exceeds this value, the wave form of the secondary-excitation current—and hence of the secondary current—begins to be distorted, owing to saturation of the CT core. This will produce unreliable results if the calculations are made assuming sinusoidal waves, the degree of unreliability increasing as the current magnitude increases. Even though one could calculate accurately the magnitude and wave shape of the secondary current, he would still have the problem of deciding how a particular relay would respond to such a current. Under such circumstances, the safest procedure is to resort to a test.

Secondary-excitation data for bushing CT's are provided by manufacturers. Occasionally, however, it is desirable to be able to obtain such data by test. This can be done accurately enough for all practical purposes merely by open-circuiting the primary circuit, applying a-c voltage of the proper frequency to the secondary, and measuring the current that flows into the secondary. The voltage should preferably be measured by a rectifier-type voltmeter. The curve of rms terminal voltage versus rms secondary current is approximately the secondary-excitation curve for the test frequency. The actual excitation voltage for such a test is the terminal voltage minus the voltage drop in the secondary resistance and leakage reactance, but this voltage drop is negligible compared with the terminal voltage until the excitation current becomes large, when the CT core begins to saturate. If a bushing CT with a completely distributed secondary winding is involved, the secondary-winding voltage drop will be due practically only to resistance, and corrections in excitation voltage for this drop can be made easily. In this way, sufficiently accurate data can be obtained up to a point somewhat beyond the knee of the secondary-excitation curve, which is usually all that is required. This method has the advantage of providing the data with the CT mounted in its accustomed place.

Secondary-excitation data for a given number of secondary turns can be made to apply to a different number of turns on the same CT by expressing the secondary-excitation voltages in "volts-per-turn" and the corresponding secondary-excitation currents in "ampere-turns." When secondary-excitation data are plotted in terms of volts-per-turn and ampere-turns, a single curve will apply to any number of turns.

The secondary-winding impedance can be found by test, but it is usually impractical to do so except in the laboratory. Briefly, it involves energizing the primary and secondary windings with equal and opposite ampere-turns, approximately equal to rated values, and measuring the voltage drop across the secondary winding.[3] This voltage divided by the secondary current is called the "unsaturated secondary-winding impedance." If we know the secondary-winding resistance, the unsaturated secondary leakage reactance can be calculated. If a bushing CT has secondary leakage flux because of an undistributed secondary winding, the CT should be tested in an enclosure of magnetic material that is the same as its pocket in the circuit breaker or transformer, or else most unreliable results will be obtained.

The most practical way to obtain the secondary leakage reactance may sometimes be to make an overcurrent ratio test, power-system current being used to get good wave form, with the CT in place, and with its secondary short-circuited through a moderate burden. The only difficulty of this method is that some means is necessary to measure the primary current accurately. Then, from the data obtained, and by using the secondary-excitation curve obtained as previously described, the secondary leakage reactance can be calculated. Such a calculation should be accurately made, taking into account the vector relations of the exciting and secondary currents and adding the secondary and burden resistance and reactance vectorially.

ASA ACCURACY CLASSIFICATION

The ASA accuracy classification[4] for current transformers used for relaying purposes provides a measure of a CT's accuracy. This method of classification assumes that the CT is supplying 20 times its rated secondary current to its burden, and the CT is classified on the basis of the maximum rms value of voltage that it can maintain at its secondary terminals without its ratio error exceeding a specified amount.

Standard ASA accuracy classifications are as shown. The letter "H" stands for "high internal secondary impedance," which is a

10H10	10L10
10H20	10L20
10H50	10L50
10H100	10L100
10H200	10L200
10H400	10L400
10H800	10L800
2.5H10	2.5L10
2.5H20	2.5L20
2.5H50	2.5L50
2.5H100	2.5L100
2.5H200	2.5L200
2.5H400	2.5L400
2.5H800	2.5L800

characteristic of CT's having concentrated secondary windings. The letter "L" stands for "low internal secondary impedance," which is a characteristic of bushing-type CT's having completely distributed secondary windings or of window type having two to four secondary coils with low secondary leakage reactance. The number before the letter is the maximum specified ratio error in percent $(= 100|\text{RCF} - 1|)$, and the number after the letter is the maximum specified secondary terminal voltage at which the specified ratio error may exist, for a secondary current of 20 times rated. For a 5-ampere secondary, which is the usual rating, dividing the maximum specified voltage by 100 amperes (20×5 amperes) gives the maximum specified burden impedance through which the CT will pass 100 amperes with no more than the specified ratio error.

At secondary currents from 20 to 5 times rated, the H class of transformer will accommodate increasingly higher burden impedances than at 20 times rated without exceeding the specified maximum ratio error, so long as the product of the secondary current times the burden impedance does not exceed the specified maximum voltage at 20 times rated. This characteristic is the deciding factor when there is a question whether a given CT should be classified as "H" or as "L." At secondary currents from rated to 5 times rated, the maximum permissible burden impedance at 5 times rated (calculated as before) must not be exceeded if the maximum specified ratio error is not to be exceeded.

At secondary currents from rated to 20 times rated, the L class of transformer may accommodate no more than the maximum specified burden impedance at 20 times rated without exceeding the maximum specified ratio error. This assumes that the secondary leakage reactance is negligible.

The reason for the foregoing differences in the permissible burden impedances at currents below 20 times rated is that in the H class of transformer, having the higher secondary-winding impedance, the voltage drop in the secondary winding decreases with reduction in secondary current more rapidly than the secondary-excitation voltage decreases with the reduction in the allowable amount of exciting current for the specified ratio error. This fact will be better understood if one will calculate permissible burden impedances at reduced currents, using the secondary-excitation method.

For the same voltage and error classifications, the H transformer is better than the L for currents up to 20 times rated.

In some cases, the ASA accuracy classification will give very conservative results in that the actual accuracy of a CT may be nearly twice as good as the classification would indicate. This is particularly true in older CT's where no design changes were made to make them conform strictly to standard ASA classifications. In such cases, a CT that can actually maintain a terminal voltage well above a certain standard classification value, but not quite as high as the next higher standard value, has to be classified at the lower value. Also, some CT's can maintain terminal voltages in excess of 800 volts, but because there is no higher standard voltage rating, they must be classified "800."

The principal utility of the ASA accuracy classification is for specification purposes, to provide an indication of CT quality. The higher the number after the letter H or L, the better is the CT. However, a published ASA accuracy classification applies only if the full secondary winding is used; it does not apply to any portion of a secondary winding, as in tapped bushing-CT windings. It is perhaps obvious that with fewer secondary turns, the output voltage will be less. A bushing CT that is superior when its full secondary winding is used may be inferior when a tapped portion of its winding is used if the partial winding has higher leakage reactance because the turns are not well distributed around the full periphery of the core. In other words, the ASA accuracy classification for the full winding is not necessarily a measure of relative accuracy if the full secondary winding is not used.

If a bushing CT has completely distributed tap windings, the ASA accuracy classification for any tapped portion can be derived from the classification for the total winding by multiplying the maximum specified voltage by the ratio of the turns. For example, assume that a given 1200/5 bushing CT with 240 secondary turns is classified as 10L400; if a 120-turn completely distributed tap is used, the applicable classification is 10L200, etc. This assumes that the CT is not actually better than its classification.

Strictly speaking, the ASA accuracy classification is for a burden having a specified power factor. However, for practical purposes, the burden power factor may be ignored.

If the information obtainable from the ASA accuracy classification indicates that the CT is suitable for the application involved, no further calculations are necessary. However, if the CT appears to be unsuitable, a more accurate study should be made before the CT is rejected.

SERIES CONNECTION OF LOW-RATIO BUSHING CT'S

It will probably be evident from the foregoing that a low-ratio bushing CT, having 10 to 20 secondary turns, has rather poor accuracy at high currents. And yet, occasionally, such CT's cannot be avoided, as for example, where a high-voltage, low-current circuit or power-transformer winding is involved where rated full-load current is only, say, 50 amperes. Then, two bushing CT's per phase are sometimes used with their secondaries connected in series. This halves the burden on each CT, as compared with the use of one CT alone, without changing the over-all ratio. And, consequently, the secondary-excitation voltage is halved, and the secondary-excitation current is considerably reduced with a resulting large improvement in accuracy. Such an arrangement may require voltage protectors to hold down the secondary voltage should a fault occur between the primaries of the two CT's.

THE TRANSIENT OR STEADY-STATE ERRORS OF SATURATED CT'S

To calculate first the transient or steady-state output of saturated CT's, and then to calculate at all accurately the response of protective relays to the distorted wave form of the CT output, are a most formidable problem. With perhaps one exception,[5] there is little in the literature that is very helpful in this respect.

Fortunately, one can get along quite well without being able to make such calculations. With the help of calculating devices, comprehensive studies[6] have been made that provide general guiding principles for applying relays so that they will perform properly even though the CT output is affected by saturation. And relaying equipments have been devised that can be properly adjusted on the basis of very simple calculations. Examples of such equipments will be described later.

We are occasionally concerned lest a CT be too accurate when extremely high primary short-circuit currents flow! Even though the CT itself may be properly applied, the secondary current may be high enough to cause thermal or mechanical damage to some element in the secondary circuit before the short-circuit current can be interrupted. One should not assume that saturation of a CT core will limit the magnitude of the secondary current to a safe value. At very high primary currents, the air-core coupling between primary and secondary of wound-type CT's will cause much more secondary current to flow than one might suspect. It is recommended that, if the short-time thermal or mechanical limit of some element of the secondary circuit would be exceeded should the CT maintain its nameplate ratio, the CT manufacturer should be consulted. Where there is such possibility, damage can be prevented by the addition of a small amount of series resistance to the existing CT burden.

OVERVOLTAGE IN SATURATED CT SECONDARIES

Although the *rms* magnitude of voltage induced in a CT secondary is limited by core saturation, very high voltage *peaks* can occur.[7] Such high voltages are possible if the CT burden impedance is high, and if the primary current is many times the CT's continuous rating. The peak voltage occurs when the rate-of-change of core flux is highest, which is approximately when the flux is passing through zero. The maximum flux density that may be reached does not affect the magnitude of the peak voltage. Therefore, the magnitude of the peak voltage is practically independent of the CT characteristics other than the nameplate ratio.

One series of tests on bushing CT's produced peak voltages whose magnitudes could be expressed empirically as follows:

$$e = 3.5ZI^{0.53}$$

where e = peak voltage in volts.
Z = unsaturated magnitude of CT burden impedance in ohms.
I = primary current divided by the CT's nameplate ratio. (Or, in other words, the rms magnitude of the secondary current if the ratio-correction factor were 1.0.)

The value of Z should include the unsaturated magnetizing impedance of any idle CT's that may be in parallel with the useful burden. If a tap on the secondary winding is being used, as with a bushing CT, the peak voltage across the full winding will be the calculated value for the tap multiplied by the ratio of the turns on the full winding to the turns on the tapped portion being used; in other words, the CT will step up the voltage as an autotransformer. Because it is the practice to ground one side of the secondary winding, the voltage that is induced in the secondary will be impressed on the insulation to ground. The standard switchgear high potential test to ground is 1500 volts rms, or 2121 volts peak; and the standard CT test voltage is 2475 volts rms or 3500 volts peak.[1] The lower of these two should not be exceeded.

Harmfully high secondary voltages may occur in the CT secondary circuit of generator differential-relaying equipment when the generator kva rating is low but when very high short-circuit kva can be supplied by the system to a short circuit at the generator's terminals. Here, the magnitude of the primary current on the system side of the generator windings may be many times the CT rating. These CT's will try to supply very high secondary currents to the operating coils of the generator differential relay, the unsaturated impedance of which may be quite high. The resulting high peak voltages could break down the insulation of the CT's, the secondary wiring, or the differential relays, and thereby prevent the differential relays from operating to trip the generator breakers.

Such harmfully high peak voltages are not apt to occur for this reason with other than motor or generator differential-relaying equipments because the CT burdens of other equipments are not usually so high. But, wherever high voltage is possible, it can be limited to safe values by overvoltage protectors.

Another possible cause of overvoltage is the switching of a capacitor bank when it is very close to another energized capacitor bank.[8]

The primary current of a CT in the circuit of a capacitor bank being energized or de-energized will contain transient high-frequency currents. With high-frequency primary and secondary currents, a CT burden reactance, which at normal frequency is moderately low, becomes very high, thereby contributing to CT saturation and high peak voltages across the secondary. Overvoltage protectors may be required to limit such voltages to safe values.

It is recommended that the CT manufacturer be consulted whenever there appears to be a need for overvoltage protectors. The protector characteristics must be coordinated with the requirements of a particular application to (1) limit the peak voltage to safe values, (2) not interfere with the proper functioning of the protective-relaying equipment energized from the CT's, and (3) withstand the total amount of energy that the protector will have to absorb.

PROXIMITY EFFECTS

Large currents flowing in a conductor close to a current transformer may greatly affect its accuracy. A designer of compact equipment, such as metal-enclosed switchgear, should guard against this effect. If one has all the necessary data, it is a reasonably simple matter to calculate the necessary spacings to avoid excessive error.[9]

Polarity and Connections

The relative polarities of CT primary and secondary terminals are identified either by painted polarity marks or by the symbols "H_1" and "H_2" for the primary terminals and "X_1" and "X_2" for the secondary terminals. The convention is that, when primary current enters the H_1 terminal, secondary current leaves the X_1 terminal, as shown by the arrows in Fig. 4. Or, when current enters the H_2 terminal, it leaves the X_2 terminal. When paint is used, the terminals corresponding to H_1 and X_1 are identified.

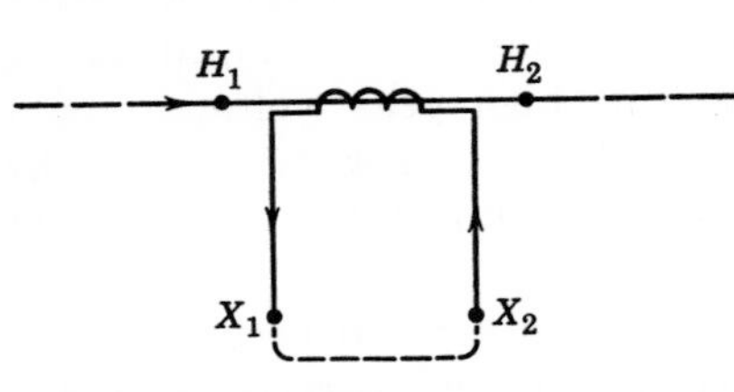

Fig. 4. The polarity of current transformers.

Standard practice is to show the corresponding terminals in connection diagrams merely by squares, as in Fig. 5.

Since a-c current is continually reversing its direction, one might well ask what the significance is of polarity marking. Its significance is in showing the direction of current flow relative to another current or to a voltage, as well as to aid in making the proper connections. If CT's were not interconnected, or if the current from one CT did

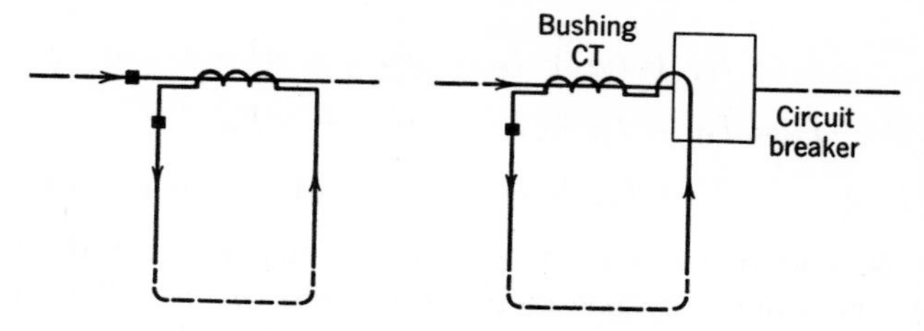

Fig. 5. Convention for showing polarity on diagrams.

not have to cooperate with a current from another CT, or with a voltage from a voltage source, to produce some desired result such as torque in a relay, there would be no need for polarity marks.

WYE CONNECTION

CT's are connected in wye or in delta, as the occasion requires. Figure 6 shows a wye connection with phase and ground relays. The currents I_a, I_b, and I_c are the vector currents, and the CT ratio is assumed to be 1/1 to simplify the mathematics. Vectorially, the

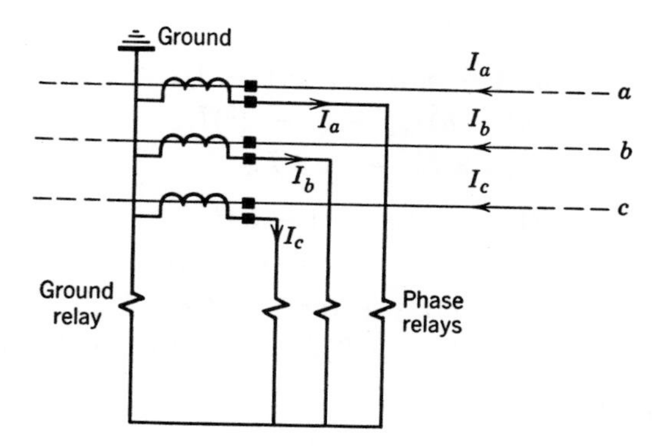

Fig. 6. Wye connection of current transformers.

primary and secondary currents are in phase, neglecting phase-angle errors in the CT's.

The symmetrical-component method of analysis is a powerful tool, not only for use in calculating the power-system currents and voltages for unbalanced faults but also for analyzing the response of protective relays. In terms of phase-sequence components of the power-system currents, the output of wye-connected CT's is as follows:

$$I_a = I_{a1} + I_{a2} + I_{a0}$$

$$I_b = I_{b1} + I_{b2} + I_{b0} = a^2 I_{a1} + a I_{a2} + I_{a0}$$

$$I_c = I_{c1} + I_{c2} + I_{c0} = a I_{a1} + a^2 I_{a2} + I_{a0}$$

$$I_a + I_b + I_c = I_{a0} + I_{b0} + I_{c0} = 3I_{a0} = 3I_{b0} = 3I_{c0}$$

where 1, 2, and 0 designate the positive-, negative-, and zero-phase-sequence components, respectively, and where "a" and "a^2" are operators that rotate a quantity counterclockwise 120° and 240°, respectively.

DELTA CONNECTION

With delta-connected CT's, two connections are possible, as shown in Fig. 7. In terms of the phase-sequence components, I_a, I_b, and I_c are the same as for the wye-connected CT's. The output currents of the delta connections of Fig. 7 are, therefore:

Connection A.

$$\begin{aligned} I_a - I_b &= (I_{a1} - I_{b1}) + (I_{a2} - I_{b2}) \\ &= (1 - a^2) I_{a1} + (1 - a) I_{a2} \\ &= (\tfrac{3}{2} + j\sqrt{3}/2) I_{a1} + (\tfrac{3}{2} - j\sqrt{3}/2) I_{a2} \end{aligned}$$

$$\begin{aligned} I_b - I_c &= (1 - a^2) I_{b1} + (1 - a) I_{b2} \\ &= a^2 (1 - a^2) I_{a1} + a(1 - a) I_{a2} \\ &= (a^2 - a) I_{a1} + (a - a^2) I_{a2} \\ &= -j\sqrt{3}\, I_{a1} + j\sqrt{3}\, I_{a2} \end{aligned}$$

$$\begin{aligned} I_c - I_a &= (1 - a^2) I_{c1} + (1 - a) I_{c2} \\ &= a(1 - a^2) I_{a1} + a^2 (1 - a) I_{a2} \\ &= (a - 1) I_{a1} + (a^2 - 1) I_{a2} \\ &= (-\tfrac{3}{2} + j\sqrt{3}/2) I_{a1} + (-\tfrac{3}{2} - j\sqrt{3}/2) I_{a2} \end{aligned}$$

Connection B.

$$\begin{aligned} I_a - I_c &= -(I_c - I_a) \\ &= (\tfrac{3}{2} - j\sqrt{3}/2) I_{a1} + (\tfrac{3}{2} + j\sqrt{3}/2) I_{a2} \end{aligned}$$

$$\begin{aligned} I_b - I_a &= -(I_a - I_b) \\ &= (-\tfrac{3}{2} - j\sqrt{3}/2) I_{a1} + (-\tfrac{3}{2} + j\sqrt{3}/2) I_{a2} \end{aligned}$$

$$\begin{aligned} I_c - I_b &= -(I_b - I_c) \\ &= j\sqrt{3}\, I_{a1} - j\sqrt{3}\, I_{a2} \end{aligned}$$

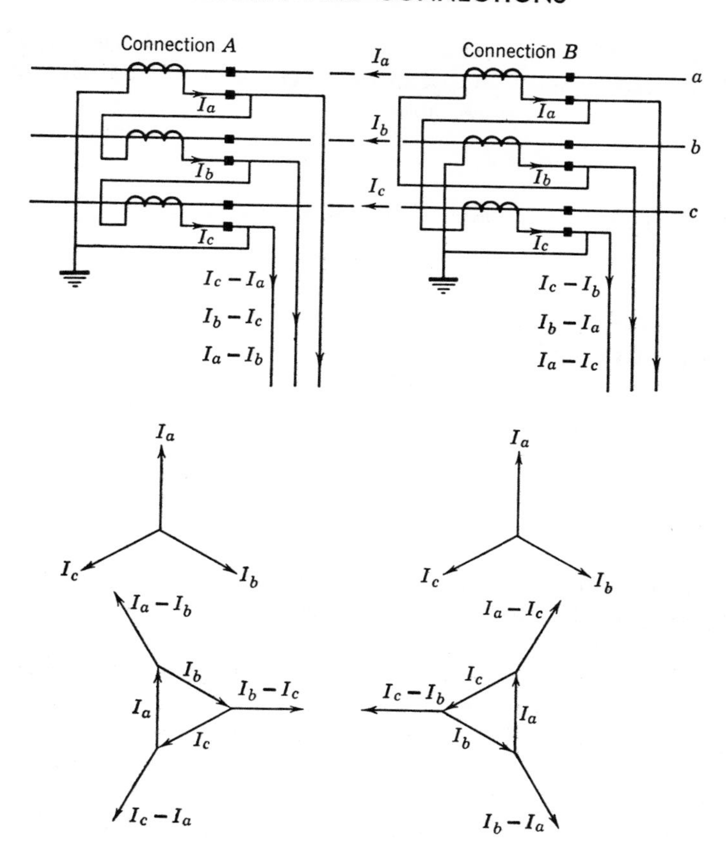

Fig. 7. Delta connections of current transformers and vector diagrams for balanced three-phase currents.

It will be noted that the zero-phase-sequence components are not present in the output circuits; they merely circulate in the delta connection. It will also be noted that connection B is merely the reverse of connection A.

For three-phase faults, only positive-phase-sequence components are present. The output currents of connection A become:

$$I_a - I_b = (\tfrac{3}{2} + j\sqrt{3}/2)I_{a1}$$

$$I_b - I_c = -j\sqrt{3}\, I_{a1}$$

$$I_c - I_a = (-\tfrac{3}{2} + j\sqrt{3}/2)I_{a1}$$

For a phase-b-to-phase-c fault, if we assume the same distribution of positive- and negative-phase-sequence currents (which is permissible if we assume that the negative-phase-sequence impedances equal the positive-phase-sequence impedances), $I_{a2} = -I_{a1}$, and the output currents of connection A become:

$$I_a - I_b = j\sqrt{3}\, I_{a1}$$

$$I_b - I_c = -j2\sqrt{3}\, I_{a1}$$

$$I_c - I_a = j\sqrt{3}\, I_{a1}$$

For a phase-a-to-ground fault, if we again assume the same distribution of positive- and negative-phase-sequence currents, $I_{a2} = I_{a1}$, and the output currents of connection A become:

$$I_a - I_b = 3I_{a1}$$

$$I_b - I_c = 0$$

$$I_c - I_a = -3I_{a1}$$

The currents for a two-phase-to-ground fault between phases b and c can be obtained in a similar manner if one knows the relation between the impedances in the negative- and zero-phase-sequence networks. It is felt, however, that the foregoing examples are sufficient to illustrate the technique involved. The assumptions that were made as to the distribution of the currents are generally sufficiently accurate, but they are not a necessary part of the technique; in any actual case, one would know the true distribution and also any angular differences that might exist, and these could be entered in the fundamental equations.

The output currents from wye-connected CT's can be handled in a similar manner.

THE ZERO-PHASE-SEQUENCE-CURRENT SHUNT

Figure 8 shows how three auxiliary CT's can be connected to shunt zero-phase-sequence currents away from relays in the secondary of wye-connected CT's. Other forms of such a shunt exist, but the one shown has the advantage that the ratio of the auxiliary CT's is not important so long as all three are alike. Such a shunt is useful in a differential circuit where the main CT's must be wye-connected but where zero-phase-sequence currents must be kept from the phase relays. Another use is to prevent misoperation of single-phase directional relays during ground faults under certain conditions. These will be discussed more fully later.

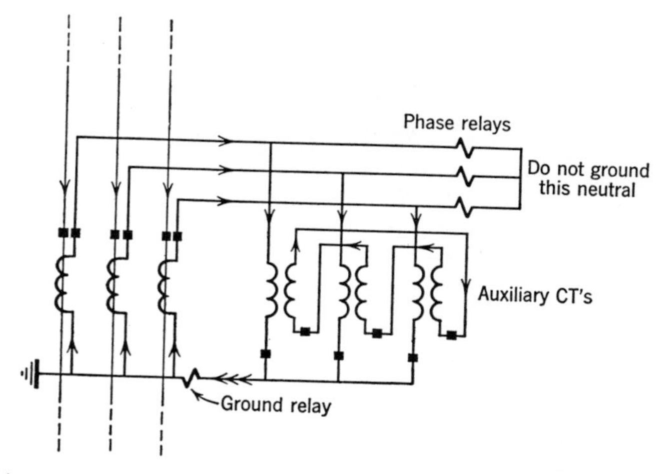

Fig. 8. A zero-phase-sequence-current shunt. Arrows show flow of zero-phase-sequence current.

Problems

1. What is the ASA accuracy classification for the full winding of the bushing CT whose secondary-excitation characteristic and secondary resistance are given on Fig. 3?

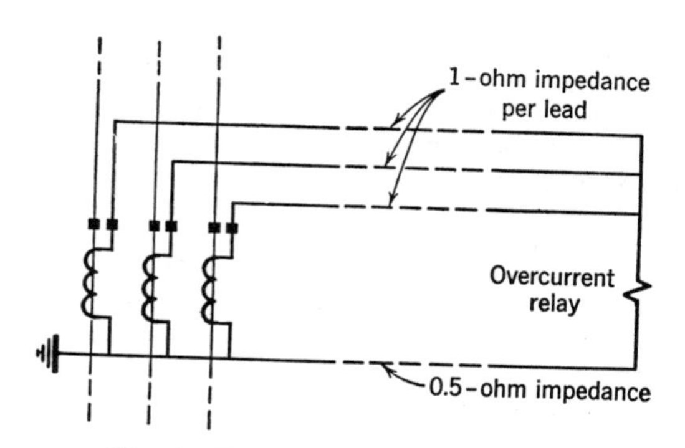

Fig. 9. Illustration for Problem 2.

2. For the overcurrent relay connected as shown in Fig. 9, determine the value of pickup current that will provide relay operation at the lowest possible value of primary current in one phase.

If the overcurrent relay has a pickup of 1.5 amperes, its coil impedance at 1.5 amperes is 2.4 ohms. Assume that the impedance at pickup current varies inversely as the square of pickup current, and that relays of any desired pickup are available to you.

The CT's are the same as the 20-turn tap of the CT whose secondary-excitation characteristic is shown in Fig. 3.

Bibliography

1. "General Requirements for Transformers, Regulators, and Reactors," *Publ. C57.11–1948;* "American Standard Requirements, Terminology, and Test Code for Instrument Transformers," *Publ. C57.13–1954;* and "Guide for Loading and Operation of Instrument Transformers," *Publ. C57.33,* American Standards Assoc., Inc., 70 East 45th St., New York 17, N. Y.

"Application Guide for Grounding of Instrument Transformer Secondary Circuits and Cases," *Publ. 52,* American Institute of Electrical Engineers, 33 West 39th St., New York 18, N. Y.

2. *ASA C57.23,* see Reference 1.

"A Simple Method for the Determination of Bushing-Current-Transformer Characteristics," by S. D. Moreton, *AIEE Trans., 62* (1943), pp. 581–585. Discussions, pp. 948–952.

"A Simple Method for Determination of Ratio Error and Phase Angle in Current Transformers," by E. C. Wentz, *AIEE Trans., 60* (1941), pp. 949–954. Discussions, p. 1369.

3. "A Proposed Method for the Determination of Current-Transformer Errors," by G. Camilli and R. L. Ten Broeck, *AIEE Trans., 59* (1940), pp. 547–550. Discussions, pp. 1138–1140.

"Overcurrent Performance of Bushing-Type Current Transformers," by C. A. Woods, Jr., and S. A. Bottonari, *AIEE Trans., 59* (1940), pp. 554–560. Discussions, pp. 1140–1144.

"Computation of Accuracy of Current Transformers," by A. T. Sinks, *AIEE Trans., 59* (1940), pp. 663–668. Discussions, pp. 1252–1253.

4. *ASA C57.13,* see Reference 1.

5. "Current Transformers and Relays for High-Speed Differential Protection, with Particular Reference to Offset Transient Currents," by W. K. Sonnemann and E. C. Wentz, *AIEE Trans., 59* (1940), pp. 481–488. Discussions, p. 1144.

6. "Transient Characteristics of Current Transformers during Faults," by C. Concordia, C. N. Weygandt, and H. S. Shott, *AIEE Trans., 61* (1942), pp. 280–285. Discussions, pp. 469–470.

"Transient Characteristics of Current Transformers during Faults, Part II," by F. S. Rothe and C. Concordia, *AIEE Trans., 66* (1947), pp. 731–734.

"The Effect of Current-Transformer Residual Magnetism on Balanced-Current or Differential Relays," by H.T. Seeley, *AIEE Trans., 62* (1943), pp. 164–168. Discussions, p. 384.

7. "Peak Voltages Induced by Accelerated Flux Reversals in Reactor Cores Operating above Saturation Density," by Theodore Specht and E. C. Wentz, *AIEE Trans., 65* (1946), pp. 254–263.

"Overvoltages in Saturable Series Devices," by A. Boyajian and G. Camilli, *AIEE Trans., 70* (1951), pp. 1845–1851. Discussions, pp. 1852–1853.

8. "Overvoltage Protection of Current-Transformer Secondary Windings and Associated Circuits," by R. H. Kaufmann and G. Camilli, *AIEE Trans., 62* (1943), pp. 467–472. Discussions, pp. 919–920.

9. "The Accuracy of Current Transformers Adjacent to High-Current Busses," by R. A. Pfuntner, *AIEE Trans., 70* (1951), pp. 1656–1661. Discussions, p. 1662.

8 VOLTAGE TRANSFORMERS

Two types of voltage transformer are used for protective-relaying purposes, as follows: (1) the "instrument potential transformer," hereafter to be called simply "potential transformer," and (2) the "capacitance potential device." A potential transformer is a conventional transformer having primary and secondary windings. The primary winding is connected directly to the power circuit either between two phases or between one phase and ground, depending on the rating of the transformer and on the requirements of the application. A capacitance potential device is a voltage-transforming equipment using a capacitance voltage divider connected between phase and ground of a power circuit.

Accuracy of Potential Transformers

The ratio and phase-angle inaccuracies of any standard ASA accuracy class[1] of potential transformer are so small that they may be neglected for protective-relaying purposes if the burden is within the "thermal" volt-ampere rating of the transformer. This thermal volt-ampere rating corresponds to the full-load rating of a power transformer. It is higher than the volt-ampere rating used to classify potential transformers as to accuracy for metering purposes. Based on the thermal volt-ampere rating, the equivalent-circuit impedances of potential transformers are comparable to those of distribution transformers.

The "burden" is the total external volt-ampere load on the secondary at rated secondary voltage. Where several loads are connected in parallel, it is usually sufficiently accurate to add their individual volt-amperes arithmetically to determine the total volt-ampere burden.

If a potential transformer has acceptable accuracy at its rated voltage, it is suitable over the range from zero to 110% of rated voltage. Operation in excess of 10% overvoltage may cause increased errors and excessive heating.

Where precise accuracy data are required, they can be obtained from ratio-correction-factor curves and phase-angle-correction curves supplied by the manufacturer.

Capacitance Potential Devices

Two types of capacitance potential device are used for protective relaying: (1) the "coupling-capacitor potential device," and (2) the "bushing potential device." The two devices are basically alike, the principal difference being in the type of capacitance voltage divider

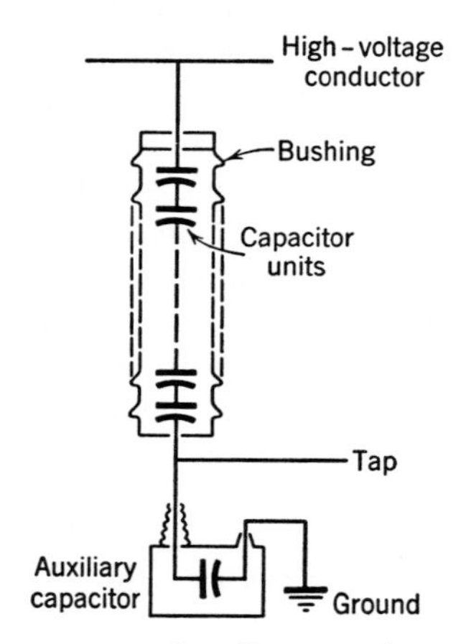

Fig. 1. Coupling-capacitor voltage divider.

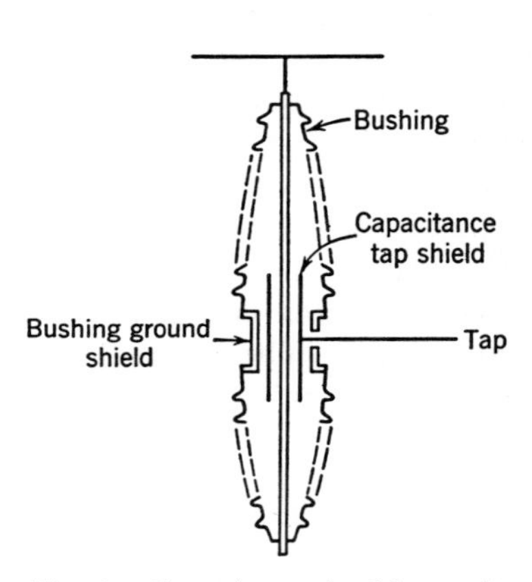

Fig. 2. Capacitance-bushing voltage divider.

used, which in turn affects their rated burden. The coupling-capacitor device uses as a voltage divider a "coupling capacitor" consisting of a stack of series-connected capacitor units, and an "auxiliary capacitor," as shown schematically in Fig. 1. The bushing device uses the capacitance coupling of a specially constructed bushing of a circuit breaker or power transformer, as shown schematically in Fig. 2.

Both of these relaying potential devices are called "Class A" devices.[2] They are also sometimes called "In-phase" or "Resonant" devices[3] for reasons that will be evident later. Other types of potential devices, called "Class C" or "Out-of-phase" or "Non-resonant," are also described in References 2 and 3, but they are not generally suitable for protective relaying, and therefore they will not be considered further here.

A schematic diagram of a Class A potential device including the capacitance voltage divider is shown in Fig. 3. Not shown are the

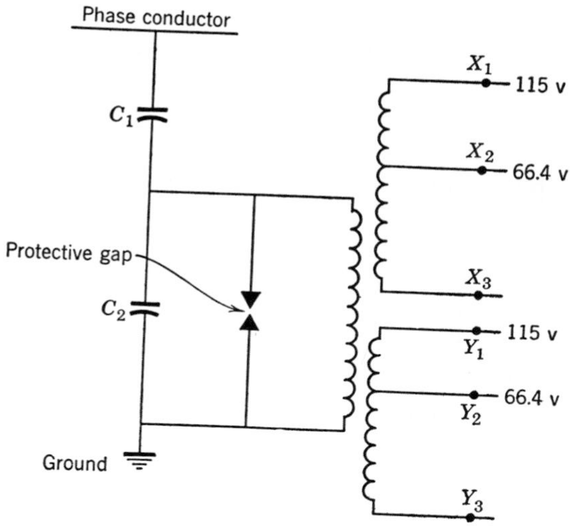

Fig. 3. Schematic diagram of a Class A potential device.

means for adjusting the magnitude and phase angle of the secondary voltage; the means for making these adjustments vary with different manufacturers, and a knowledge of them is not essential to our present purposes. The Class A device has two secondary windings as shown. Both windings are rated 115 volts, and one must have a 66.4-volt tap. These windings are connected in combination with the windings of the devices of the other two phases of a three-phase power circuit. The connection is "wye" for phase relays and "broken delta" for ground relays. These connections will be illustrated later.

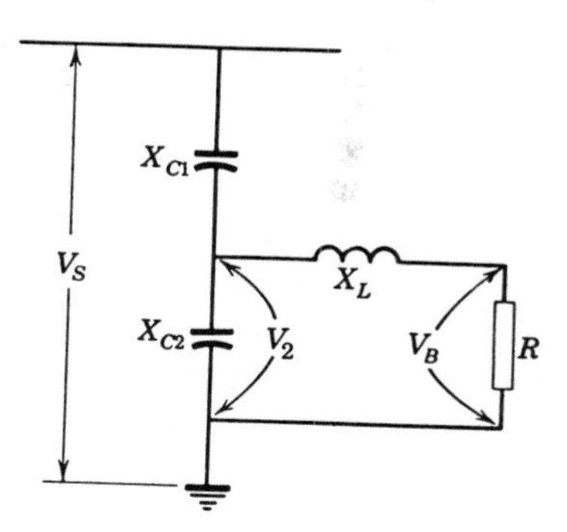

Fig. 4. Equivalent circuit of a Class A potential device.

The equivalent circuit of a Class A device is shown in Fig. 4. The equivalent reactance X_L is adjustable to make the burden voltage V_B be in phase with the phase-to-ground voltage of the system V_S. The burden is shown as a resistor because, so far as it is possible, it is the practice to correct the power factor of the burden approximately to unity by the use of auxiliary capacitance burden. When the device is properly adjusted,

$$X_L = \frac{X_{C1}X_{C2}}{X_{C1} + X_{C2}} \tag{1}$$

which explains why the term "Resonant" is applied to this device. Actually, X_{C2} is so small compared with X_{C1} that X_L is practically equal to X_{C2}. Therefore, X_L and X_{C2} would be practically in parallel resonance were it not for the presence of the burden impedance.

The gross input in watts from a power circuit to a capacitance potential-device network is:[3]

$$W = 2\pi f C_1 V_S V_2 \sin \alpha \quad \text{watts} \tag{2}$$

where f = power-system frequency.
α = phase angle between V_S and V_2.
C_1 = capacitance of main capacitor (see Fig. 3) in farads.

V_S and V_2 are volts defined as in Fig. 4. If the losses in the network are neglected, equation 2 will give the output of the device. For special applications, this relation is useful for estimating the rated burden from the known rated burden under standard conditions; it is only necessary to compare the proportions in the two cases, remembering that, for a given rating of equipment, the tap voltage V_2 varies directly as the applied voltage V_S.

For a given group of coupling-capacitor potential devices, the product of the capacitance of the main capacitor C_1 and the rated circuit-voltage value of V_S is practically constant; in other words, the number of series capacitor units that comprise C_1 is approximately directly proportional to the rated circuit voltage. The capacitance of the auxiliary capacitor C_2 is the same for all rated circuit voltages, so as to maintain an approximately constant value of the tap voltage V_2 for all values of rated circuit voltage.

For bushing potential devices, the value of C_1 is approximately constant over a range of rated voltages, and the value of C_2 is varied by the use of auxiliary capacitance to maintain an approximately constant value of the tap voltage V_2 for all values of rated circuit voltage.

STANDARD RATED BURDENS OF CLASS A POTENTIAL DEVICES

The rated burden of a secondary winding of a capacitance potential device is specified in watts at rated secondary voltage when rated phase-to-ground voltage is impressed across the capacitance voltage divider. The rated burden of the device is the sum of the watt burdens that may be impressed on both secondary windings simultaneously.

Adjustment capacitors are provided in the device for connecting in parallel with the burden on one secondary winding to correct the total-burden power factor to unity or slightly leading.

The standard[2] rated burdens of bushing potential devices are given in Table 1.

Table 1. Rated Burdens of Bushing Potential Devices

Rated Circuit Voltage, kv		Rated Burden, watts
Phase-to-Phase	Phase-to-Ground	
115	66.4	25
138	79.7	35
161	93.0	45
230	133.0	80
287	166.0	100

The rated burden of coupling-capacitor potential devices is 150 watts for any of the rated circuit voltages, including those of Table 1.

STANDARD ACCURACY OF CLASS A POTENTIAL DEVICES

Table 2 gives the standard maximum deviation in voltage ratio and phase angle for rated burden and for various values of primary voltage, with the device adjusted for the specified accuracy at rated primary voltage.

Table 2. Ratio and Phase-Angle Error versus Voltage

Primary Voltage, percent of rated	Maximum Deviation	
	Ratio, percent	Phase Angle, degrees
100	± 1.0	± 1.0
25	± 3.0	± 3.0
5	± 5.0	± 5.0

Table 3 gives the standard maximum deviation in voltage ratio and phase angle for rated voltage and for various values of burden with the device adjusted for the specified accuracy at rated burden.

Table 3. Ratio and Phase-Angle Error versus Burden

Burden, percent of rated	Maximum Deviation	
	Ratio, percent	Phase Angle, degrees
100	± 1.0	± 1
50	± 6.0	± 4
0	± 12.0	± 8

Table 3 shows that for greatest accuracy, the burden should not be changed without readjusting the device.

EFFECT OF OVERLOADING

As the burden is increased beyond the rated value, the errors will increase at about the rate shown by extrapolating the data of Table 3, which is not very serious for protective relaying. Apart from the possibility of overheating, the serious effect is the accompanying increase of the tap voltage (V_2 of Fig. 4). An examination of the equivalent circuit, Fig. 4, will show why the tap voltage increases with increasing burden. It has been said that X_L is nearly equal to X_{C2}, and therefore these two branches of the circuit will approach parallel resonance as R is decreased (or, in other words, as the burden is increased). Hence, the tap voltage will tend to approach V_S. As the burden is increased above the rated value, the tap voltage will increase approximately proportionally.

The objection to increasing the tap voltage is that the protective gap must then be adjusted for higher-than-normal arc-over voltage. This lessens the protection afforded the equipment. The circuit elements protected by the gap are specified[2] to withstand 4 times the normal tap voltage for 1 minute. Ordinarily, the gap is adjusted to arc over at about twice normal voltage. This is about as low an arc-over as the gap may be adjusted to have in view of the fact that for some ground faults the applied voltage (and hence the tap voltage) may rise to $\sqrt{3}$ times normal. Obviously, the gap must not be permitted to arc over for any voltage for which the protective-relaying equipment must function. Since the ground-relay burden loads the devices only when a ground fault occurs, gap flashover may be a problem when thermal overloading is not a problem. Before purposely overloading a capacitance potential device, one should consult the manufacturer.

As might be suspected, short-circuiting the secondary terminals of the device (which is extreme overloading) will arc over the gap continuously while the short circuit exists. This may not cause any damage to the device, and hence it may not call for fusing, but the gap will eventually be damaged to such an extent that it may no longer protect the equipment.

Even when properly adjusted, the protective gap might arc over during transient overvoltages caused by switching or by lightning. The duration of such arc-over is so short that it will not interfere with the proper operation of protective relays. The moment the overvoltage ceases, the gap will stop arcing over because the impedance of the main capacitor C_1 is so high that normal system voltage cannot maintain the arc.

It is emphasized that the standard rated burdens are specified as though a device were connected and loaded as a single-phase device. In practice, however, the secondary windings of three devices are interconnected and loaded jointly. Therefore, to determine the actual loading on a particular device under unbalanced voltage conditions, as when short circuits occur, certain conversions must be made. This is described later in more detail for the broken-delta burden. Also, the effective burden on each device resulting from the phase-to-phase and phase-to-neutral burdens should be determined if the loading is critical; this is merely a circuit problem that is applicable to any kind of voltage transformer.

NON-LINEAR BURDENS

A "non-linear" burden is a burden whose impedance decreases because of magnetic saturation when the impressed voltage is increased. Too much non-linearity in its burden will let a capacitance potential device get into a state of ferroresonance,[4] during which steady overvoltages of highly distorted wave form will exist across the burden. Since these voltages bear no resemblance to the primary voltages, such a condition must be avoided.

If one must know the maximum tolerable degree of non-linearity, he should consult the manufacturer. Otherwise, the ferroresonance condition can be avoided if all magnetic circuits constituting the burden operate at rated voltage at such low flux density that any possible momentary overvoltage will not cause the flux density of any magnetic circuit to go beyond the knee of its magnetization curve (or, in other words, will not cause the flux density to exceed about 100,000 lines per square inch). Since the potential-device secondary-winding voltage may rise to $\sqrt{3}$ times rated, and the broken-delta voltage may rise to 3 times rated, the corresponding phase-to-neutral and broken-delta burdens may be required to have no more than $1/\sqrt{3}$ and $1/3$, respectively, of the maximum allowable flux density at rated voltage.

If burdens with closed magnetic circuits, such as auxiliary potential transformers, are not used, there is no likelihood of ferroresonance. Class A potential devices are provided with two secondary windings purposely to avoid the need of an auxiliary potential transformer. The relays, meters, and instruments generally used have air gaps in their magnetic circuits, or operate at low enough flux density to make their burdens sufficiently linear.

THE BROKEN-DELTA BURDEN AND THE WINDING BURDEN

The broken-delta burden is usually composed of the voltage-polarizing coils of ground directional relays. Each relay's voltage-coil circuit contains a series capacitor to make the relay have a lagging angle of maximum torque. Consequently, the voltage-coil circuit has a leading power factor. The volt-ampere burden of each relay is expressed by the manufacturer in terms of the rated voltage of the relay. The broken-delta burden must be expressed in terms of the rated voltage of the potential-device winding or the tapped portion of the winding—whichever is used for making up the broken-delta connection. If the relay- and winding-voltage ratings are the same, the broken-delta burden is the sum of the relay burdens. If the voltage ratings are different, we must re-express the relay burdens in terms of the voltage rating of the broken-delta winding before adding them, remembering that the volt-ampere burden will vary as the square of the voltage, assuming no saturation.

The actual volt-ampere burdens imposed on the individual windings comprising the broken-delta connection are highly variable and are only indirectly related to the broken-delta burden. Normally, the three winding voltages add vectorially to zero. Therefore, no current flows in the circuit, and the burden on any of the windings is zero. When ground faults occur, the voltage that appears across the broken-delta burden corresponds to 3 times the zero-phase-sequence component of any one of the three phase-to-ground voltages at the potential-device location. We shall call this voltage "$3V_0$." What the actual magnitude of this voltage is depends on how solidly the system neutrals are grounded, on the location of the fault with respect to the potential device in question, and on the configuration of the transmission circuits so far as it affects the magnitude of the zero-phase-sequence reactance. For faults at the potential-device location, for which the voltage is highest, $3V_0$ can vary approximately from 1 to 3 times the rated voltage of each of the broken-delta windings. (This voltage can go even higher in an ungrounded-neutral system should a state of ferroresonance exist, but this possibility is not considered here because it must not be permitted to exist.) If we assume no magnetic saturation in the burden, its maximum current magnitude will vary with the voltage over a 1 to 3 range.

The burden current flows through the three broken-delta windings in series. As shown in Fig. 5, the current is at a different phase angle with respect to each of the winding voltages. Since a ground fault can occur on any phase, the positions of any of the voltages of Fig. 5

relative to the burden current can be interchanged. Consequently, the burden on each winding may have a wide variety of characteristics under different circumstances.

Another peculiarity of the broken-delta burden is that the load is really carried by the windings of the unfaulted phases, and that the

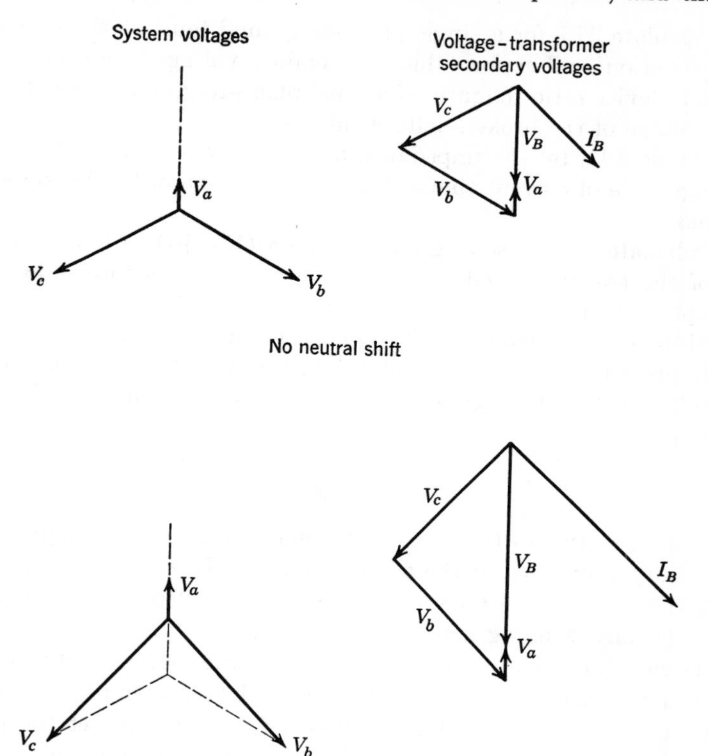

Fig. 5. Broken-delta voltages and current for a single-phase-to-ground fault on phase a some distance from the voltage transformer.

voltages of these windings do not vary in direct proportion to the voltage across the broken-delta burden. The voltages of the unfaulted-phase windings are not nearly as variable as the broken-delta-burden voltage. The winding voltages of the unfaulted phases vary from approximately rated voltage to $\sqrt{3}$ times rated, while the broken-delta-burden voltage, and hence the current, is varying from less than rated to approximately 3 times rated.

As a consequence of the foregoing, on the basis of rated voltage, the burden on any winding can vary from less than the broken-delta burden to $\sqrt{3}$ times it. For estimating purposes, the $\sqrt{3}$ multiplier would be used, but, if the total burden appeared to be excessive, one would want to calculate the actual burden. To do this, the following steps are involved:

1. Calculate $3V_0$ for a single-phase-to-ground fault at the potential-device location, and express this in secondary-voltage terms, using as a potential-device ratio the ratio of normal phase-to-ground voltage to the rated voltage of the broken-delta windings.
2. Divide $3V_0$ by the impedance of the broken-delta burden to get the magnitude of current that will circulate in each of the broken-delta windings.
3. Calculate the phase-to-ground voltage ($V_{b1} + V_{b2} + V_{b0}$, etc.) of each of the two unfaulted phases at the voltage-transformer location, and express it in secondary-voltage terms as for $3V_0$.
4. Multiply the current of (2) by each voltage of (3).
5. Express the volt-amperes of (4) in terms of the rated voltage of the broken-delta windings by multiplying the volt-amperes of (4) by the ratio:

$$\left[\frac{V_{\text{rated}}}{\text{Voltage of 3}}\right]^2$$

It is the practice to treat the volt-ampere burden as though it were a watt burden on each of the three windings. It will be evident from Fig. 5 that, depending on which phase is grounded, the volt-ampere burden on any winding could be practically all watts.

It is not the usual practice to correct the power factor of the broken-delta burden to unity as is done for the phase burden. Because this burden usually has a leading power factor, to correct the power factor to unity would require an adjustable auxiliary burden that had inductive reactance. Such a burden would have to have very low resistance and yet it would have to be linear. In the face of these severe requirements, and in view of the fact that the broken-delta burden is usually a small part of the total potential-device burden, such corrective burden is not provided in standard potential devices.

COUPLING-CAPACITOR INSULATION COORDINATION AND ITS EFFECT ON THE RATED BURDEN

The voltage rating of a coupling capacitor that is used with protective relaying should be such that its insulation will withstand the flashover voltage of the circuit at the point where the capacitor is connected. Table 4 lists the standard[2] capacitor withstand test

Table 4. Standard Withstand Test Voltages for Coupling Capacitors

Rated Circuit Voltage, kv		Withstand Test Voltages		
			Low Frequency	
Phase-to-Phase	Phase-to-Ground	Impulse, kv	Dry 1-Min, kv	Wet 10-Sec, kv
115	66.4	550	265	230
138	79.7	650	320	275
161	93.0	750	370	315
230	133.0	1050	525	445
287	166.0	1300	655	555

voltages for some circuit-voltage ratings for altitudes below 3300 feet.

The flashover voltage of the circuit at the capacitor location will depend not only on the line insulation but also on the insulation of other terminal equipment such as circuit breakers, transformers, and lightning arresters. However, there may be occasions when these other terminal equipments may be disconnected from the line, and the capacitor will then be left alone at the end of the line without benefit of the protection that any other equipment might provide. For example, a disconnect may be opened between a breaker and the capacitor, or a breaker may be opened between a transformer or an arrester and the capacitor. If such can happen, the capacitor must be able to withstand the voltage that will flash over the line at the point where the capacitor is connected.

Some lines are overinsulated, either because they are subjected to unusual insulator contamination or because they are insulated for a future higher voltage than the present operating value. In any event, the capacitor should withstand the actual line flashover voltage unless there is other equipment permanently connected to the line that will hold the voltage down to a lower value.

At altitudes above 3300 feet, the flashover value of air-insulated equipment has decreased appreciably. To compensate for this decrease, additional insulation may be provided for the line and for the other terminal equipment. This may require the next higher standard voltage rating for the coupling capacitor, and it is the practice to specify the next higher rating if the altitude is known to be over 3300 feet.

When a coupling-capacitor potential device is to be purchased for operation at the next standard rated circuit voltage below the coupling-capacitor rating, the manufacturer should be so informed. In such a case, a special auxiliary capacitor will be furnished that will provide normal tap voltage even though the applied voltage is one step less than rated. This will give the device a rated burden of 120 watts. If a special auxiliary capacitor were not furnished, the rated burden would be about 64% of 150 watts instead of 80%. The foregoing will become evident on examination of equation 2 and on consideration of the fact that each rated insulation class is roughly 80% of the next higher rating.

The foregoing applies also to bushing potential devices, except that sometimes a non-standard transformer unit may be required to get 80% of rated output when the device is operating at the next standard rated circuit voltage below the bushing rating.

COMPARISON OF INSTRUMENT POTENTIAL TRANSFORMERS AND CAPACITANCE POTENTIAL DEVICES

Capacitance potential devices are used for protective relaying only when they are sufficiently less expensive than potential transformers. Potential devices are not as accurate as potential transformers, and also they may have undesirable transient inaccuracies unless they are properly loaded.[5] When a voltage source for the protective relays of a single circuit is required, and when the circuit voltage is approximately 69 kv and higher, coupling-capacitor potential devices are less costly than potential transformers. Savings may be realized somewhat below 69 kv if carrier current is involved, because a potential-device coupling capacitor can be used also, with small additional expense, for coupling the carrier-current equipment to the circuit. Bushing potential devices, being still less costly, may be even more economical, provided that the devices have sufficiently high rated-burden capacity. However, the main capacitor of a bushing potential device cannot be used to couple carrier-current equipment to a power circuit. When compared on a dollars-per-volt-ampere basis, potential transformers are much cheaper than capacitance potential devices.

When two or more transmission-line sections are connected to a common bus, a single set of potential transformers connected to the bus will generally have sufficient capacity to supply the protective-relaying equipments of all the lines, whereas one set of capacitance potential devices may not. The provision of additional potential devices will quickly nullify the difference in cost. In view of the foregoing, one should at least consider bus potential transformers, even for a single circuit, if there is a likelihood that future requirements might involve additional circuits.

Potential transformers energized from a bus provide a further slight advantage where protective-relaying equipment is involved in which dependence is placed on "memory action" for reliable operation.

When a line section protected by such relaying equipment is closed in on a nearby fault, and if potential transformers connected to the bus are involved, the relays will have had voltage on them before the line breaker was closed, and hence the memory action can be effective. If the voltage source is on the line side of the breaker, as is usually true with capacitance potential devices, there will have been no voltage on the relays initially, and memory action will be ineffective. Consequently, the relays may not operate if the voltage is too low owing to the presence of a metallic fault with no arcing, thereby requiring back-up relaying at other locations to clear the fault from the system. However, the likelihood of the voltage being low enough to prevent relay operation is quite remote, but the relays may be slow.

Some people object to bus potential transformers on the basis that trouble in a potential transformer will affect the relaying of all the lines connected to the bus. This is not too serious an objection, particularly if the line relays are not allowed to trip on loss of voltage during normal load, and if a voltage-failure alarm is provided.

Where ring buses are involved, there is no satisfactory location for a single set of bus potential transformers to serve the relays of all circuits. In such cases, capacitance potential devices on the line side of the breakers of each circuit are the best solution when they are cheaper.

The Use of Low-Tension Voltage

When there are step-down power transformers at a location where voltage is required for protective-relaying equipment, the question naturally arises whether the relay voltage can be obtained from the low-voltage side of the power transformers, and thereby avoid the expense of a high-voltage source. Such a low-voltage source can be used under certain circumstances.

The first consideration is the reliability of the source. If there is only one power transformer, the source will be lost if this power transformer is removed from service for any reason. If there are two or more power transformers in parallel, the source is probably sufficiently reliable if the power transformers are provided with separate breakers.

The second consideration is whether there will be a suitable source for polarizing directional-ground relays if such relays are required. If the power transformers are wye-delta, with the high-voltage side connected in wye and the neutral gounded, the neutral current can be used for polarizing. Of course, the question of whether a single power transformer can be relied on must be considered as in the preceding paragraph. If the high-voltage side is not a grounded wye, then a high-voltage source must be provided for directional-ground relays, and it may as well be used also by the phase relays.

Finally, if distance relays are involved, the desirability of "transformer-drop compensation" must be investigated. This subject will be treated in more detail when we consider the subject of transmission-line protection.

The necessary connections of potential transformers for obtaining the proper voltages for distance relays will be discussed later in this chapter. Directional-overcurrent relays can use any conventional potential-transformer connection.

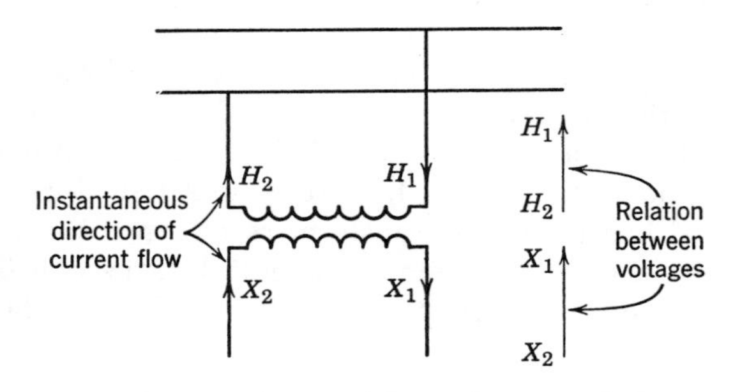

Fig. 6. Significance of potential-transformer polarity marks.

Polarity and Connections

The terminals of potential transformers are marked to indicate the relative polarities of the primary and secondary windings. Usually, the corresponding high-voltage and low-voltage terminals are marked "H_1" and "X_1," respectively (and "Y_1" for a tertiary). In capacitance potential devices, only the X_1 and Y_1 terminals are marked, the H_1 terminal being obvious from the configuration of the equipment.

The polarity marks have the same significance as for current transformers, namely, that, when current enters the H_1 terminal, it leaves the X_1 (or Y_1) terminal. The relation between the high and low voltages is such that X_1 (or Y_1) has the same instantaneous polarity as H_1, as shown in Fig. 6. Whether a transformer has additive or subtractive polarity may be ignored because it has absolutely no effect on the connections.

Distance relays for interphase faults must be supplied with voltage corresponding to primary phase-to-phase voltage, and any one of the three connections shown in Fig. 7 may be used. Connection A is chosen when polarizing voltage is required also for directional-ground relays; this will be discussed later in this chapter. The equivalent of connection A is the only one used if capacitance potential devices are involved. Connections B and C do not provide means for polariz-

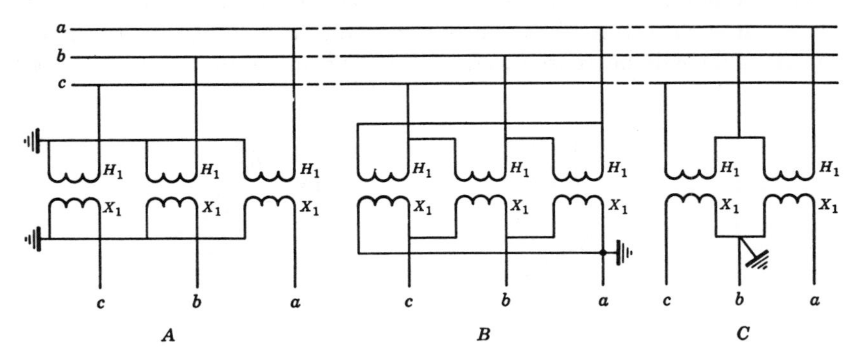

Fig. 7. Connections of potential transformers for distance relays.

ing directional-ground relays; of these two, connection C is the one generally used because it is less expensive since it employs only two potential transformers. The burden on each potential transformer is less in connection B, which is the only reason it would ever be chosen.

The voltages between the secondary leads for all three connections of Fig. 7 are the same, and in terms of symmetrical components are:

$$\begin{aligned} V_{ab} &= V_a - V_b \\ &= V_{a1} + V_{a2} + V_{a0} - V_{b1} - V_{b2} - V_{b0} \\ &= (1 - a^2)V_{a1} + (1 - a)V_{a2} \\ &= (\tfrac{3}{2} + j\sqrt{3}/2)V_{a1} + (\tfrac{3}{2} - j\sqrt{3}/2)V_{a2} \end{aligned}$$

Similarly,

$$\begin{aligned} V_{bc} &= (1 - a^2)V_{b1} + (1 - a)V_{b2} \\ &= a^2(1 - a^2)V_{a1} + a(1 - a)V_{a2} \\ &= (a^2 - a)V_{a1} + (a - a^2)V_{a2} \\ &= -j\sqrt{3}\,V_{a1} + j\sqrt{3}\,V_{a2} \end{aligned}$$

$$\begin{aligned} V_{ca} &= (1 - a^2)V_{c1} + (1 - a)V_{c2} \\ &= a(1 - a^2)V_{a1} + a^2(1 - a)V_{a2} \\ &= (a - 1)V_{a1} + (a^2 - 1)V_{a2} \\ &= (-\tfrac{3}{2} + j\sqrt{3}/2)V_{a1} + (-\tfrac{3}{2} - j\sqrt{3}/2)V_{a2} \end{aligned}$$

It will be observed that these relations are similar to those obtained for the output currents of the delta-connected CT's of Chapter 7, Fig. 7.

LOW-TENSION VOLTAGE FOR DISTANCE RELAYS

The potential transformers must be connected to the low-voltage source in such a way that the phase-to-phase voltages on the high-voltage side will be reproduced. The connection that must be used

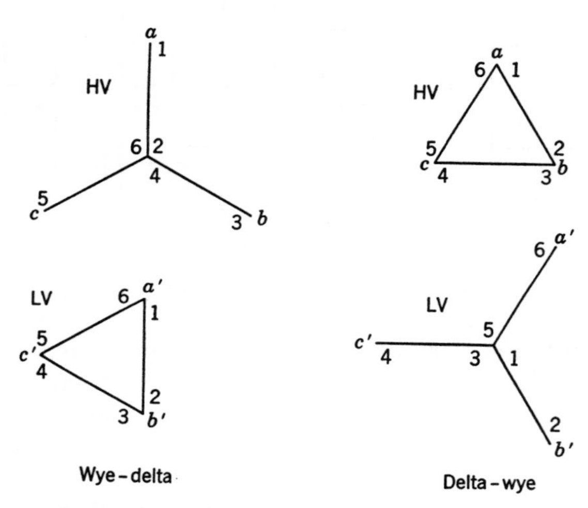

Fig. 8. Three-phase voltages for standard connection of power transformers.

will depend on the power-transformer connections. If, as is not usually the case, the power-transformer bank is connected wye-wye or delta-delta, the potential-transformer connections would be the same as though the potential transformers were on the high-voltage side. Usually, however, the power transformers are connected wye-delta or delta-wye.

First, let us become acquainted with the standard method of connecting wye-delta or delta-wye power transformers. Incidentally, in stating the connections of a power-transformer bank, the high-voltage connection is stated first; thus a wye-delta transformer bank has its high-voltage side connected in wye, etc. The standard method of connecting power transformers does not apply to potential transformers (which are connected as required), but the technique involved in making the desired connections will apply also to potential transformers. The standard connection for power transformers is that, with balanced three-phase load on the transformer bank, the current in each phase on the high-voltage side will lead by 30° the current in each corresponding phase on the low-voltage side. Also, the no-load phase-to-phase voltages on the high-voltage side will lead the corresponding low-voltage phase-to-phase voltages by 30°. For this to be true, the three-phase voltages must be as in Fig. 8, where a' corresponds to a, b' to b, and c' to c. The numbers on the voltage vectors of Fig. 8 designate the corresponding ends of the transformer windings, 1–2 designating the primary and secondary windings of one transformer, etc. Now, consider three single-phase transformers as in Fig. 9 with their primary and secondary windings designated 1–2, etc. If we assume that the transformers are rated for either phase-to-phase or phase-to-

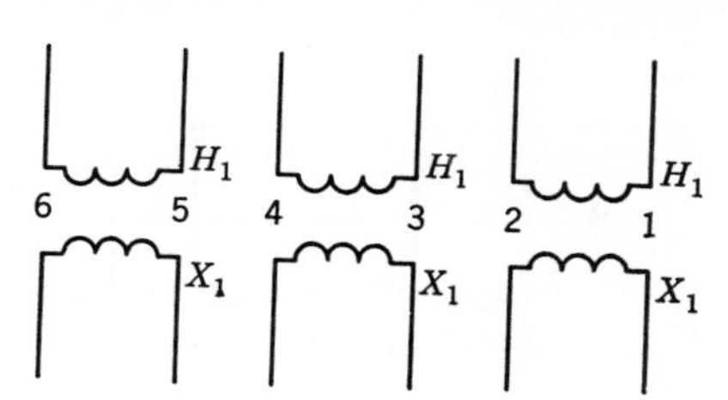

Fig. 9. Numbering the ends of the transformer windings preparatory to making three-phase connections.

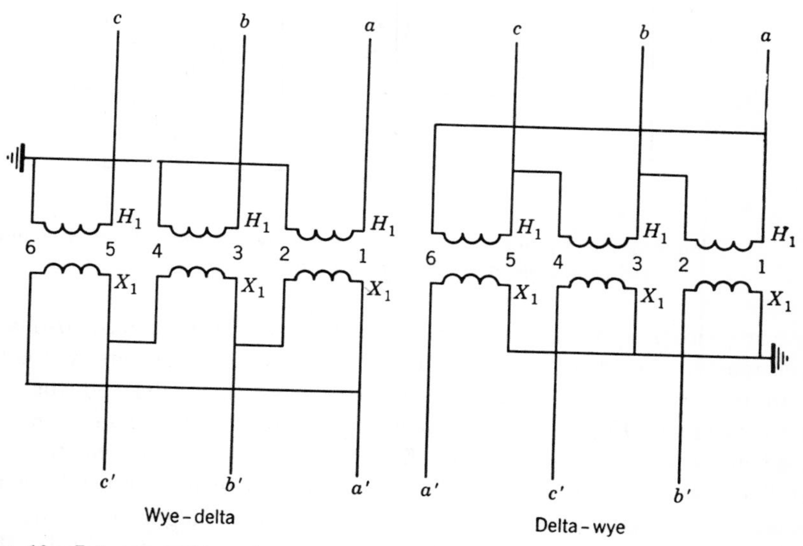

Fig. 10. Interconnecting the transformers of Fig. 9 according to Fig. 8 to get standard connections.

ground connection, it is only necessary to connect together the numbered ends that are shown connected in Fig. 8, and the connections of Fig. 10 will result.

We can now proceed to examine the connections of potential

transformers on the low-voltage side that are used for the purpose of supplying voltage to distance relays. Figure 11 shows the connections if the power transformers are connected wye-delta. Figure 12 shows the connections if the power transformers are connected delta-wye.

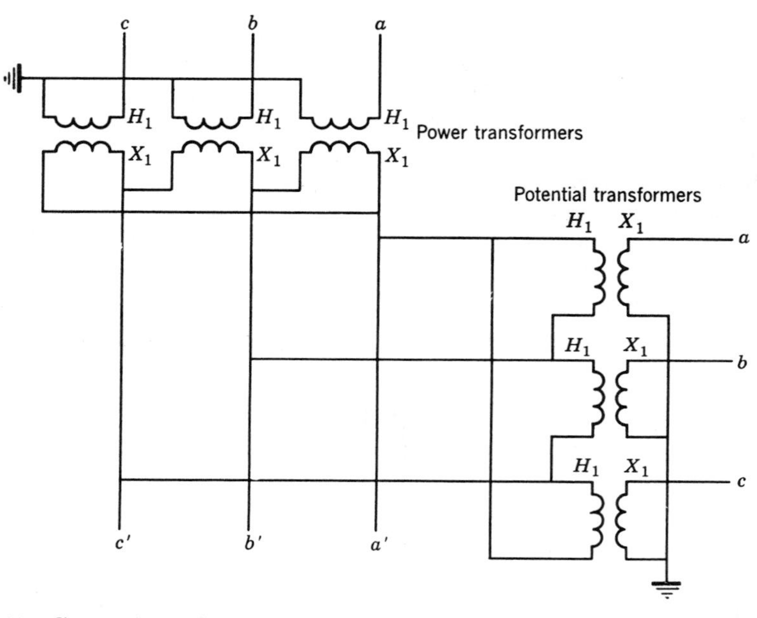

Fig. 11. Connections of potential transformers on low-voltage side of wye-delta power transformer for use with distance relays.

For either power-transformer connection, the phase-to-phase voltages on the secondary side of the potential transformers will contain the same phase-sequence components as those derived for the connections of Fig. 7, if we neglect the voltage drop or rise owing to load or fault currents that may flow through the power transformer. If, for one reason or another, the potential transformers must be connected delta-delta or wye-wye, or if the voltage magnitude is incorrect, auxiliary potential transformers must be used to obtain the required voltages for the distance relays.

The information given for making the required connections for distance relays should be sufficient instruction for making any other desired connections for phase relays. Other application considerations involved in the use of low-voltage sources for distance and other relays will be discussed later.

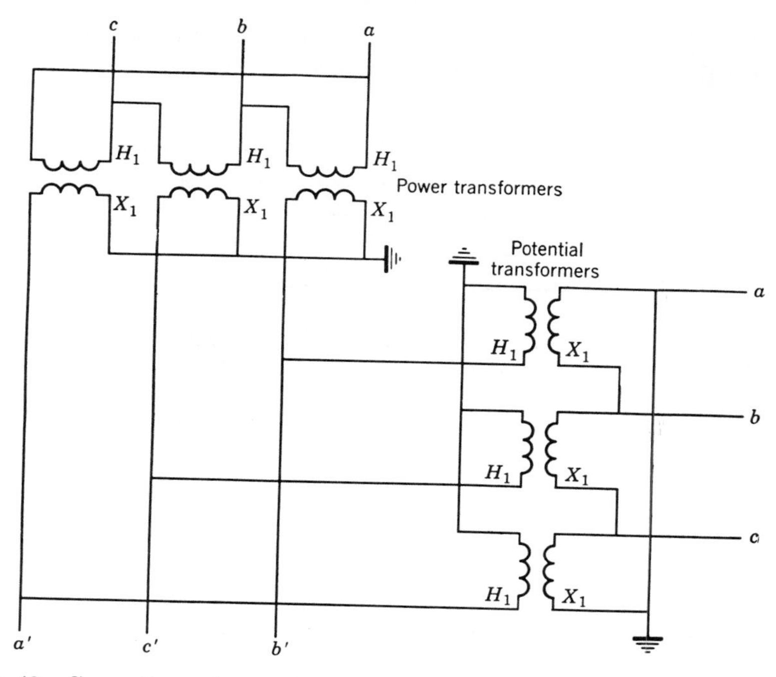

Fig. 12. Connections of potential transformers on low-voltage side of delta-wye power transformer for use with distance relays.

CONNECTIONS FOR OBTAINING POLARIZING VOLTAGE FOR DIRECTIONAL-GROUND RELAYS

The connections for obtaining the required polarizing voltage are shown in Fig. 13. This is called the "broken-delta" connection. The voltage that will appear across the terminals *nm* is as follows:

$$
\begin{aligned}
V_{nm} &= V_a + V_b + V_c \\
&= (V_{a1} + V_{a2} + V_{a0}) + (V_{b1} + V_{b2} + V_{b0}) + (V_{c1} + V_{c2} + V_{c0}) \\
&= V_{a0} + V_{b0} + V_{c0} = 3V_{a0} = 3V_{b0} = 3V_{c0}
\end{aligned}
$$

In other words, the polarizing voltage is 3 times the zero-phase-sequence component of the voltage of any phase.

The actual connections in a specific case will depend on the type of voltage transformer involved and on the secondary voltage required for other than ground relays. If voltage for distance relays must also be supplied, the connections of Fig. 14 would be used.

If voltage is required only for polarizing directional-ground relays, three coupling capacitors and one potential device, connected as in

Fig. 15 would suffice. The voltage obtained from this connection is 3 times the zero-phase-sequence component.

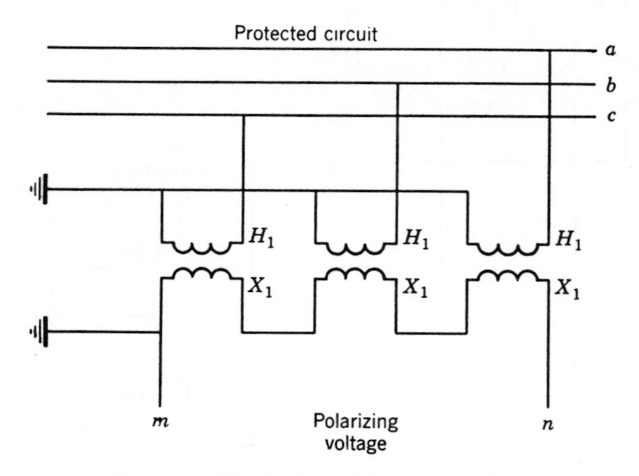

Fig. 13. The broken-delta connection.

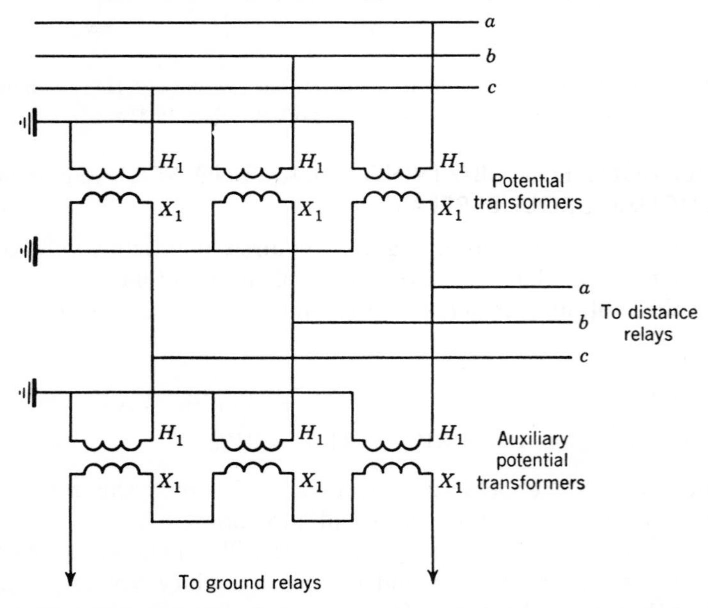

Fig. 14. Potential-transformer connections for distance and ground relays.

The connection of Fig. 15 cannot always be duplicated with bushing potential devices because at least some of the capacitance corresponding to the auxiliary capacitor C_2 might be an integral part of the bushing and could not be separated from it. The capacitance to ground of interconnecting cable may also have a significant effect.

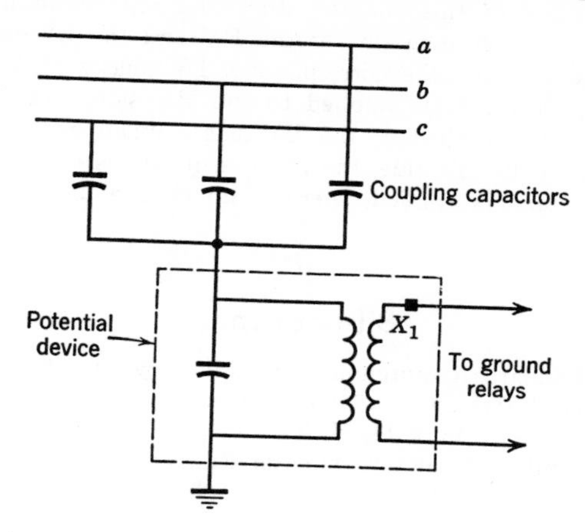

Fig. 15. Connection of three coupling capacitors and one potential device for providing polarizing voltage for directional-ground relays.

The three capacitance taps may be connected together, and a special potential device may be connected across the tap voltage as shown in Fig. 16, but the rated burden may be less than that of Table 1.

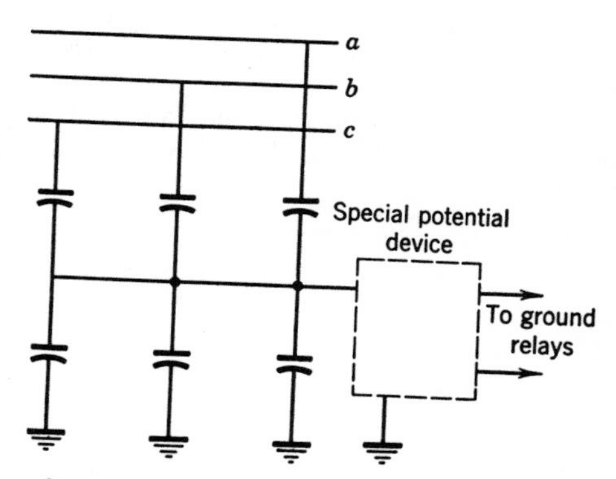

Fig. 16. Use of one potential device with three capacitance bushings.

Incidentally, a capacitance bushing cannot be used to couple carrier current to a line because there is no way to insert the required carrier-current choke coil in series with the bushing capacitance between the tap and ground, to prevent short-circuiting the output of the carrier-current transmitter.

Problem

1. Given a wye-delta power transformer with standard connections. According to the definitions of this chapter, draw the connection diagram and the three-phase voltage vector diagram for the HV and the LV sides, labeling the HV phases *a*, *b*, and *c* and the corresponding LV phases *a*′, *b*′, and *c*′, (1) for positive-phase-sequence voltage applied to the HV side, and (2) for negative-phase-sequence voltage applied to the HV side. When positive-phase-sequence voltage is applied to the HV side, the phase sequence is *a-b-c*. When negative-phase-sequence voltage is applied, there is no change in connections, but the phase sequence is *a-c-b*.

Bibliography

1. "American Standard Requirements, Terminology, and Test Code for Instrument Transformers," *Publ. C57.13–1954*, American Standards Association, Inc., 70 East 45th Street, New York 17, N. Y.

2. "Outdoor Coupling Capacitors and Capacitance Potential Device," *Publ. 31*, American Institute of Electrical Engineers, 33 West 39th Street, New York 18, N. Y.

"Standards for Coupling Capacitors and Potential Devices," *Publ. 47–123*, National Electrical Manufacturers Association, 155 East 44th Street, New York 17, N. Y.

3. "Application of Capacitance Potential Devices," by J. E. Clem, *AIEE Trans., 48* (1939), pp. 1–8. Discussions, pp. 8–10.

4. *Transient Performance of Electric Power Systems,* by R. Rudenberg, McGraw-Hill Book Co., New York, 1950.

5. "The Effects of Coupling-Capacitor Potential-Device Transients on Protective-Relay Operation," *AIEE Committee Report, AIEE Trans., 70* (1951), pp. 2089–2096. Discussions, p. 2096.

"Transient and Steady-State Performance of Potential Devices," by E. L Harder, P. O. Langguth, and C. A. Woods, Jr., *AIEE Trans., 59* (1940), pp. 91–99 Discussions, pp. 99–102.

9 METHODS FOR ANALYZING, GENERALIZING, AND VISUALIZING RELAY RESPONSE

The material that has been presented thus far will enable one to translate power-system currents and voltages into protective-relay response in any given case. From that standpoint, the material of this chapter is unnecessary. Nor is this chapter intended to teach one how to determine these currents and voltages by the methods of symmetrical components,[1,2] since it is assumed that this is known. The purpose of this chapter is best explained by a simple example.

In Chapter 7, we learned that a relay coil connected in the neutral lead of three wye-connected current transformers would have a current in it equal to $3I_{a0}$. Assuming that this is an overcurrent relay, we can immediately say that this relay will respond only to zero-phase-sequence current. This is important and useful knowledge, because we then know that the relay will respond only to faults involving ground. Furthermore, we do not have to calculate the positive- and negative-phase-sequence components of current in the circuit protected by the relay; all we need to know is the zero-phase-sequence component. Moreover, merely by looking at the phase-sequence diagram for any fault, we can tell whether this relay will receive zero-phase-sequence current, and how the magnitude and direction of this current will change with a change in fault location. Therefore, it is evident that we have at our disposal a much broader conception of the response of this relay than merely knowing that it will operate whenever it receives more than a certain magnitude of current. The value of being able to visualize and generalize relay response will become even more evident in the case of any relay that responds to certain combinations of voltage, current, and phase angle.

The *R*-*X* Diagram

The *R*-*X* diagram was introduced in Chapter 4 to show the operating characteristics of distance relays. Now, we are about to use it to study the response of distance-type relays to various abnormal system conditions. With this diagram, the operating characteristic of any distance relay can be superimposed on the same graph with any system characteristic, making the response of the relay immediately apparent.

A distance relay operates for certain relations between the magnitudes of voltage, current, and the phase angle between them. For any type of system-operating condition, there are certain characteristic relations between the voltage, current, and phase angle at a given distance-relay location in the system. Thus, the procedure is to construct a graph showing the relations between these three quantities (1) as supplied from the system, and (2) as required for relay operation.

PRINCIPLE OF THE *R-X* DIAGRAM

As described in Chapter 4, the basis of the R-X diagram is the resolution of the three variables—voltage, current, and phase angle—into two variables. This is done by dividing the rms magnitude of voltage by the rms magnitude of current and calling this an impedance "Z." (For the moment, let us not be concerned with the significance of this impedance.) Then, the resistance and reactance components of Z are derived, by means of the familiar relations $R = Z \cos \theta$ and $X = Z \sin \theta$. We shall call θ positive when I lags V, assuming certain relay connections and references. These values of R and X are the coordinates of a point on the R-X diagram representing a given combination of V, I, and θ. R and X may be positive or negative, but Z must always be positive; any negative values of Z obtained by substituting certain values of θ in an equation should be ignored since they have no significance.

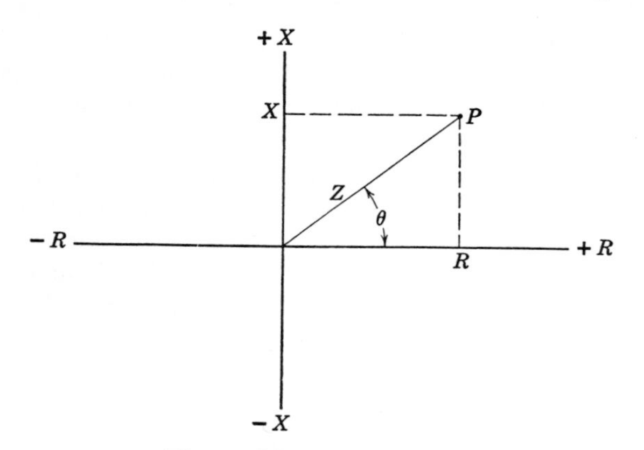

Fig. 1. The R-X diagram.

Figure 1 shows the R-X diagram and a point P representing fixed values of V, I, and θ, when I is assumed to lag V by an angle less than 90°. A straight line drawn from the origin to P represents Z, and θ is the angle measured counterclockwise from the $+R$ axis to Z.

It is probably obvious that P can be located from a knowledge of Z and θ without deriving the R and X components. Or, by calculating the complex ratio of V to I, the values of R and X can be obtained immediately without consideration of θ. If V, I, and θ vary, several points can be plotted, and a curve can be drawn through these points to represent the characteristic.

CONVENTIONS FOR SUPERIMPOSING RELAY AND SYSTEM CHARACTERISTICS

In order to superimpose the plot of a relay characteristic on the plot of a system characteristic to determine relay operation, both plots must be on the same basis. A given relay operates in response to voltage and current obtained from certain phases.[3] Therefore, the system characteristic must be plotted in terms of these same phase quantities as they exist at the relay location. Also, the coordinates must be in the same units. The per unit or percent system is generally employed for this purpose. If actual ohms are used, both the power-system and the relay characteristics must be on either a primary or a secondary basis, taking into account the current- and voltage-transformation ratios, as follows:

$$\text{Secondary ohms} = \text{Primary ohms} \times \frac{\text{CT ratio}}{\text{VT ratio}}$$

Finally, both coordinates must have the same scale, because certain characteristics are circular if the scales are the same.

It is necessary to establish a convention for relating a relay characteristic to a system characteristic on the R-X diagram. The convention must satisfy the requirement that, for a system condition requiring relay operation, the system characteristic must lie in the operating region of the relay characteristic. The convention is to make the signs of R and X positive when power and lagging reactive power flow in the tripping direction of the distance relays under balanced three-phase conditions. Lagging reactive power is here considered to flow in a certain direction when current flows in that direction as though into a load whose reactance is predominantly inductive. It is the practice to assume "delta" (defined later) currents and voltages as the basis for plotting both system and relay characteristics when phase distance relays are involved, or phase current and the corresponding phase-to-ground voltage (called "wye" quantities) when ground distance relays are involved. In either case, three relays are involved, each receiving current and voltage from different phases. Either the delta or the wye voltage-and-current combinations will give the same point on the R-X diagram under balanced three-phase conditions.

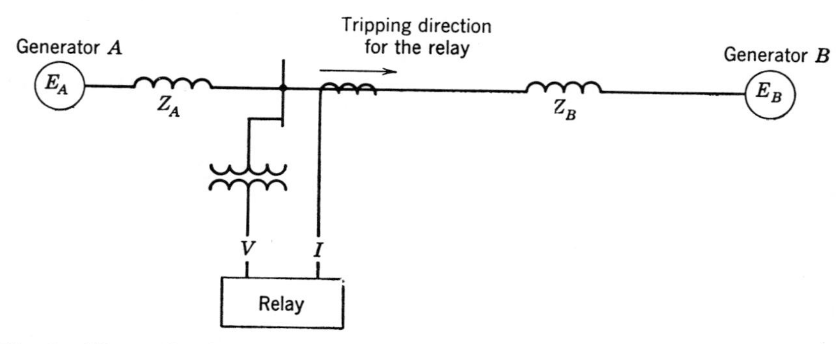

Fig. 2. Illustration for the convention for relating relay and system characteristics on the R-X diagram.

To illustrate this convention, refer to Fig. 2 where a distance relay is shown to be energized by voltage and current at a given location in a system. The coordinates of the impedance point on the R-X diagram representing a balanced three-phase system condition as viewed in the tripping direction of the relay will have the signs as shown in Table 1. Leading reactive power is here considered to flow in a certain direction when current flows in that direction as though into a load whose reactance is predominantly capacitive.

Table 1. Conventional Signs of *R and X*

Condition	Sign of R	Sign of X
Power from A toward B	+	
Power from B toward A	−	
Lagging reactive power from A toward B		+
Lagging reactive power from B toward A		−
Leading reactive power from A toward B		−
Leading reactive power from B toward A		+

The following relations give the numerical values of R and X for any balanced three-phase condition:

$$R = \frac{V^2 W}{W^2 + RVA^2}$$

$$X = \frac{V^2 RVA}{W^2 + RVA^2}$$

where V is the phase-to-phase voltage, W is the three-phase power, and RVA is the three-phase reactive power. R and X are components of the positive-phase-sequence impedance which could be obtained under balanced three-phase conditions by dividing any phase-to-neutral voltage by the corresponding phase current. All the quantities in the formulae must be expressed in actual values (i.e., ohms, volts, watts, and reactive volt-amperes), or all in percent or per unit.

By applying the proper signs to R and X, one can locate the point on the R-X diagram representing the impedance for any balanced three-phase system condition. For example, the point P of Fig. 1 would represent a condition where power and lagging reactive power were being supplied from A toward B in the tripping direction of the relay.

For a relay in the system of Fig. 2 whose tripping direction is opposite to that shown, interchange A and B in the designation of the generators of Fig. 2 and in Table 1; in other words, follow the rule already given that the signs of R and X are positive when power and lagging reactive power flow in the tripping direction of the relay. For example, if the point P of Fig. 1 represents a given condition of power and reactive power flow as it appears to the relay of Fig. 2, then to a relay with opposite tripping direction the same condition appears as a point diametrically opposite to P on Fig. 1. Occasionally, it may be desired to show on the same diagram the characteristics of relays facing in opposite directions. Then the rule cannot be followed, and care must be taken to avoid confusion.

Because it is customary to think of impedance in terms of combinations of resistance and reactance of circuit elements, one may wonder what significance the Z of an R-X diagram has. By referring to Fig. 2, it can easily be shown that the ratio of V to I is as follows:

$$\frac{V}{I} = Z = \frac{E_A Z_B + E_B Z_A}{E_A - E_B}$$

where all quantities are complex numbers, and where E_A and E_B are the generated voltages of generators A and B, respectively. Therefore, in general, Z is not directly related to any actual impedance of the system. From this general equation, system characteristics can be developed for loss of synchronism between the generators or for loss of excitation in either generator.

For normal load, loss of synchronism, loss of excitation, and three-phase faults—all balanced three-phase conditions—a system characteristic has the same appearance to each of the three distance relays that are energized from different phases. For unbalanced short circuits, the characteristic has a different appearance to each of the three relays, as we shall see shortly.

Other conventions involved in the use of the R-X diagram will be described as it becomes necessary.

By using distance-type relay units individually or in combination, any region of the R-X diagram can be encompassed or set apart from another region by one or more relay characteristics. With the knowledge of the region in which any system characteristic will lie or through which it will progress, one can place distance-relay characteristics in such a way that a desired kind of relay operation will be obtained only for a particular system characteristic.

Short Circuits

For general studies, it is the practice to think of a power system in terms of a two-generator equivalent, as in Fig. 3. The generated voltages of the two generators are assumed to be equal and in phase. The equivalent impedances to the left of the relay location and to

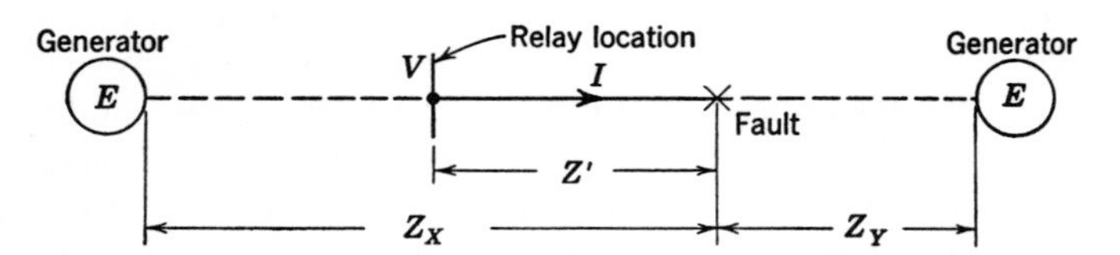

Fig. 3. Equivalent-system diagram for defining relay quantities during faults.

the right of the short circuit are those that will limit the magnitudes of the short-circuit currents to the actual known values. The short circuit is assumed to lie in the tripping direction of the relay.

The possible effect of mutual induction from a circuit paralleling the portion of the system between the relay and the fault will be neglected. Also, load and charging current will be neglected; however, they may not be negligible if the fault current is very low.

Nomenclature to identify specific values or combinations of the quantities indicated on Fig. 3 will be as follows:

Z = System impedance viewed both ways from the fault $= \dfrac{Z_X Z_Y}{Z_X + Z_Y}$

C = Ratio of the relay current I to the total current in the fault $= \dfrac{Z_Y}{Z_X + Z_Y}$

Subscripts a, b, and c denote phases a, b, and c, respectively. Throughout this book, positive phase sequence is assumed to be a-b-c. Subscripts 1, 2, and 0 denote positive, negative, and zero phase sequence, respectively.

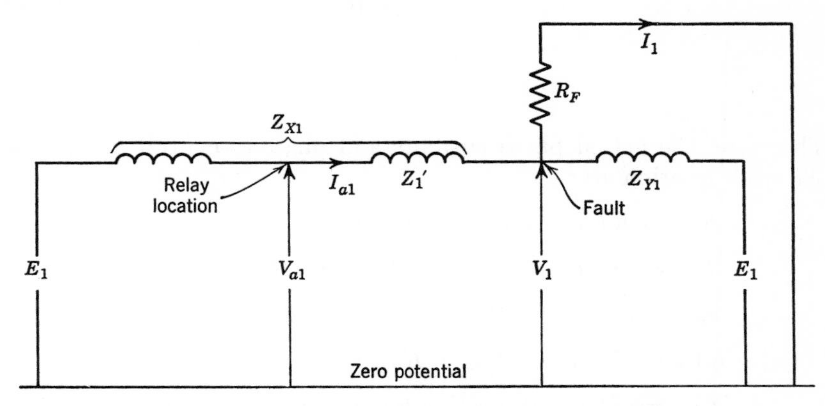

Fig. 4. Positive-phase-sequence network for a three-phase fault.

THREE-PHASE SHORT CIRCUITS

For a three-phase fault, the positive-phase-sequence network is as shown in Fig. 4 for the quantities of phase a. Whenever the term "three-phase" fault is used, it will be assumed that the fault is balanced, i.e., that only positive-phase-sequence quantities are involved. The quantity R_F is the resistance in the short circuit, assumed to be from phase to neutral of each phase.

By inspection, we can write:

$$I_1 = \frac{E_1}{Z_1 + R_F}$$

$$I_{a1} = \frac{Z_{Y1}I_1}{Z_{Y1} + Z_{X1}} = C_1I_1 = \frac{C_1E_1}{Z_1 + R_F}$$

$$V_1 = I_1R_F = \frac{E_1R_F}{Z_1 + R_F}$$

$$V_{a1} = V_1 + I_{a1}Z_1' = \frac{E_1R_F}{Z_1 + R_F} + \frac{C_1E_1Z_1'}{Z_1 + R_F}$$

Let

$$\frac{E_1}{Z_1 + R_F} = \frac{1}{K}$$

Then

$$KI_{a1} = C_1$$

$$KV_{a1} = R_F + C_1Z_1'$$

Since there are no negative- or zero-phase-sequence quantities for a three-phase fault, we can write:

$$KI_{a2} = 0$$

$$KI_{a0} = 0$$

$$KV_{a2} = 0$$

$$KV_{a0} = 0$$

Therefore, the actual phase currents and phase-to-neutral voltages at the relay location are:

$$KI_a = KI_{a1} + KI_{a2} + KI_{a0} = C_1$$

$$KI_b = a^2KI_{a1} + aKI_{a2} + KI_{a0} = a^2C_1$$

$$KI_c = aKI_{a1} + a^2KI_{a2} + KI_{a0} = aC_1$$

$$KV_a = KV_{a1} + KV_{a2} + KV_{a0} = R_F + C_1Z_1'$$

$$KV_b = a^2KV_{a1} + aKV_{a2} + KV_{a0} = a^2(R_F + C_1Z_1')$$

$$KV_c = aKV_{a1} + a^2KV_{a2} + KV_{a0} = a(R_F + C_1Z_1')$$

If delta-connected CT's are involved,

$$K(I_a - I_b) = (1 - a^2)C_1$$

$$K(I_b - I_c) = (a^2 - a)C_1$$

$$K(I_c - I_a) = (a - 1)C_1$$

The phase-to-phase voltages are:

$$KV_{ab} = K(V_a - V_b) = (1 - a^2)(R_F + C_1Z_1')$$

$$KV_{bc} = K(V_b - V_c) = (a^2 - a)(R_F + C_1Z_1')$$

$$KV_{ca} = K(V_c - V_a) = (a - 1)(R_F + C_1Z_1')$$

PHASE-TO-PHASE SHORT CIRCUITS

Figure 5 shows the phase a phase-sequence networks for a phase-b-to-phase-c fault. By inspection, we can write:

$$I_1 = \frac{E_1}{Z_1 + Z_2 + R_F} = \frac{E_1}{2Z_1 + R_F}$$

(assuming $Z_2 = Z_1$). We shall let $E_1/(2Z_1 + R_F) = 1/K$ for the same reason as for the three-phase short-circuit calculations, realizing that the values of K are not the same in both cases. It should also be realized that the values of R_F will probably be different in both cases.

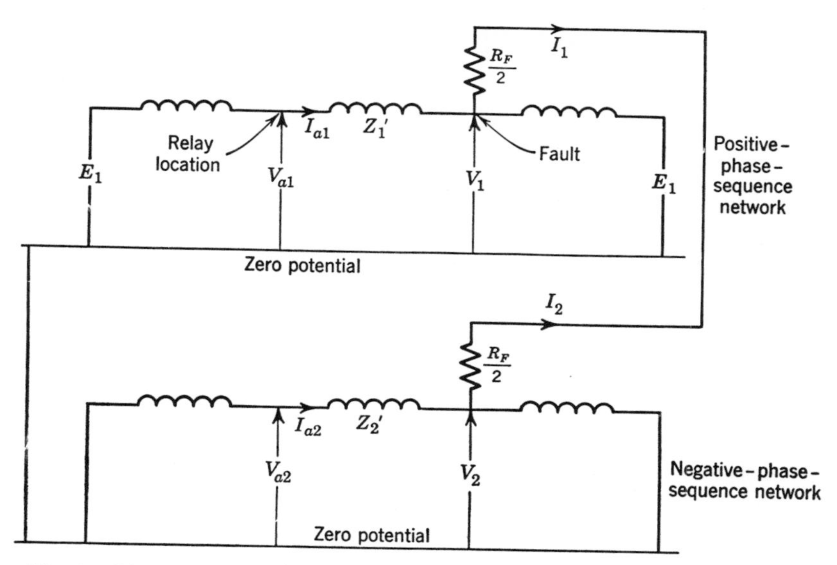

Fig. 5. Phase *a* phase-sequence networks for a phase-*b*-to-phase-*c* fault.

Here, R_F is the resistance between the faulted phases. Figure 5 shows $R_F/2$ in series with each network, in order to be consistent with certain references for this chapter (1,2,5). This is simply a caution not to attach too much significance to apparent differences in the R_F term. Continuing, we can write:

$$V_1 = (I_1 - I_2)\frac{R_F}{2} - I_2Z_2$$

$$= I_1R_F + I_1Z_1 \quad \text{since} \quad I_2 = -I_1 \text{ and } Z_2 = Z_1 \text{ (assumed)}$$

$$V_{a1} = V_1 + I_{a1}Z_1'$$

$$= I_1(R_F + Z_1 + C_1Z_1') \quad \text{since} \quad I_{a1} = C_1I_1 \text{ by definition}$$

$$= \frac{1}{K}(R_F + Z_1 + C_1Z_1')$$

$$KV_{a1} = R_F + Z_1 + C_1Z_1'$$

$$V_2 = -I_2Z_2 = I_1Z_1$$

$$V_{a2} = V_2 + I_{a2}Z_2'$$

But $Z_2' = Z_1'$ for transmission lines, and $I_{a2} = C_2I_2$ by definition, and hence:

$$V_{a2} = I_1Z_1 + C_2I_2Z_1'$$

We shall further assume that $C_2 I_2 = -C_1 I_1$, and hence:

$$V_{a2} = I_1 Z_1 - C_1 I_1 Z_1'$$
$$= I_1 (Z_1 - C_1 Z_1')$$
$$KV_{a2} = Z_1 - C_1 Z_1'$$

By definition, $I_{a1} = C_1 I_1$,

$$KI_{a1} = C_1$$

Since $I_{a2} = -I_{a1}$,

$$KI_{a2} = -KI_{a1} = -C_1$$

Since there are no zero-phase-sequence quantities,

$$KV_{a0} = 0$$
$$KI_{a0} = 0$$

We could continue, as for the three-phase fault, and write the actual currents and voltages at the relay location, but the technique is the same as before. The final values are given in Tables 2 and 3. Also,

Table 2. Currents during Faults

Quantity at Relay Location	Value of Quantity for Various Types of Fault		
	Three-Phase	Phase b to Phase c	Phase a to Ground
KI_{a1}	C_1	C_1	C_1
KI_{a2}	0	$-C_1$	C_1
KI_{a0}	0	0	C_0
KI_a	C_1	0	$C_0 + 2C_1$
KI_b	$a^2 C_1$	$(a^2 - a)C_1$	$C_0 - C_1$
KI_c	aC_1	$-(a^2 - a)C_1$	$C_0 - C_1$
$K(I_a - I_b)$	$(1 - a^2)C_1$	$-(a^2 - a)C_1$	$3C_1$
$K(I_b - I_c)$	$(a^2 - a)C_1$	$2(a^2 - a)C_1$	0
$K(I_c - I_a)$	$(a - 1)C_1$	$-(a^2 - a)C_1$	$-3C_1$
$K(I_a + I_b + I_c)$	0	0	$3C_0$
K	$\dfrac{Z_1 + R_F}{E_1}$	$\dfrac{2Z_1 + R_F}{E_1}$	$\dfrac{2Z_1 + Z_0 + 3R_F}{E_1}$

the calculations for a phase-a-to-ground fault will be omitted, but the values are given in the tables. The tables do not include the values for a phase-b-to-phase-c-to-ground fault because they are too complicated, and are not very significant. The purpose here is not to present all the data, but to show the technique involved in determining it, and it is felt that in this respect enough data will have been given. Complete data will be found in two different AIEE papers.[4,5]

Table 3. Voltages during Faults

Quantity at Relay Location	Value of Quantity for Various Types of Fault		
	Three-Phase	Phase b to Phase c	Phase a to Ground
KV_{a1}	$C_1Z_1'+R_F$	$C_1Z_1'+Z_1+R_F$	$C_1Z_1'+Z_1+Z_0+3R_F$
KV_{a2}	0	$Z_1-C_1Z_1'$	$C_1Z_1'-Z_1$
KV_{a0}	0	0	$C_0Z_0'-Z_0$
KV_a	$C_1Z_1'+R_F$	$2Z_1+R_F$	$2C_1Z_1'+C_0Z_0'+3R_F$
KV_b	$a^2(C_1Z_1'+R_F)$	$(a^2-a)C_1Z_1'-Z_1+a^2R_F$	$-C_1Z_1'+(a^2-a)Z_1+(a^2-1)Z_0+C_0Z_0'+3a^2R_F$
KV_c	$a(C_1Z_1'+R_F)$	$(a-a^2)C_1Z_1'-Z_1+aR_F$	$-C_1Z_1'+(a-a^2)Z_1+(a-1)Z_0+C_0Z_0'+3aR_F$
$K(V_a-V_b)$	$(1-a^2)(C_1Z_1'+R_F)$	$(a-a^2)C_1Z_1'+3Z_1+(1-a^2)R_F$	$3C_1Z_1'-(a^2-a)Z_1-(a^2-1)(Z_0+3R_F)$
$K(V_b-V_c)$	$(a^2-a)(C_1Z_1'+R_F)$	$2(a^2-a)C_1Z_1'+(a^2-a)R_F$	$2(a^2-a)Z_1+(a^2-a)(Z_0+3R_F)$
$K(V_c-V_a)$	$(a-1)(C_1Z_1'+R_F)$	$(a-a^2)C_1Z_1'-3Z_1+(a-1)R_F$	$-3C_1Z_1'+(a-a^2)Z_1+(a-1)(Z_0+3R_F)$
$K(V_a+V_b+V_c)$	0	0	$3(C_0Z_0'-Z_0)$
K	$\dfrac{Z_1+R_F}{E_1}$	$\dfrac{2Z_1+R_F}{E_1}$	$\dfrac{2Z_1+Z_0+3R_F}{E_1}$

DISCUSSION OF ASSUMPTIONS

The error in assuming the positive- and negative-phase-sequence impedances to be equal depends on the operating speed of the relay involved and on the location of the fault and relay relative to a generator. All the error is in the assumed generator impedances. Whereas the negative-phase-sequence reactance of a generator is constant, the positive-phase-sequence reactance will grow larger from subtransient to synchronous within a very short time after a fault occurs. However, unless the fault is at the terminals of a generator, the constant impedance of transformers and lines between a generator and the fault will tend to lessen the effect of changes in the generator's positive-phase-sequence reactance.

The assumption that positive- and negative-phase-sequence impedances are equal is sufficiently accurate for analyzing the response of high-speed distance-type relay units. For the short time that it takes a high-speed relay to operate after a fault has occurred, the equivalent positive-phase-sequence reactance of a generator is nearly enough equal to the negative-phase-sequence reactance so that there is negligible over-all error in assuming them to be equal. Actually, it is usually the practice to use the rated-voltage direct-axis transient reactance for both.

When time-delay distance-type units are involved, such as for back-up relaying, the assumption of equal positive- and negative-phase-sequence impedances could produce significant errors. This is especially true for a fault near a generator and with the relay located between the generator and the fault, in which event the relay's operation would be largely dependent on the generator characteristics alone. On the other hand, the "reach" of a back-up relay does not have to be as precise as that of high-speed primary relaying, which permits more error in its adjustment. Suffice it to say that this possibility of error, even with time-delay relays, is generally ignored for distance relaying except in very special cases.

Actually, this consideration of the possible error resulting from assuming equal positive- and negative-phase-sequence impedances is somewhat academic where distance relays are involved. Whatever error there may be will generally affect the response of only those relays on whose operation we do not ordinarily rely anyway. An exception to this is for faults on the other side of a wye-delta or delta-wye transformer, which we shall consider later.

Neglecting the effect of mutual induction, where mutual induction exists, would noticeably affect the response of only ground relays.

It is the practice to ignore mutual induction when dealing with phase relays. The effect of mutual induction on the response of ground distance relays may have to be considered; this is treated in detail in Reference 3.

DETERMINATION OF DISTANCE-RELAY OPERATION FROM THE DATA OF TABLES 2 AND 3

Modern distance relays are single-phase types. The three relays used for phase-fault protection are supplied with the following combinations of current and voltage:

Current	Voltage
$I_a - I_b$	$V_{ab} = V_a - V_b$
$I_b - I_c$	$V_{bc} = V_b - V_c$
$I_c - I_a$	$V_{ca} = V_c - V_a$

These quantities are often called "delta currents" and "delta voltages." Each relay is intended to provide protection for faults involving the phases between which its voltage is obtained.

It was shown in Chapter 4 that all distance relays operate according to some function of the ratio of the voltage to the current supplied to the relay. We are now able to determine what the ratio of these quantities is for each phase relay for any type of fault, by using the data of Tables 2 and 3. These ratios are given in Table 4.

Table 4. Impedances "Seen" by Phase Distance Relays for Various Kinds of Short Circuits

	Value of Ratio for Various Types of Fault		
Ratio	3-Phase	Phase b to Phase c	Phase a to Ground
$\dfrac{K(V_a - V_b)}{K(I_a - I_b)}$	$Z_1' + \dfrac{R_F}{C_1}$	$Z_1' - j\sqrt{3}\,Z_{X1} - \dfrac{a}{C_1} R_F$	$Z_1' + j\dfrac{\sqrt{3}}{3} Z_{X1} + \dfrac{(1 - a^2)(Z_0 + 3R_F)}{3C_1}$
$\dfrac{K(V_b - V_c)}{K(I_b - I_c)}$	$Z_1' + \dfrac{R_F}{C_1}$	$Z_1' + \dfrac{R_F}{2C_1}$	∞
$\dfrac{K(V_c - V_a)}{K(I_c - I_a)}$	$Z_1' + \dfrac{R_F}{C_1}$	$Z_1' + j\sqrt{3}\,Z_{X1} - \dfrac{a^2}{C_1} R_F$	$Z_1' - j\dfrac{\sqrt{3}}{3} Z_{X1} - \dfrac{(a - 1)(Z_0 + 3R_F)}{3C_1}$

For a three-phase fault, all three relays "see" the positive-phase-sequence impedance of the circuit between the relays and the fault, plus a multiple of the arc resistance. This multiple depends on the fraction of the total fault current that flows at the relay location and is larger foı smaller fractions.

For a phase-to-phase fault at a given location, the relay energized by the voltage between the faulted phases sees the same impedance as for three-phase faults at that location, except, possibly, for differences in the R_F term. As already noted, the value of R_F may be different for the two types of fault. The subject of fault resistance will be treated later in more detail when we consider the application of distance relays. The other two relays see other impedances.

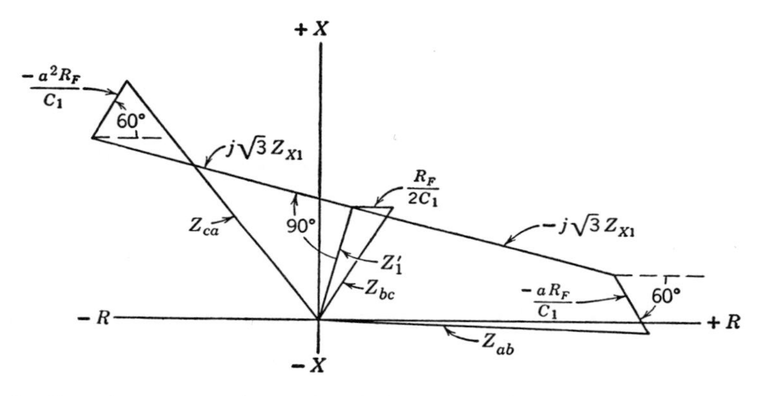

Fig. 6. Impedances seen by each of the three phase distance relays for a phase-b-to-phase-c fault.

These values of impedance seen by the three relays can be shown on an R-X diagram, as in Fig. 6. The terms Z_{bc}, Z_{ab}, and Z_{ca} identify the impedances seen by the relays obtaining voltage between phases bc, ab, and ca, respectively. If we were to follow rigorously the conventions already described for the R-X diagram, we should draw three separate diagrams, one each for the constructions for obtaining Z_{bc}, Z_{ab}, and Z_{ca}. This is because each one involves the ratio of different quantities. However, since the relay characteristics would be the same on all three diagrams, it is more convenient to put all impedance characteristics on the same diagram, and also it reveals certain interesting interrelations, as we shall see shortly.

For phase-b-to-phase-c faults, with or without arcs, and located anywhere on a line section from the relay location out to a certain distance, the heads of the three impedance radius vectors will lie on or within the boundaries of the shaded areas of Fig. 7. These areas would be generated if we were to let Z_1' and R_F of Fig. 6 increase from zero to the value shown.

To use the data shown by Fig. 7, it is only necessary to superimpose the characteristic of any distance relay using one of the combinations of delta current and voltage in order to determine its operating tendencies. This has been done on Fig. 7 for an impedance-type distance relay adjusted to operate for all faults having any impedance within the shaded area Z_{bc}. Had we shown the three fault areas Z_{ab}, Z_{bc}, and Z_{ca} on three different R-X diagrams, the relay characteristic would still have looked the same on all three

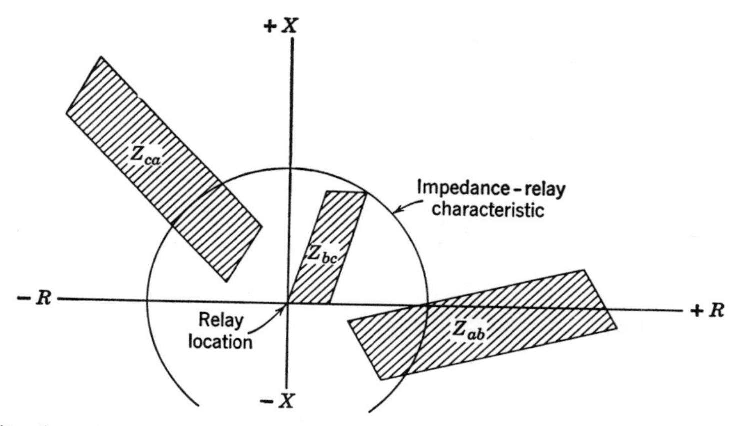

Fig. 7. Impedance areas seen by the three phase distance relays for various locations of a phase-b-to-phase-c fault with and without an arc.

diagrams since the practice is to adjust all three relays alike. Therefore, the relay characteristic of Fig. 7 may be thought of as that of any one of the three relays. For any portion of shaded area lying inside the relay characteristic, it is thereby indicated that for certain locations of the phase-b-to-phase-c fault, the relay represented by that area will operate.

For the adjustment of Fig. 7, all three relays will operate for nearby faults, represented by certain values of Z_{ca} and Z_{ab}, where the shaded areas fall within the operating characteristic of the impedance relay. Such operation is not objectionable, but the target indications might lead one to conclude that the fault was three-phase instead of phase-to-phase.

We may generalize the picture of Fig. 7, and think of the Z_{bc} area as representing the appearance of a phase-to-phase fault to the distance relay that is supposed to operate for that fault. Then, the Z_{ca} area shows the appearance of the fault to the relay using the voltage lagging the faulted phase voltage (sometimes called the "lagging" relay); and the Z_{ab} area shows the appearance of the fault to the relay using the voltage leading the faulted phase voltage (sometimes called the "leading" relay).

We can construct the diagram of Fig. 6 graphically, as in Fig. 8, neglecting the effect of arc resistance. Draw the line OF equal to Z_1' of Fig. 6. Extend OF to A, making FA equal to Z_{X1}, the positive-phase-sequence impedance from the fault to the end of the system back

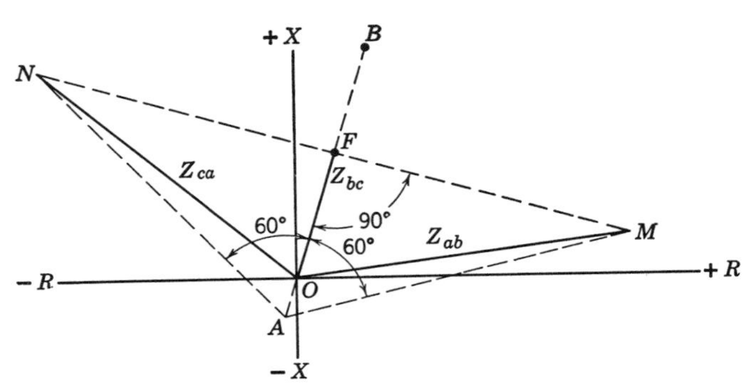

Fig. 8. Graphical construction of Fig. 6, neglecting the effect of arc resistance.

of the relay location. Actually, FA is Z_{X2}, but we are assuming the positive- and negative-phase-sequence impedances to be equal. Draw a line through F perpendicular to AF. From A, draw lines at 60° to AF until they intersect the perpendicular to AF at M and N. Then:

$$OF = Z_{bc}$$

$$OM = Z_{ab}$$

$$ON = Z_{ca}$$

The proof that this construction is valid is that the tangent of the angle FAM is equal to:

$$\frac{\sqrt{3}\, Z_{X1}}{Z_{X1}} = \sqrt{3}$$

which is the tangent of 60°. The construction for showing the effect of fault resistance can be added to Fig. 8 by the method used for Fig. 6.

It should be noted that the line segments FO and OA would not lie on the same straight line as in Fig. 8 if their X/R ratios were different. Then a straight line FA that did not go through the origin would be drawn, and the rest of the construction would be based on it, the perpendicular to it being drawn through F, and the lines AM and AN being drawn at 60° to it.

The appearance of a phase-a-to-ground fault to phase distance relays is shown in Fig. 9. Except for the last term in Table 4, this diagram can be constructed graphically by drawing the two construction lines at 30° to FA.

Similar constructions can be made for distance relays used for ground-fault protection. However, these constructions will not be described here. A paper containing a complete treatment of this subject is listed in the Bibliography.[5]

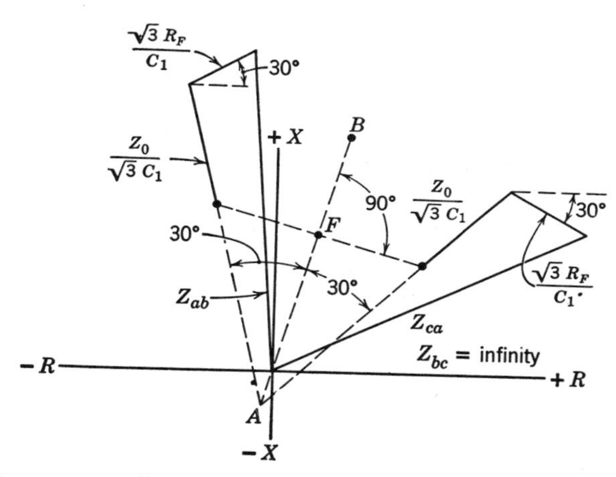

Fig. 9. Appearance of a phase-a-to-ground fault to phase distance relays.

The foregoing constructions, showing what different kinds of faults look like to distance relays, are of more academic than practical value. In other words, although these constructions are extremely helpful for understanding how distance relays respond to different kinds of faults, one seldom has to make such constructions to apply distance relays. It is usually only necessary to locate the point representing the appearance of a fault to the one relay that should operate for the fault. In other words, only the positive-phase-sequence impedance between the relay and the fault is located. The information gained from such constructions explains why relay target indications cannot always be relied on for determining what kind of fault occurred; in other words, three targets (apparently indicating a three-phase fault) might show for a nearby phase-to-phase fault. Or a phase relay might show a target for a nearby single-phase-to-ground fault, etc. The construction has also been useful for explaining a tendency of certain ground relays to "overreach" for phase faults;[5] because of this tendency it is customary to provide means for blocking tripping by ground distance relays when a fault involves two or more phases, or at least to block tripping by the ground relays that can overreach.

The principal use of the foregoing type of construction for practical application purposes is when one must know how distance relays will respond to faults on the other side of a power transformer. This will now be considered.

EFFECT OF A WYE-DELTA OR A DELTA-WYE POWER TRANSFORMER BETWEEN DISTANCE RELAYS AND A FAULT

For other than three-phase faults, the presence of a wye-delta or delta-wye transformer between a distance relay and a fault changes the complexion of the fault as viewed from the distance-relay location,[6] because of the phase shift and the recombination of the currents and voltages from one side to the other of the power transformer. In passing through the transformer from the fault to the relay location, the positive-phase-sequence currents and voltages of the corresponding phases are shifted 30° in one direction, and the negative-phase-sequence quantities are shifted 30° in the other direction. The zero-phase-sequence quantities are not transmitted through such a power transformer. The 30° shift described here is not at variance with the 90° shift described in some textbooks on symmetrical components. The 90° shift is a simpler mathematical manipulation, but it does not apply to what is usually considered the *corresponding* phase quantities.

In terms of only the magnitude of per unit quantities in an equivalent system diagram, the only effect of the presence of a power transformer is its impedance in the phase-sequence circuits. But, to combine the per unit phase-sequence quantities and convert them to volts and amperes at the relay location, one must first shift the per unit quantities by the proper phase angle from their positions on the fault side of the power transformer. If the power transformer has the standard connections described in Chapter 8 whereby the high-voltage phase currents lead the corresponding low-voltage phase currents by 30° under balanced three-phase conditions, the positive-phase-sequence currents and voltages on the HV side lead the corresponding positive-phase-sequence components on the LV side by 30°. (Under balanced three-phase conditions, only positive-phase-sequence quantities exist, and the vector diagram for this condition is a positive-phase-sequence diagram; this is a good way to determine the direction and amount of shift for any connection.) The negative-phase-sequence quantities on the HV side lag the corresponding LV quantities by 30°, or, in other words, they are shifted by the same amount as the positive-phase-sequence quantities, but in the opposite direction.

For three-phase faults where there are only positive-phase-sequence currents and voltages, the fact that these quantities are shifted 30° in going through the transformer may be ignored because all of them are shifted in the same direction. On the R-X diagram, a three-phase fault on the other side of a power transformer is represented merely by adding to the positive-phase-sequence impedance between the relay and the transformer the positive-phase-sequence impedance of the transformer and of the line between the transformer and the fault. If the impedances are in ohms, it is only necessary to be sure that the impedances of the transformer and the line between it and the fault are expressed in terms of the rated voltage of the transformer on the relay side; if the impedances are in percent or per unit, each impedance should be based on the base voltage of the portion of the circuit in which it exists, exactly the same as for any short-circuit study. It is probably evident that all three relays will operate alike for a three-phase fault.

The net effect of the shift in the positive- and negative-phase-sequence components on the impedance appearing to a distance relay on the HV side for a fault on the LV side is compared in Table 5 with data from Table 4 for a phase-b-to-phase-c fault. For the purposes of comparison, the same impedance values between the relay and the fault are assumed for both fault locations, and the effect of fault resistance is neglected.

Table 5. Comparison of Appearance of Phase-*b*-to-Phase-*c* Faults on Either Side of a Wye-Delta or Delta-Wye Power Transformer to Distance Relays on the HV Side

	HV Fault (From Table 4)	LV Fault
Z_{ab}	$Z_1' - j\sqrt{3}\,Z_{X1}$	$Z_1' - j\dfrac{\sqrt{3}}{3}Z_{X1}$
Z_{bc}	Z_1'	$Z_1' + j\dfrac{\sqrt{3}}{3}Z_{X1}$
Z_{ca}	$Z_1' + j\sqrt{3}\,Z_{X1}$	∞

The effect of the LV fault on the graphical construction for showing these impedances on an R-X diagram is to shift the construction lines

AM, *AN*, and *AF* by 30° in the counterclockwise direction from their positions in Fig. 8 to the new positions *AM′*, *AN′*, and *AF′* as illustrated in Fig. 10. In other words, $Z_{ab} = OM'$, $Z_{bc} = OF'$, and Z_{ca} = infinity

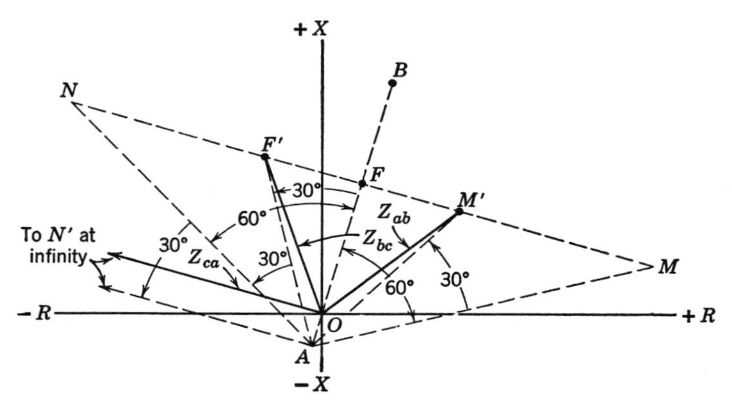

Fig. 10. Appearance to phase distance relays of a phase-*b*-to-phase-*c* fault on the low-voltage side of a wye-delta or delta-wye power transformer.

because the construction line is parallel to the line *MN*. Table 5 and Fig. 10 apply whether the power transformer is connected wye-delta or delta-wye, so long as it has the standard connections described in

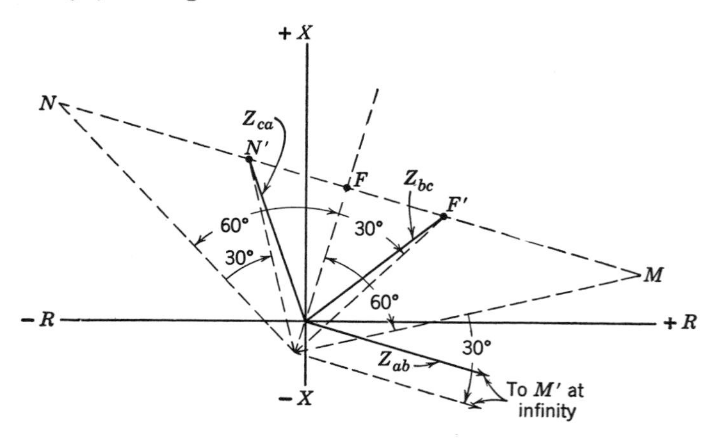

Fig. 11. Appearance, to phase distance relays on the LV side, of a phase-*b*-to-phase-*c* fault on the HV side of a wye-delta or delta-wye power transformer.

Chapter 8. It will be noted that the construction of Fig. 10 also uses as a starting point the line *OF* which represents the positive-phase-sequence impedance for a three-phase fault at the fault location, including the transformer impedance.

If the relay is on the LV side of the power transformer, and the phase-*b*-to-phase-*c* fault is on the HV side, the construction of Fig. 11 applies. It will be noted that here the construction lines are shifted 30° clockwise from their positions for the fault on the same side as the relay.

Comparing the foregoing *R-X* diagrams, we note that as we move the fault from the relay side of the power transformer to the other side, the construction lines shift 30° in the same direction that the negative-phase-sequence components shift in going through the transformer from the relay to the other side. This assumes that the transformer has the standard connections.

Space does not permit the consideration of the appearance to phase distance relays of single-phase-to-ground faults on the other side of a transformer, or of the appearance to ground distance relays of various types of faults on the other side of a transformer. Suffice it to say, a single-phase-to-ground fault on one side looks like a phase-to-phase fault on the other side, and vice versa, when wye-delta or delta-wye power transformers are involved.

The significant information that we get from studies of this kind may be summarized as follows:

1. If we want to be sure that any kind of distance relay on one side will not operate for any kind of fault on the other side, we must be sure that it will not operate for a three-phase fault.
2. If we want to be sure that a phase distance relay on one side will operate for any kind of fault on the other side, we must be sure that it will operate for a phase-to-phase fault.
3. If we want to be sure that a ground distance relay on one side will operate for any kind of fault on the other side, we must be sure that it will operate for a single-phase-to-ground fault. This assumes that the system neutral is grounded in such a way that a single-phase-to-ground fault at the location under consideration will cause short-circuit current to flow at the relay location.

After one has tried to adjust distance relays to provide back-up protection for certain faults on the other side of a transformer bank, it will become evident that it would be more practical either to use back-up units connected to measure distance correctly for such faults or to use the "reversed-third-zone" principle which will be described in Chapter 14. The connection of back-up units to measure distance correctly is the same principle as that when low-tension current and voltage are used, except that the high-tension quantities must be so combined as to duplicate the low-tension quantities.

Power Swings and Loss of Synchronism

Power swings are surges of power such as those after the removal of a short circuit, or those resulting from connecting a generator to its system at an instant when the two are out of phase. The charac-

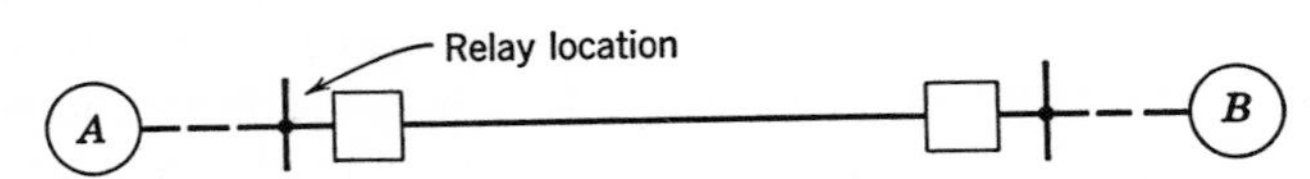

Fig. 12. One-line diagram of a system, illustrating loss-of-synchronism characteristics.

teristic of a power swing is the same as the early stages of loss of synchronism, and hence the loss-of-synchronism characteristic can describe both phenomena.

Consider the one-line diagram of Fig. 12 where a section of transmission line is shown with generating sources beyond either end of the line section. As for short-circuit studies, it is the practice, when possible, to represent the system by its two-generator equivalent.

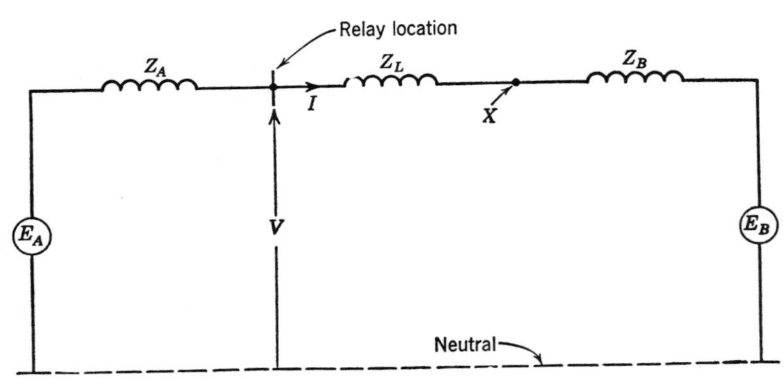

Fig. 13. System constants and relay current and voltage for Fig. 12.

The location of a relay is shown whose response to loss of synchronism between the two generating sources is to be studied. Each generating source may be either an actual generator or an equivalent generator representing a group of generators that remain in synchronism. (If generators within the same group lose synchronism with each other, this simple approach to the problem cannot be used, and a network-analyzer study may be required to provide the desired data.) The effects of shunt capacitance and of shunt loads are neglected.

Figure 13 indicates the pertinent phase-to-neutral (positive-phase-sequence) impedances and the generated voltages of the system of Fig. 12, and also the phase current and phase-to-neutral voltage at the relay location. An equivalent generator reactance of 90% of the direct-axis rated-current transient reactance most nearly represents a generator during the early stages of a power swing, and should be used for calculating the impedances in such an equivalent circuit.[7] In practice the generator reactance and the generated voltage are assumed to remain constant.

We can derive the relay quantities as follows:

$$I = \frac{E_A - E_B}{Z_A + Z_L + Z_B}$$

$$V = E_A - IZ_A = E_A - \frac{(E_A - E_B)Z_A}{Z_A + Z_L + Z_B}$$

$$\frac{V}{I} = Z = \frac{E_A}{E_A - E_B}(Z_A + Z_L + Z_B) - Z_A$$

If we let E_B be the reference, and let E_A advance in phase ahead of E_B by the angle θ, and if we let the magnitude of E_A be equal to nE_B, where n is a scalar, then

$$\frac{E_A}{E_A - E_B} = \frac{n(\cos\theta + j\sin\theta)}{n(\cos\theta + j\sin\theta) - 1}$$

This equation will resolve into the form:

$$\frac{E_A}{E_A - E_B} = \frac{n[(n - \cos\theta) - j\sin\theta]}{(n - \cos\theta)^2 + \sin^2\theta}$$

If we take the special case where $n = 1$, the equation becomes:

$$\frac{E_A}{E_A - E_B} = \frac{1}{2}\left(1 - j\cot\frac{\theta}{2}\right)$$

Therefore, Z becomes:

$$Z = \frac{Z_A + Z_L + Z_B}{2}\left(1 - j\cot\frac{\theta}{2}\right) - Z_A$$

This value of Z is shown on the R-X diagram of Fig. 14 for a particular value of θ less than 180°. Point P is thereby seen to be a point on the loss-of-synchronism characteristic. Further thought will reveal that all other points on the loss-of-synchronism characteristic will lie on the dashed line through P. This line is the perpendicular bisector of the straight line connecting A and B.

The loss-of-synchronism characteristic has been expressed in terms of the ratio of a phase-to-neutral voltage to the corresponding phase current. Under balanced three-phase conditions that exist during loss of synchronism, this ratio is exactly the same as the ratio of

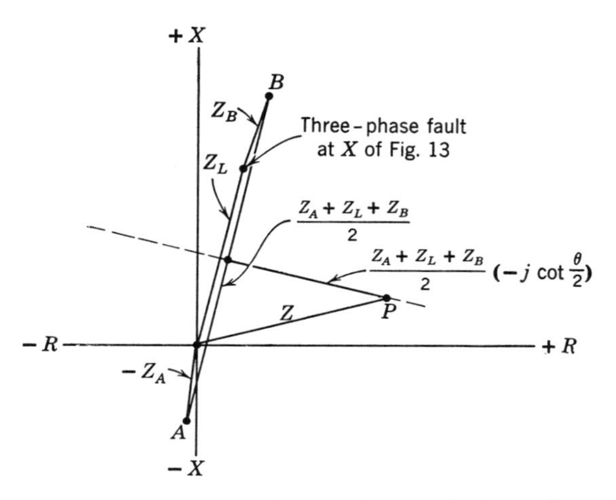

Fig. 14. Construction for locating a point on the loss-of-synchronism characteristic.

delta voltage to delta current that was used earlier in this chapter for describing the appearance of short circuits to phase distance relays. Therefore, it is permissible to superimpose such loss-of-synchronism characteristics, short-circuit characteristics, and distance-relay characteristics on the same R-X diagram. For example, a three-phase fault (X of Fig. 13) at the far end of the line section from the relay location appears to distance relays as a point at the end of Z_L, as shown on Fig. 14.

The foregoing observation will lead to the further observation that the point where the loss-of-synchronism characteristic intersects the impedance Z_L would also represent a three-phase fault at that point. In other words, at one instant during loss of synchronism, the conditions are exactly the same as for a three-phase fault at a point approximately midway electrically between the ends of the system. This point is called the "electrical center" or the "impedance center" of the system. This point would be exactly midway if the various impedances had the same X/R ratio, or, in other words, if all of them were in line on the diagram. The point where the loss-of-synchronism characteristic intersects the total-impedance line AB is reached when generator A has advanced to 180° leading generator B.

As shown in Fig. 15, the location of P for any angle θ between the generators can be found graphically by drawing a straight line from either end of the total-impedance line AB at the angle $\left(90 - \frac{\theta}{2}\right)$ to AB. The point P is the intersection of this straight line with the loss-of-synchronism characteristic, which is the perpendicular bisector of the line AB. When θ is 90°, P lies on the circle whose diameter is the total-impedance line AB. This fact is useful to remember because it provides a simple method to locate a point corresponding to about the maximum load transfer.

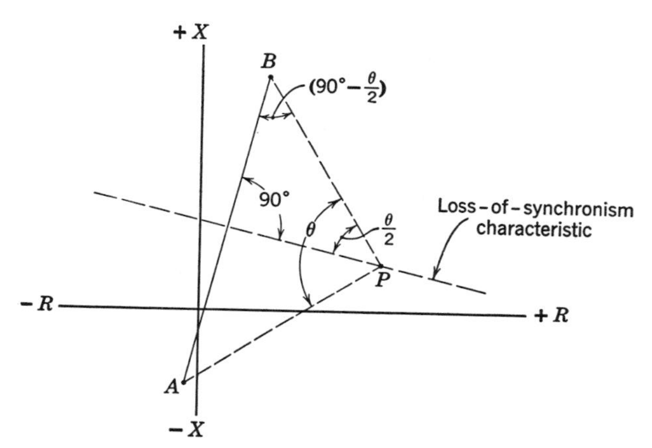

Fig. 15. Locating any point on the loss-of-synchronism characteristic corresponding to any value of θ.

The preceding development of the loss-of-synchronism characteristic was for the special case of $n = 1$. For most purposes, the characteristic resulting from this assumption is all that one needs to know in order to understand distance-type-relay response to loss of synchronism. However, without going into too much detail, let us at least obtain a qualitative picture of the general case where n is greater or less than unity.

All loss-of-synchronism characteristics are circles with their centers on extensions of the total-impedance line AB of Fig. 15. The characteristic when $n = 1$ is a circle of infinite radius. Any of these characteristics could be derived by successive calculations, if we assumed a value for n and then let θ vary from 0 to 360° in the general formula:

$$Z = (Z_A + Z_L + Z_B)n\,\frac{(n - \cos\theta) - j\sin\theta}{(n - \cos\theta)^2 + \sin^2\theta} - Z_A$$

Or one might manipulate the formula mathematically so as to get expressions for the diameter and the location of the center of the circle for any value of n. Figure 16 shows three loss-of-synchronism characteristics for $n > 1$, $n = 1$, and $n < 1$. The total-system-impedance line is again shown as AB.

The dashed circle through A, B, P', P, and P'' is an interesting aspect because all points on this circle to the right of the line AB, such as P', P, and P'' are for the same angle θ by which generator A has advanced ahead of generator B. This might be expected in view of the fact that the angle between a pair of lines drawn from any

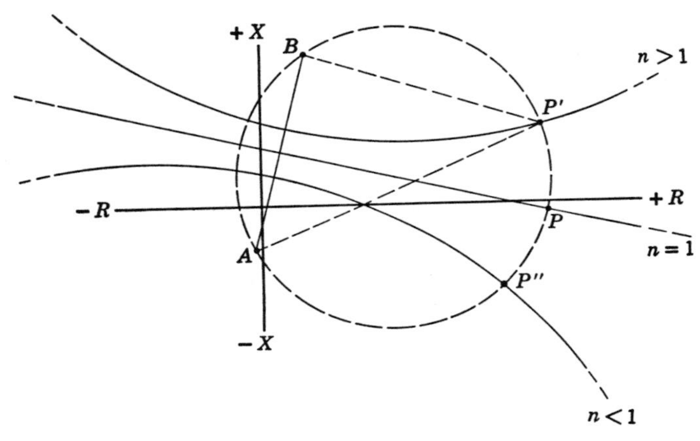

Fig. 16. General loss-of-synchronism characteristics.

point on this part of the circle to A and B is equal to the angle between another pair of lines drawn from any other point on this part of the circle to A and B.

Another interesting aspect of the diagram of Fig. 16 is that the ratio of the lengths of a pair of straight lines drawn from any point on the right-hand part of the circle to A and B will be equal to n. In other words, $P'A/P'B = n$. This suggests a simple method by which the loss-of-synchronism characteristic can be constructed graphically for any value of n. With a compass, one can easily locate three such points, all of which satisfy this same relation; having three points, one can then draw the circle.

This also suggests how to derive mathematical expressions for the radius of the circle and the location of its center. It is only necessary to assume the two possible locations of P' on the line AB and its extension, as shown on Fig. 17, and to obtain the characteristics of the circle that satisfy the relation $P'A/P'B = n$ for both locations.

According to this suggestion, it can be shown that, if we let Z_T be the total system impedance, then, for $n > 1$:

$$\text{Distance from } B \text{ to center of circle} = \frac{Z_T}{n^2 - 1}$$

$$\text{Radius of the circle} = \frac{nZ_T}{n^2 - 1}$$

These are illustrated in Fig. 17.

The circles for $n < 1$ are symmetrical to those for $n > 1$, but with their centers beyond A; the same formulae can be used if $1/n$ is inserted in place of n.

The construction of the loss-of-synchronism characteristic is more complicated if the effects of shunt capacitance and loads are taken

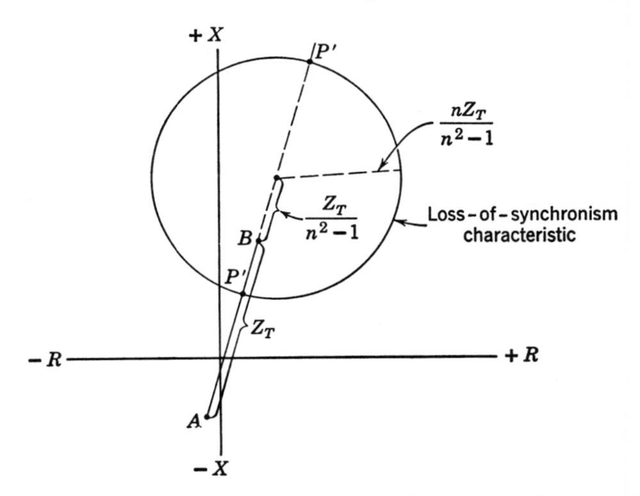

Fig. 17. Graphical construction of loss-of-synchronism characteristic.

into account. Also, the presence of a fault during a power swing or loss of synchronism further complicates the problem. However, seldom if ever will one need any more information than has been given here. The entire subject is treated most comprehensively in references 5 and 8 of the Bibliography.

EFFECT ON DISTANCE RELAYS OF POWER SWINGS OR LOSS OF SYNCHRONISM

To determine the response of any distance relay, it is only necessary to superimpose its operating characteristic on the R-X diagram of the loss-of-synchronism characteristic. This will show immediately whether any portion of the loss-of-synchronism characteristic enters the relay's operating region.

The fact that the loss-of-synchronism characteristic passes through or near the point representing a three-phase short circuit at the electrical center of the system indicates that any distance relay whose range of operation includes this point will have an operating tendency. Whether the relay will actually complete its operation and trip its breaker depends on the operating speed of the relay and the length of time during which the loss-of-synchronism conditions will produce

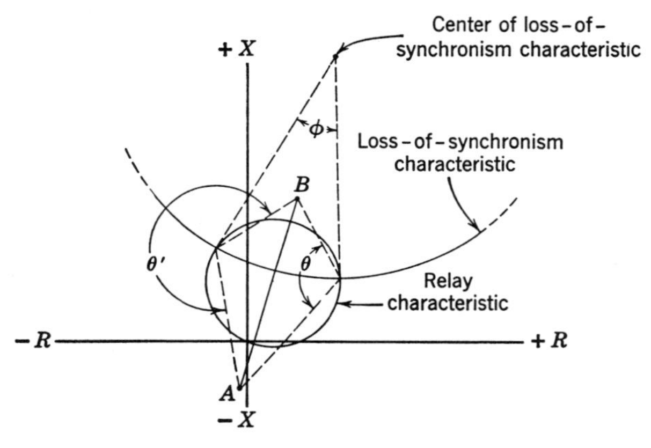

Fig. 18. Determination of relay operating tendency during loss of synchronism.

an operating tendency. Only the back-up-time step of a distance relay is likely to involve enough time delay to avoid such tripping. For a given rate of slip S in cycles per second, one can determine how long the operating tendency will last. In Fig. 18, $(\theta' - \theta)$ gives the angular change of the slipping generator in degrees relative to the other generator. If we assume a constant rate of slip, the time during which an operating tendency will last is $t = (\theta' - \theta)/360S$ seconds. Note that the angular change is not given by the angle ϕ of Fig. 18; once around the loss-of-synchronism circle is one slip cycle, but the rate of movement around the circle is not constant for a constant rate of slip.

The effect of changes in system configuration or generating capacities on the location of the loss-of-synchronism characteristic should be taken into account in determining the operating tendencies of distance relays. Such changes will shift the position of the electrical center with respect to adjoining line sections, and hence may at one time put the electrical center within reach of relays at one location, and at another time put it within reach of relays at another location.

This subject will be treated in more detail when we consider the application of distance relays to transmission-line protection.

Response of Polyphase Directional Relays to Positive- and Negative-Phase-Sequence Volt-Amperes

It has been shown[9] that the torque of polyphase directional relays can be expressed in terms of positive- and negative-phase-sequence volt-amperes. The proof of this fact involves the demonstration that only currents and voltages of the same phase sequence produce net

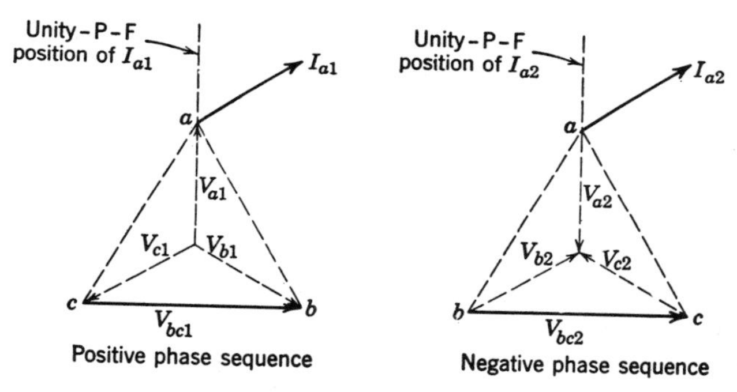

Fig. 19. Corresponding positive- and negative-phase-sequence torque-producing quantities for the quadrature connection.

torque in a polyphase directional relay. Moreover, when delta voltages or delta currents energize a polyphase directional relay, zero-phase-sequence currents produce no net torque because there are no zero-phase-sequence components in the delta quantities.

Making use of the foregoing facts, let us examine briefly the information that can be derived regarding the torques produced by the various conventional connections described in Chapter 3. Consider first the quadrature connection.

Figure 19 shows the corresponding positive- and negative-phase-sequence currents and voltages that produce net torque in one element of a polyphase directional relay; each of the other two elements will produce the same net torque since the positive- and negative-phase-sequence currents and voltages are balanced three phase. Additional torques are produced in each of the three elements by currents and voltages of opposite phase sequence, and these torques are not necessarily equal in each element, but they add to zero in the three elements, and hence can be neglected. (If single-phase directional relays were involved, these additional torques would have to be considered.)

Before proceeding to develop the torque relations, let us note that in Fig. 19 the arrows are on opposite ends of the phase-to-neutral voltages for the two phase sequences. The reason for this is that negative-phase-sequence voltages are voltage drops, whereas positive-phase-sequence voltages are voltage rises minus voltage drops; therefore, if corresponding positive- and negative-phase-sequence currents are in phase, their voltages are 180° out of phase, if we assume positive-, negative-, and zero-phase-sequence impedances to have the same X/R ratio. Another observation is that I_{a2} being shown in phase with I_{a1} has no significance; the important fact is that I_{a2} is assumed to lag the indicated unity-power-factor position by the same angle as I_{a1}. (The justification for this assumption is given later.)

If I_{a1} is β degrees from its maximum-torque position, the total torque is:

$$T \propto (V_{bc1}I_{a1} + V_{bc2}I_{a2}) \cos \beta$$

(Note that the voltage and current values in this and in the following torque equations are positive rms magnitudes, and should not be expressed as complex quantities. Note also that β may be positive or negative because the current may be either leading or lagging its maximum-torque position.) We can simplify the foregoing relation to:

$$T \propto (V_1I_1 + V_2I_2) \cos \beta$$

where the subscripts 1 and 2 denote positive- and negative-phase-sequence components, respectively, of the quantities supplied to the relay.

The vector diagrams for the 60° connection are shown in Fig. 20. The torque relations for Fig. 20 are:

$$T \propto V_1I_1 \cos \beta + V_2I_2 \cos (60° + \beta)$$

Likewise, the torque for the 30° connection is:

$$T \propto V_1I_1 \cos \beta + V_2I_2 \cos (120° + \beta)$$

From the foregoing relations, the difference in response for the various connections depends on the relative magnitude of the negative-phase-sequence voltage-current product to the positive-phase-sequence product. If there are no negative-phase-sequence quantities, the response is the same for any connection. The negative-phase-sequence product can vary from zero to a value equal to the positive-phase-sequence product. Equality of the products occurs for a phase-to-phase fault at the relay location with no fault resistance. For such a fault, our assumption of the same X/R ratio is most legitimate since we are dealing only with the positive- and negative-phase-sequence networks where this ratio is practically the same. Figure 21 shows the torque components and total torques versus β for this extreme condition where both products are equal. The so-called "zero-degree" connection will be discussed later.

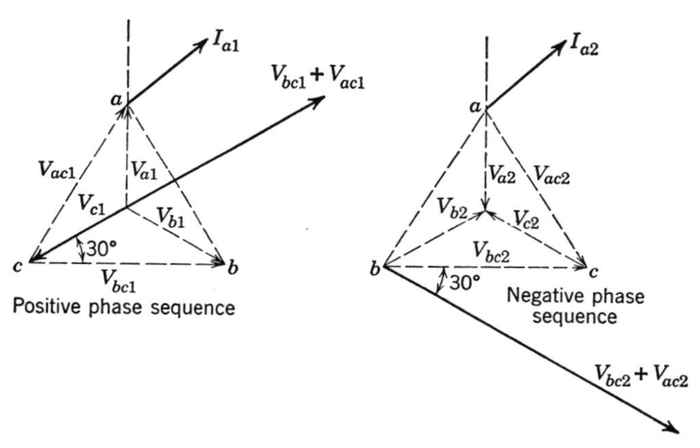

Fig. 20. Corresponding torque-producing quantities for the 60° connection.

For the limiting conditions of Fig. 21, if we assume that we want to develop positive net torque over the largest possible range of β within $\pm 90°$, the 90° connection is seen to be the best, and the 30° connection is the poorest. Remember that we may have balanced three-phase conditions alone, for which the torque is simply $V_1 I_1 \cos \beta$, or we may have unbalanced conditions for which the total torque will be produced. Therefore, under the limiting conditions of a phase-to-phase fault at the relay location, and with no fault resistance, the three connections will produce positive torque over the total range of β shown in the accompanying table.

Connection	Range of β
Quadrature	180°
60°	150°
30°	120°

The ranges for the 60° and 30° connections will increase and approach 180° as the negative-phase-sequence product becomes smaller and smaller relative to the positive-phase-sequence product, or, in other words, as the fault is farther and farther away from the relay location. The fact that under short-circuit conditions the current does not range over the maximum theoretical angular limits (current limited only by resistance or only by inductive reactance) makes all

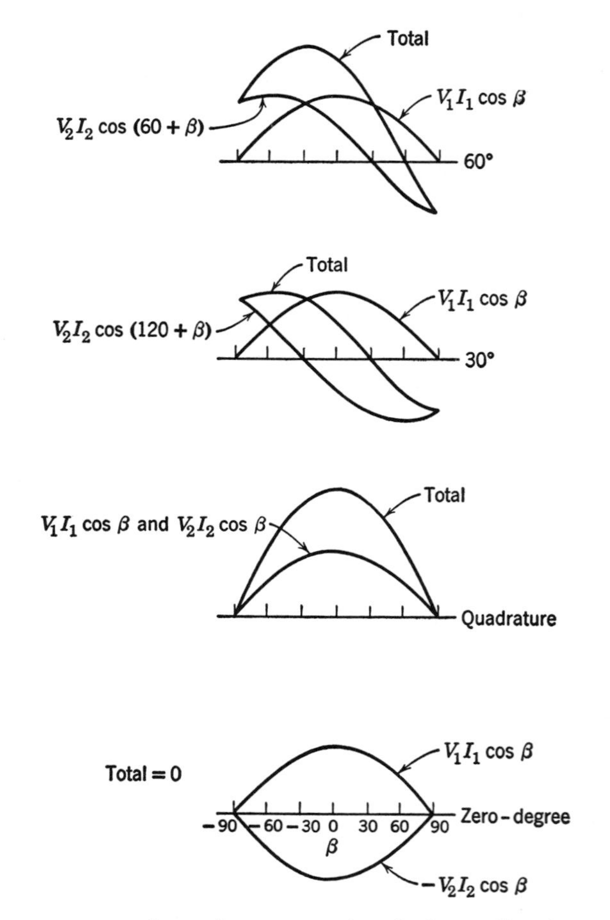

Fig. 21. Component and total torques of polyphase directional relays for conventional connections.

three connections usable, but the quadrature connection provides the largest margin of safety for correct operation. (The possibility that capacity reactance may limit the current is not considered here, but it is an important factor when series capacitors are used in lines.)

The zero-degree connection is that for which the current and voltage supplied to each relay element are in phase under balanced three-phase unity-power-factor conditions, as, for example, $I_a - I_b$ and V_{ab}. If there are no zero-phase-sequence components present, I_a and V_a could also be used. The equation for the torque produced with this connection is:

$$T \propto V_1 I_1 \cos\beta + V_2 I_2 \cos(180° + \beta)$$

$$\propto (V_1 I_1 - V_2 I_2) \cos\beta$$

The zero-degree connection will produce positive net torque over the same range of β as the quadrature connection, but the magnitude of the net torque over most of the range is less than for the other connections. For the limiting condition assumed for Fig. 21, the net torque is zero. The zero-degree connection is mentioned here to show that, by adding or subtracting the torques of a 90° polyphase relay and a zero-degree polyphase relay, operation can be obtained from either positive- or negative-phase-sequence volt-amperes. However, a most precise balance of the quantities must be effected if the net torque is to be truly representative of the desired quantity.

It follows from the foregoing that the response of a polyphase directional relay will be the same for a given type of fault whether or not there is a wye-delta or delta-wye power transformer between the relay and the fault. Although such a power transformer causes an angular shift in the phase-sequence quantities, the positive-phase-sequence quantities being shifted in one direction and the negative-phase-sequence quantities in the other, the interacting currents and voltages of the same phase sequence are shifted in the same direction and by the same amount. Hence, there is no net effect aside from the current-limiting-impedance effect of the power transformer itself.

Response of Single-Phase Directional Relays to Short Circuits

The impedances seen by single-phase directional relays for different kinds of faults can be constructed by the methods used for distance relays. Reference 10 is an excellent contribution in this respect. Unfortunately the torques of single-phase relays cannot be expressed as simply as for polyphase relays, in terms of positive- and negative-phase-sequence volt-amperes. In other words, simple generalizations cannot be made for single-phase directional relays. It is more fruitful to examine the types of application where misoperation is known to be most likely. This will be done later when we consider the application of directional relays for transmission-line protection.

Phase-Sequence Filters

It is sometimes desirable to operate protective-relaying equipment from a particular phase-sequence component of the three-phase system currents or voltages. Although the existence of phase-sequence components may be considered a mathematical concept, it is possible, nevertheless, to separate out of the three-phase currents or voltages actual quantities that are directly proportional to any of the phase-sequence components. The clue to the method by which this can be done is given by the three following equations from symmetrical-component theory:

$$I_{a1} = \tfrac{1}{3}(I_a + aI_b + a^2I_c)$$

$$I_{a2} = \tfrac{1}{3}(I_a + a^2I_b + aI_c)$$

$$I_{a0} = \tfrac{1}{3}(I_a + I_b + I_c)$$

These equations give the phase-sequence components of the current in phase a in terms of the actual three-phase currents. Knowing the phase-sequence components for phase a, we can immediately write down the components for the other two phases. The components of voltage are similarly expressed.

A phase-sequence filter does electrically what these three equations describe graphically. We can quickly dismiss the problem of obtaining the zero-phase-sequence component because we have already seen in Chapter 7 that the current in the neutral of wye-connected CT's is 3 times the zero-phase-sequence component. To obtain a quantity proportional to the positive-phase-sequence component of the current in phase a, we must devise a network that (1) shifts I_b 120° counterclockwise, (2) shifts I_c 240° counterclockwise, and (3) adds I_a vectorially to the vector sum of the other two shifted quantities. I_{a2} can be similarly obtained except for the amount of shift in I_b and I_c. Since we are seeking quantities that are only *proportional* to the actual phase-sequence quantities, we shall not be concerned by changes in magnitude of the shifted quantities so long as all three are compensated alike for magnitude changes in any of them. We should remember also that a 120° shift of a quantity in the counterclockwise direction is the same as reversing the quantity and shifting it 60° in the clockwise direction, etc.

Before considering actual networks that have been used for obtaining the various phase-sequence quantities, let us first see some other relations, derived from those already given, that show some other manipulations of the actual currents for getting the desired components. The manipulations indicated by these relations are used in some filters where a zero-phase-sequence quantity does not exist or where it has first been subtracted from the actual quantities (in other words, where $I_a + I_b + I_c = 0$). One combination is:

$$I_{a1} = \frac{1 - a^2}{3}(I_b - I_c) - I_b$$

$$I_{a2} = \frac{1 - a}{3}(I_b - I_c) - I_b$$

Another combination is:

$$I_{a1} = \frac{1 - a^2}{3}(I_a - a^2 I_b)$$

$$I_{a2} = \frac{1 - a}{3}(I_a - a I_b)$$

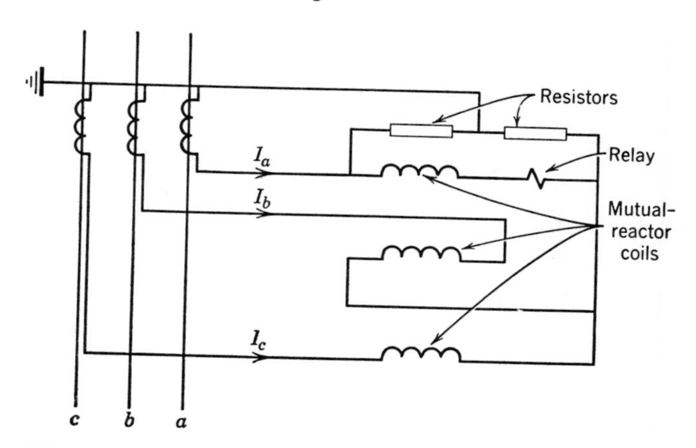

Fig. 22. A positive-phase-sequence-current filter and relay.

Still another combination is:

$$I_{a1} = \frac{a - a^2}{3}(I_b - a I_a)$$

$$I_{a2} = \frac{a^2 - a}{3}(I_b - a^2 I_a)$$

Figure 22 shows a positive-phase-sequence-current filter that has been used;[11] by interchanging the leads in which I_b and I_c flow, the filter becomes a negative-phase-sequence-current filter. A filter that provides a quantity proportional to the negative-phase-sequence component alone, or the negative plus an adjustable proportion of the zero, is shown in Fig. 23;[12] by interchanging the leads in which I_b and I_c flow, the filter becomes a positive-plus-zero-phase-sequence-current filter.

The references given for the filters of Figs. 22 and 23 contain proof of the capabilities of the filters. The purpose here is not to consider details of filter design but merely to indicate the technique involved. Variations on the filters shown are possible. Excellent descriptions of the many possible types of filters are given in References

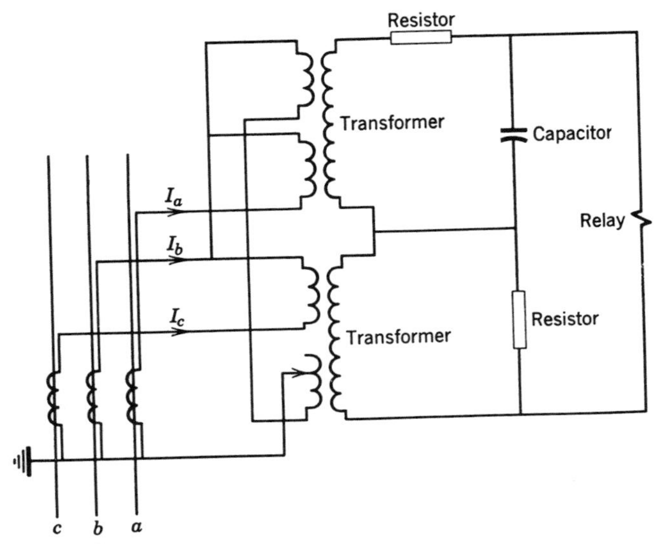

Fig. 23. A combination negative- and zero-phase-sequence-current filter and relay.

1 and 13 of the Bibliography. These books show voltage as well as current filters.

A consideration that should be observed in the use of phase-sequence filters is the effect of saturation in any of the filter coil elements on the segregating ability of the filter. Also, frequency variations and harmonics of the input currents will affect the output.[14] It will be appreciated that the actual process of shifting quantities and combining them, as indicated by symmetrical-component theory, requires a high degree of precision if the derived quantity is to be truly representative of the desired quantity.

Problems

1. Given the system shown in Fig. 24, including the transmission-line sections *AB* and *BC*, and mho-type phase distance relays for protecting these lines. On an *R-X* diagram, show quantitatively the characteristics of the three mho units

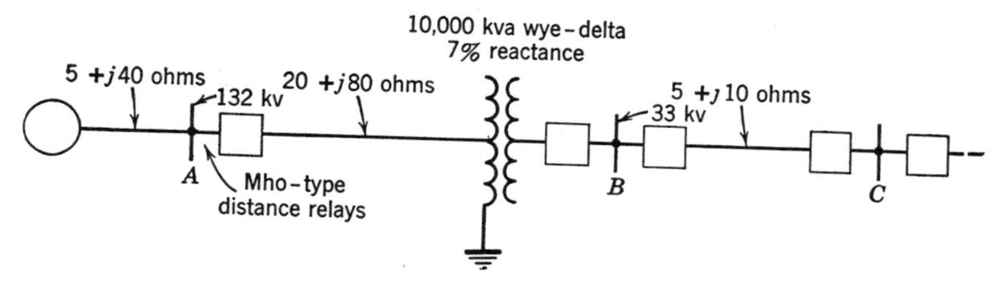

Fig. 24. Illustration for Problem 1.

of each relay for providing primary protection for the line *AB* and back-up protection for the line *BC*. Assume that the centers of the mho characteristics are at 60° lag on the *R-X* diagram.

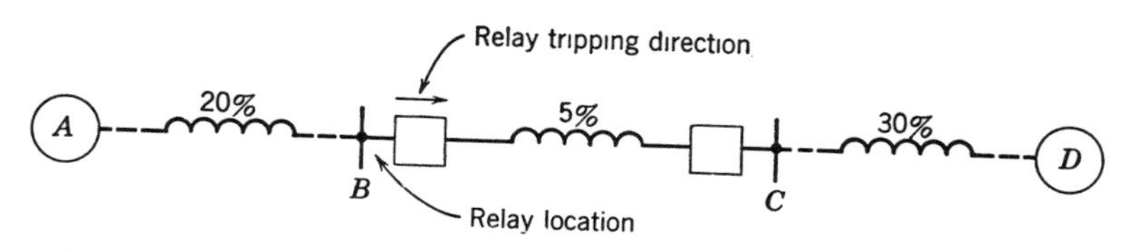

Fig. 25. Illustration for Problem 2.

2. Given a system and a distance relay located as shown in Fig. 25. Assume that the relay is a mho type having distance and time settings as follows:

Zone	Distance Setting	Time Setting
1st	90% of line *BC*	High speed
2nd	120% of line *BC*	0.3 sec
3rd	300% of line *BC*	0.5 sec

Assume also that the centers of the mho circles lie at 60° lag on the *R-X* diagram. The values on the system diagram are percent impedance on a given kva base. Assume that the X/R ratio of each impedance is tan 60°. As viewed from the relay location, the system is transmitting synchronizing power equal to base kva at base voltage over the line *BC* toward *D* at 95% lagging PF, when generator *A* starts to lose synchronism with generator *D* by speeding up with respect to *D*. Assume that the rate of slip is constant at 0.2 slip-cycle per second. Will the relay operate to trip its breaker? Illustrate by showing the relay and the loss-of-synchronism characteristics on an *R-X* diagram.

3. Given a polyphase directional relay, the three elements of which use the combinations of current and voltage shown on p. 192.

Element No.	I	V
1	$I_a - I_b$	$V_{bc} - V_{ac}$
2	$I_b - I_c$	$V_{ca} - V_{ba}$
3	$I_c - I_a$	$V_{ab} - V_{cb}$

Write the torque equations in terms of the positive- and negative-phase-sequence volt-amperes. What do you think of this as a possible connection for such relays?

Bibliography

1. *Symmetrical Components,* by C. F. Wagner and R. D. Evans, McGraw-Hill Book Co., New York, 1933.

2. *Circuit Analysis of A-C Power Systems,* Vols. I and II, by Edith Clarke, John Wiley & Sons, New York, 1943, 1950.

3. "Neuerungen im Anschlutz von Impedanz und Richtungsrelais," by R. Bauch, *Siemens Z., 9* (January, 1929), pp. 13–18.

"Zeitstufen Reaktanzschutz für Hochspannungsfreileitungen," by H. Poleck and J. Sorge, *Siemens Z., 8* (1928), pp. 694–706.

"Fundamental Basis for Distance Relaying on Three-Phase Systems," by W. A. Lewis and L. S. Tippett, *AIEE Trans., 66* (1947), pp. 694–708. Discussions, pp. 708–709.

4. "A Comprehensive Method of Determining the Performance of Distance Relays," by J. H. Neher, *AIEE Trans., 56* (1937), pp. 833–844. Discussions, p. 1515.

5. "Graphical Method for Estimating the Performance of Distance Relays during Faults and Power Swings," by A. R. van C. Warrington, *AIEE Trans., 68* (1949), pp. 608–620. Discussions, pp. 620–621.

6. "Performance of Distance Relays," by Giuseppe Calabrese, *AIEE Trans., 55* (1936), pp. 660–672. Discussions, pp. 1254–1255.

7. *Power System Stability,* Vol. II, by S. B. Crary, John Wiley & Sons, New York, 1947.

8. "Impedances Seen by Relays during Power Swings with and without Faults," by Edith Clarke, *AIEE Trans., 64* (1945), pp. 372–384.

9. "A Single Element Polyphase Relay," by A. J. McConnell, *AIEE Trans., 56* (1937), pp. 77–80, 113. Discussions, pp. 1025–1028.

"Factors Which Influence the Behavior of Directional Relays," by T. D. Graybeal, *AIEE Trans., 61* (1942), pp. 942–952.

"An Analysis of Polyphase Directional Relay Torques," by C. J. Baldwin, Jr., and B. N. Gafford, *AIEE Trans., 72,* Part III (1953), pp. 752–757. Discussions, pp. 757–759.

10. "A Study of Directional Element Connections for Phase Relays," by W. K. Sonnemann, *AIEE Trans., 69,* Part II (1950), pp. 1438–1450. Discussions, pp. 1450–1451.

11. "Thévenin's Theorem," by E. L. Harder, *Elec. J.,* October, 1938, pp. 397–401.

12. "Phase-Comparison Carrier-Current Relaying," by A. J. McConnell, T. A. Cramer, and H. T. Seeley, *AIEE Trans., 64* (1945), pp. 825–833. Discussions, pp. 973–975.

13. *Electric Transmission and Distribution Reference Book,* published by Westinghouse Electric Corp., East Pittsburgh, Pa.

14. "Frequency Errors in Negative-Sequence Segregating Networks," by E. A. Livingston, *Proc. Inst. Elec. Engrs., 99,* Part 4 (1952), pp. 390–396.

10 A-C GENERATOR AND MOTOR PROTECTION

The remaining chapters deal with the application of protective relays to each of the several elements that make up the electric power system. Although there is quite good agreement among protection engineers as to what constitutes the necessary protection and how to provide it, there are still many differences of opinion in certain areas. This book describes the general practice, giving the pros and cons where there are differences of opinion. Four standard-practice publications deal with the application of protective relays.[1,2,3,4] Manufacturers' publications are also available.[5,6,20] Bibliographies of relaying literature prepared by an AIEE committee provide convenient reference to a wealth of information for more detailed study.[7] Frequent reference will be made here to publications that have been found most informative.

The fact that this book recognizes differences of opinion should not be interpreted as complete approval of the various parallel practices. Although it is recognized that there may sometimes be special economic and technical considerations, nevertheless, much can still be done in the way of standardization.

Generator Protection

Except where specifically stated otherwise, the following will deal with generators in *attended* stations, including the generators of frequency converters.

The protection of generators involves the consideration of more possible abnormal operating conditions than the protection of any other system element. In unattended stations, automatic protection against all harmful abnormal conditions should be provided.[1] But much difference of opinion exists as to what constitutes sufficient protection of generators in attended stations. Such difference of opinion is mostly concerning the protection against abnormal operating conditions, other than short circuits, that do not necessarily require the immediate removal from service of a machine and that might be left to the control of an attendant.

The arguments that are advanced in favor of a minimum amount of automatic protective equipment are as follows: (*a*) the more automatic equipment there is to maintain, the poorer maintenance it will get, and hence it will be less reliable; (*b*) automatic equipment might operate incorrectly and trip a generator undesirably; (*c*) an operator can sometimes avoid removing a generator from service when its removal would be embarrassing. Most of the objection to automatic protective equipment is not so much that a relay will fail to operate when it should, but that it might remove a generator from service unnecessarily. Part of the basis for this attitude is simply fear. Each additional device adds another contact that can trip the generator. The more such contacts there are, the greater is the possibility that one might somehow close when it should not. There is some justification for such fears. Relays have operated improperly. Such improper operation is most likely in new installations before the installation "kinks" have been straightened out. Occasionally, an abnormal operating condition arises that was not anticipated in the design or application of the equipment, and a relay operates undesirably. Cases are on record where cleaning or maintenance personnel accidentally caused a relay to trip a generator. But, if something is known to be basically wrong with a protective relay so that it cannot be relied on to operate properly, it should not be applied or it should be corrected one way or another. Otherwise, fear alone is not a proper basis for omitting needed protection.

Admittedly, an alert and skillful operator can sometimes avoid removing a generator from service. In general, however, and with all due respect to operators, the natural fear of removing a machine from service unnecessarily could result in serious damage. Operators have been known to make mistakes during emergencies and to trip generators unnecessarily as well as to fail to trip when necessary.[8] Furthermore, during an emergency, an operator has other important things to do for which he is better fitted.

An unnecessary generator outage is undesirable, but one should not try to avoid it by the omission of otherwise desirable automatic protection. It is generally agreed that any well-designed and well-operated system should be able to withstand a short unscheduled outage of the largest generating unit.[9] It is realized that sometimes it may take several hours to make sure that there is nothing wrong with the unit and to return it to service. Nevertheless, if this is the price one has to pay to avoid the possibility of a unit's being out of service several months for repair, it is worth it. The protection of certain generators against the possibility of extensive damage may be more important than the protection of the service of the system.[9]

The practice is increasing of using centralized control, which requires more automatic equipment and less manual "on the spot" supervision, in order to provide higher standards of service with still greater efficiency.[10] Such practice requires more automatic protective-relaying equipment to provide the protection that was formerly the responsibility of attendants.[8]

SHORT-CIRCUIT PROTECTION OF STATOR WINDINGS BY PERCENTAGE-DIFFERENTIAL RELAYS

It is the standardized practice of manufacturers to recommend differential protection for generators rated 1000 kva or higher,[2] and most of such generators are protected by differential relays.[11] Above 10,000 kva, it is almost universally the practice to use differential relays.[9] Percentage-differential relaying is the best for the purpose, and it should be used wherever it can be justified economically. It is not necessarily the size of a generator that determines how good the protection should be; the important thing is the effect on the rest of the system of a prolonged fault in the generator, and how great the hardship would be if the generator was badly damaged and was out of service for a long time.

The arrangement of CT's and percentage-differential relays is shown in Fig. 1 for a wye-connected machine, and in Fig. 2 for a delta machine. If the neutral connection is made inside the generator and only the neutral lead is brought out and grounded through low impedance, percentage-differential relaying for ground faults only can be provided, as in Fig. 3. The connections for a so-called "unit" generator-transformer arrangement are shown in Fig. 4; notice that the CT's on the neutral side may be used in common by the differential-relaying equipments of the generator and the transformer.

For greatest sensitivity of differential relaying, the CT primary-current rating would have to be equal to the generator's rated full-load current. However, in practice the CT primary-current rating is as much as about 25% higher than full load, so that if ammeters are connected to the CT's their deflections will be less than full scale at rated load. It may be impossible to abide by this rule in Fig. 5; here, the primary-current rating of the CT's may have to be considerably higher than the generator's rated current, because of the higher system current that may flow through the CT's at the breakers.

The way in which the generator neutral is grounded does not influence the choice of percentage-differential relaying equipment when both ends of all windings are brought out. But, if the neutral is not grounded, or if it is grounded through high enough impedance, the differential relays should be supplemented by sensitive ground-fault relaying, which will be described later. Such supplementary

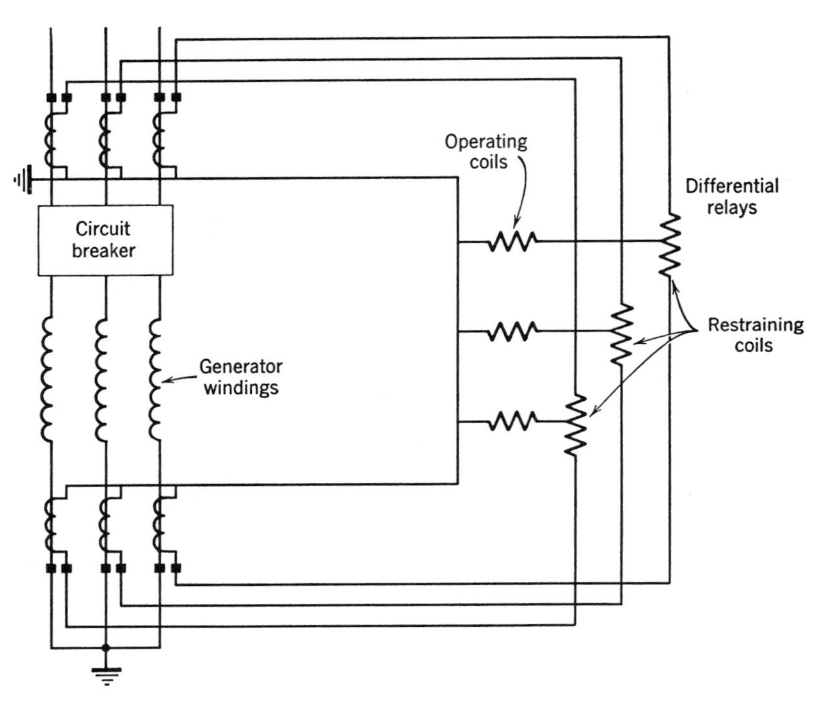

Fig. 1. Percentage-differential relaying for a wye-connected generator.

equipment is generally provided when the ground-fault current that the generator can supply to a single-phase-to-ground fault at its terminals is limited to less than about rated full-load current. Otherwise, the differential relays are sensitive enough to operate for ground faults anywhere from the terminals down to somewhat less than about 20% of the winding away from the neutral, depending on the magnitude of fault current and load current, as shown in Fig. 6, which was obtained from calculations for certain assumed equipment. This is generally considered sensitive enough because, with less than 20% of rated voltage stressing the insulation, a ground fault is most unlikely; in the rare event that a fault did occur, it would simply have to spread until it involved enough of the winding to operate a relay. To make the percentage-differential relays much more sensitive than they are would make them likely to operate undesirably on transient CT errors during external disturbances.

The foregoing raises the question of CT accuracy and loading. It is generally felt that CT's having an ASA accuracy classification of

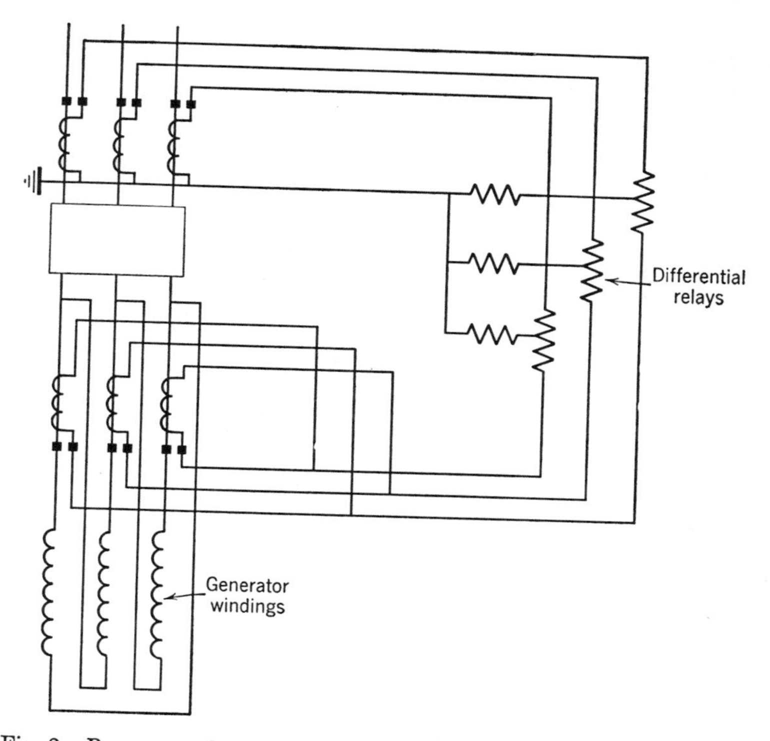

Fig. 2. Percentage-differential relaying for a delta-connected generator.

10H200 or 10L200 are satisfactory if the burdens imposed on the CT's during external faults are not excessive. If variable-percentage-differential relays (to be discussed later) are used, CT's of even lower accuracy classification may be permissible, or higher burdens may be applied. It is the *difference* in accuracy between the CT's (usually of the same type) at opposite ends of the windings that really counts. The difference between their ratio errors should not exceed about one-half of the percent slope of the differential relays for any external fault beyond the generator terminals. Such things as unequal CT secondary lead lengths, or the addition of other burdens in the leads on one side or the other, tend to make the CT's have different errors.

A technique for calculating the steady-state errors of CT's in a differential circuit will be described for the circuit of Fig. 7, where a single-phase-to-ground external fault is assumed to have occurred on the

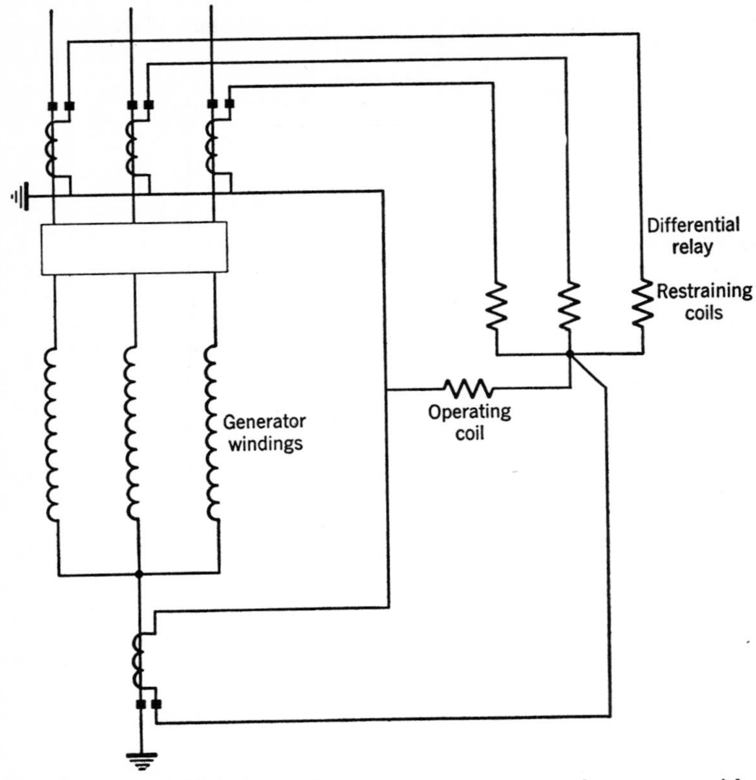

Fig. 3. Percentage-differential relaying for a wye-connected generator with only four leads brought out.

phase shown. The equivalent circuit of each CT is shown in order to illustrate the method of solution. The fact that $(I_{S1} - I_{S2})$ is flowing through the relay's operating coil in the direction shown is the result of assuming that CT_1 is more accurate than CT_2, or in other words that I_{S1} is greater than I_{S2}. We shall assume that I_{S1} and I_{S2} are in phase, and, by Kirchhoff's laws, we can write the voltages for the circuit *a-b-c-d-a* as follows:

$$E_1 - I_{S1}(Z_{S1} + 2Z_{L1} + Z_R) - (I_{S1} - I_{S2})Z_0 = 0$$

or

$$(I_{S1} - I_{S2})Z_0 = E_1 - I_{S1}(Z_{S1} + 2Z_{L1} + Z_R) \qquad (1)$$

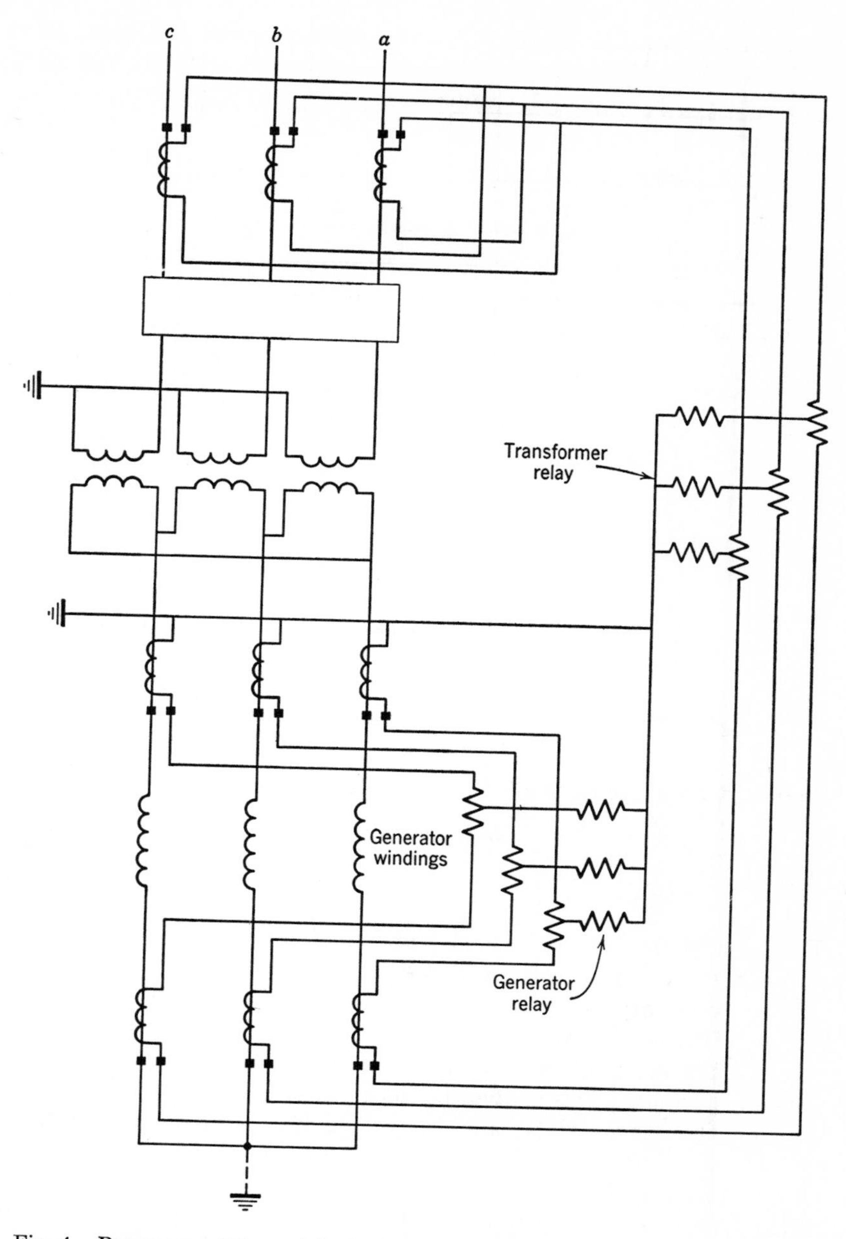

Fig. 4. Percentage-differential relaying for a unit generator and transformer. Note: phase sequence is *a-b-c*.

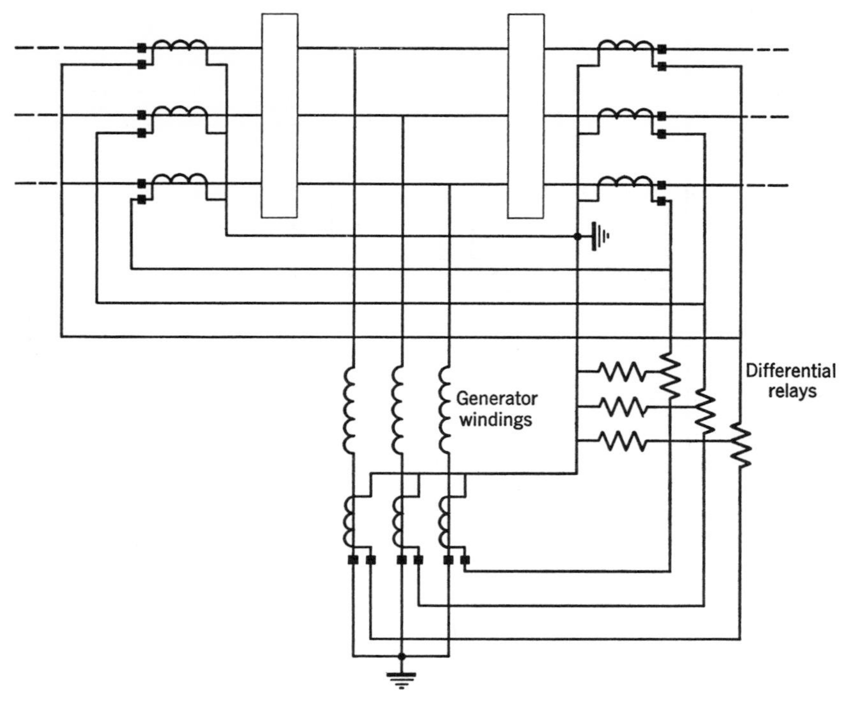

Fig. 5. Generator differential relaying with a double-breaker bus.

Similarly, for the circuit e-f-d-c-e, we can write:

$$E_2 - I_{S2}(Z_{S2} + 2Z_{L2} + Z_R) + (I_{S1} - I_{S2})Z_0 = 0$$

or

$$(I_{S1} - I_{S2})Z_0 = I_{S2}(Z_{S2} + 2Z_{L2} + Z_R) - E_2 \qquad (2)$$

For each of the two equations 1 and 2, if we assume a value of the secondary-excitation voltage E, we can obtain a corresponding value of the secondary-excitation current I_e from the secondary-excitation curve. Having I_e, we can get I_S from the relation $I_S = I/N - I_e$, where I is the initial rms magnitude of the fundamental component of primary current. This enables us to calculate the value of $(I_{S1} - I_{S2})$. Finally, the curves of I_{S1} and I_{S2} versus $(I_{S1} - I_{S2})$ for each of the two CT's is plotted on the same graph, as in Fig. 8. For only one value of the abscissa $(I_{S1} - I_{S2})$ will the difference between the two ordinates I_{S1} and I_{S2} be equal to that value of the abscissa, and this is the point that gives us the solution to the problem. Once we know the values of I_{S1} and I_{S2}, we can quickly determine whether the differential relay will operate for the maximum external fault current.

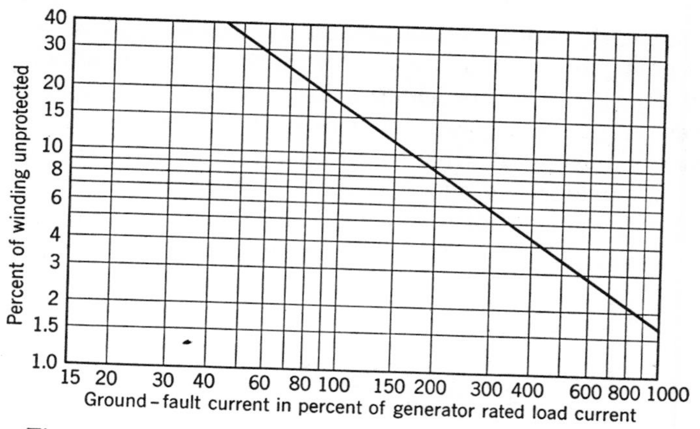

Fig. 6. Percent of winding unprotected for ground faults.

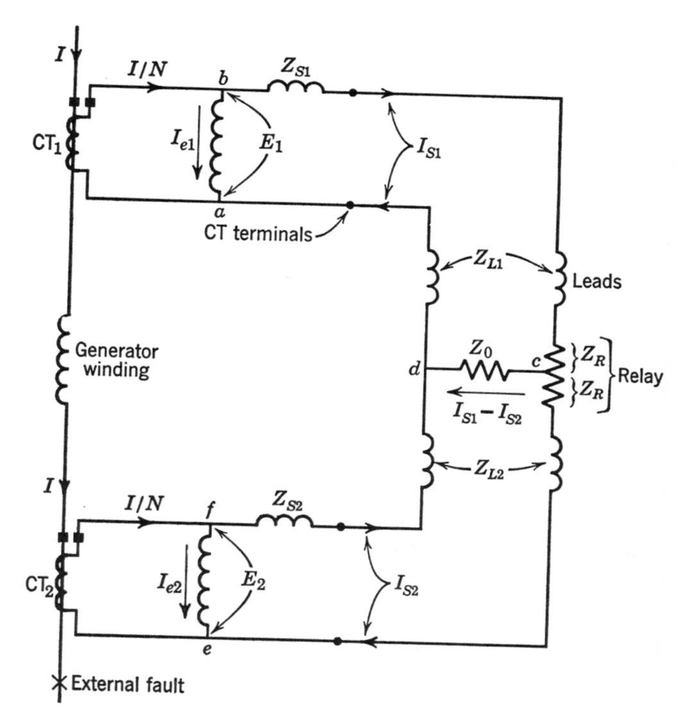

Fig. 7. Equivalent circuit for calculating CT errors in a generator differential circuit.

From the example of Fig. 7, it will become evident that, for an external fault, if there is a tendency for one CT to be more accurate than the other, any current that flows through the operating coil of the relay imposes added burden on the more accurate CT and reduces the burden on the less accurate CT. Thus, there is a natural tendency in a current-differential circuit to resist CT unbalances, and this tendency is greater the more impedance there is in the relay's operating coil. This is not to say, however, that there may not sometimes be enough unbalance to cause incorrect differential-relay opera-

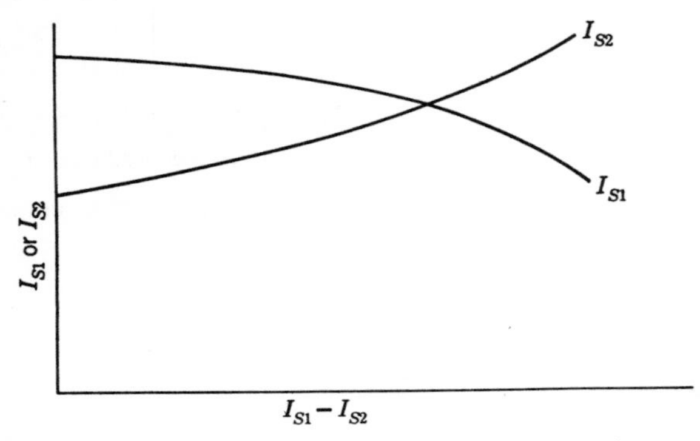

Fig. 8. Plot of the relations of the currents of Fig. 7 for various assumed values of E_1 and E_2.

tion when the burden of leads or other devices in series with one CT is sufficiently greater than the burden in series with the other. If so, it becomes necessary to add compensating burden on one side to more nearly balance the burdens. If the CT's on one side are inherently considerably more accurate than on the other side, shunt burden, having saturation characteristics more or less like the secondary-excitation curve of the less accurate CT's, can be connected across the terminals of the more accurate CT's; this has the effect of making the two sets of CT's equally poor, but, at any rate, more nearly alike.

Another consequence of widely differing CT secondary-excitation characteristics may be "locking-in" for internal faults. In such a case, the inferior CT's may be incapable of inducing sufficient rms voltage in their secondaries to keep the good CT's from forcing current through the inferior CT's secondaries in opposition to their induced voltage, thereby providing a shunt around the differential relay's operating coil and preventing operation. Adding a shunt burden across the good CT's, as was described in the foregoing paragraph, is a solution to this difficulty. The larger the impedance of the operating coil, the more likely locking-in is to happen. However, the circumstances that make it possible are rare.

Chapter 7 described the possibility of harmfully high overvoltages in CT secondary circuits of generator-differential relays when the system is capable of supplying to a generator fault short-circuit current whose magnitude is many times the rating of the CT's. In such cases, it is necessary to use overvoltage limiters, as treated in more detail in Chapter 7.

Whether to use high-speed relays or only the somewhat slower "instantaneous" relays is sometimes a point of contention. If system stability is involved, there may be no question but that high-speed relays must be used. Otherwise, the question is how much damage will be prevented by high-speed relays. The difference in the damage caused by the current supplied by the generator will probably be negligible in view of the continuing flow of fault current because of the slow decay of the field flux. But, if the system fault-current contribution is very large, considerable damage may be prevented by the use of high-speed relays and main circuit breaker. It is easy enough to compare the capabilities of doing damage in terms of I^2t, but the cost of repair is not necessarily directly proportional, and there are no good data in this respect. The savings to be made in the cost of slower-speed relays are insignificant compared to the cost of generators, and there cannot fail to be some benefits with high-speed relays. Except for "match and line-up" considerations, the slower-speed relays might well be eliminated.

Generally, the practice is to have the percentage-differential relays trip a hand-reset multicontact auxiliary relay. This auxiliary relay simultaneously initiates the following: (1) trip main breaker, (2) trip field breaker, (3) trip neutral breaker if provided, (4) shut down the prime mover, (5) turn on CO_2 if provided, (6) operate an alarm and/or annunciator. The auxiliary relay may also initiate the transfer of station auxiliaries from the generator terminals to the reserve source, by tripping the auxiliary breaker. Whether to provide a main-field breaker or only an exciter-field breaker is a point of contention. The tripping of a main-field breaker instead of only an exciter-field breaker will minimize the damage, but there is insufficient evidence to prove whether it is worth the additional expense where other important factors urge the omission of a main-field breaker. Consequently, the practice is divided.

THE VARIABLE-PERCENTAGE-DIFFERENTIAL RELAY

High-speed percentage-differential relays having variable ratio—or percent-slope—characteristics are preferred. At low values of through current, the slope is about 5%, increasing to well over 50% at the high values of through current existing during external faults. This characteristic permits the application of sensitive high-speed relaying equipment using conventional current transformers, with no danger of undesired tripping because of transient inaccuracies in the CT's. To a certain extent, poorer CT's may be used—or higher burdens may be applied—than with fixed-percent-slope relays.

Two different operating principles are employed to obtain the variable characteristic. In both, saturation of the operating element is responsible for a certain amount of increase in the percent slope. In one equipment,[13] saturation alone causes the slope to increase to about 20%; further increase is caused by the effect on the relay response of angular differences between the operating and restraining currents that occur owing to CT errors at high values of external short-circuit current. The net effect of both saturation and phase angle is to increase the slope to more than 50%.

The other equipment[14] obtains a slope greater than 50% for large values of through current entirely by saturation of the operating element. A principle called "product restraint" is used to assure operation for internal short circuits. Product restraint provides restraint sufficient to overcome the effect of any CT errors for external short circuits; for internal short circuits when the system supplies very large currents to a fault, there is no restraint.

PROTECTION AGAINST TURN-TO-TURN FAULTS IN STATOR WINDINGS

Differential relaying, as illustrated in Figs. 1–5, will not respond to faults between turns because there is no difference in the currents at the ends of a winding with shorted turns; a turn fault would have to burn through the major insulation to ground or to another phase before it could be detected. Some of the resulting damage would be prevented if protective-relaying equipment were provided to function for turn faults.

Turn-fault protection has been devised for multicircuit generators, and is used quite extensively, particularly in Canada.[15] In the United States, the government-operated hydroelectric generating stations are the largest users. Because the coils of modern large steam-turbine generators usually have only one turn, they do not need turn-fault protection because turn faults cannot occur without involving ground.

Even though the benefits of turn-fault protection would apply equally well to single-circuit generators, equipment for providing this protection for such generators has not been used, although methods have been suggested;[15,18] as will be seen later, the equipment used for multicircuit generators is not applicable to single-circuit generators.

To justify turn-fault protection, apart from what value it may have as duplicate protection, one must evaluate the savings in damage and outage time that it will provide. In a unit generator-transformer arrangement, considerable saving is possible where the generator operates ungrounded, or where high-resistance grounding or ground-fault-neutralizer grounding is used; if the ground-detecting equipment is not permitted to trip the generator breakers, a turn fault could burn much iron before the fault could spread to another phase and operate the differential relay. Even if the ground-detecting equipment is arranged to trip the generator breakers, it would probably be too slow to prevent considerable iron burning. (The foregoing leads to the further conclusion that if the generator has single-circuit windings with no turn-fault protection, the ground-fault detector should operate as quickly as possible to trip the generator breakers.)

For other than unit generator-transformer arrangements, and where the generator neutral is grounded through low impedance, the justification for turn-fault protection is not so apparent. The amount of iron burning that it would save would not be significant because conventional differential relaying will prevent excessive iron burning.[19] Consequently, the principal saving would be in the cost of the coil-repair job, and it is questionable whether there would be a significant saving there.

The conventional method for providing turn-fault protection is called "split-phase" relaying, and is illustrated in Fig. 9. If there are more than two circuits per phase, they are divided into two equal groups of parallel circuits with a CT for each group. If there is an odd number of circuits, the number of circuits in each of the two groups will not be equal, and the CT's must have different primary-current ratings so that under normal conditions their secondary currents will be equal. Split-phase relaying will operate for any type of short circuit in the generator windings, although it does not provide as good protection as differential relaying for some faults. The split-phase relays should operate the same hand-reset auxiliary tripping relay that is operated by the differential relays.

An inverse-time overcurrent relay is used for split-phase relaying rather than an instantaneous percentage-differential relay, in order to get the required sensitivity. For its use to be justified, split-phase relaying must respond when a single turn is short-circuited. Moreover, the relay equipment must not respond to any transient unbalance that there may be when external faults occur. If percentage restraint were used to prevent such undesired operation, the restraint caused by load current would make the relay too insensitive at full load.

Consequently, time delay is relied on to prevent operation on transients.

Time delay tends somewhat to nullify the principal advantage of turn-fault protection, namely, that of tripping the generator breakers before the fault has had time to develop serious proportions. A supplementary instantaneous overcurrent unit is used together with the

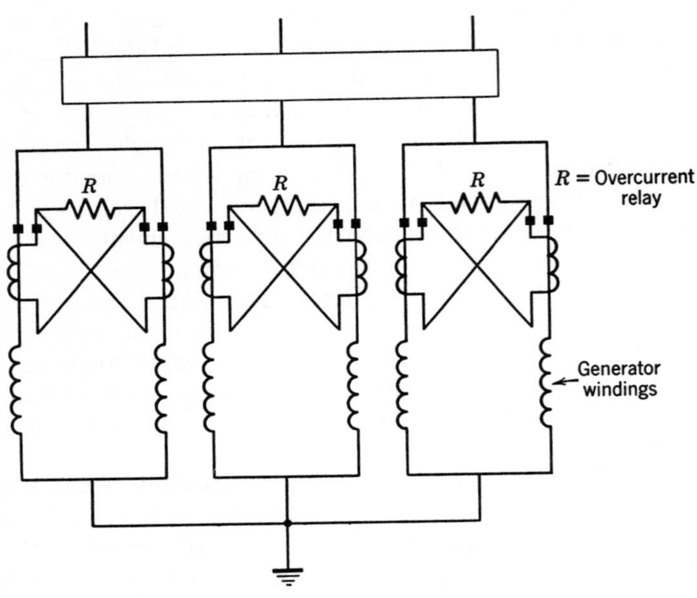

Fig. 9. Split-phase relaying for a multicircuit generator.

inverse-time unit, but the pickup of the instantaneous unit has to be so high to avoid undesired operation on transients that it will not respond unless several turns are short-circuited.

Faster and more sensitive protection can be provided if a double-primary, single-secondary CT, as shown in Fig. 10, is used rather than the two separate CT's shown in Fig. 9.[16] Such a double-primary CT eliminates all transient unbalances except those existing in the primary currents themselves. With such CT's and with close attention to the generator design to minimize normal unbalance, very sensitive instantaneous protection is possible.[17] Such practices have been limited to Canada.

Split-phase relaying at its best could not completely replace over-all differential relaying which is required for protection of the generator circuit beyond the junctions of the paralleled windings. However, some people feel that if split-phase relaying is used with unit generator-transformer arrangements, it is not necessary to have separate generator-differential relaying if the transformer differential relaying includes the generator in its protective zone. This would be true if the split-phase relaying was instantaneous. The greater sensitivity

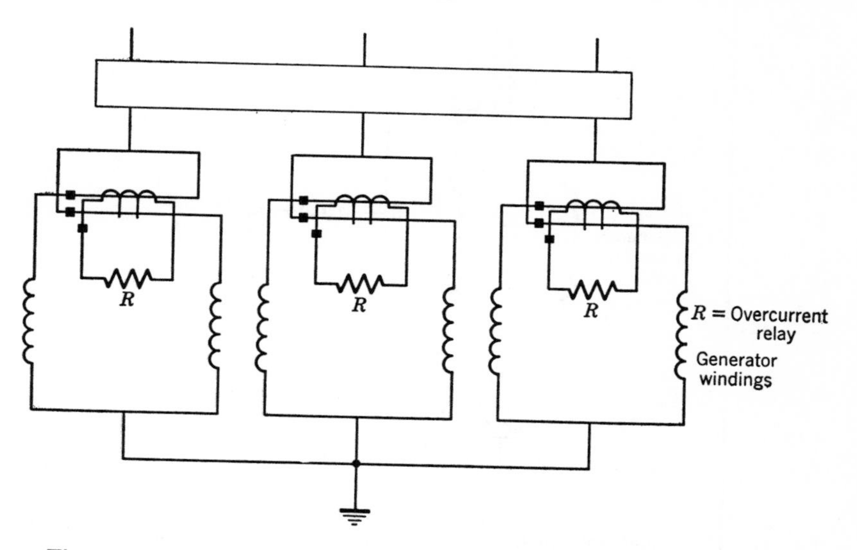

Fig. 10. Split-phase relaying using double-primary current transformers.

of generator-differential relaying is not needed for faults beyond the generator windings. The principal advantage of retaining the generator-differential relaying, apart from the duplicate protection that it affords, is the value of its target indication in helping to locate a fault. However, if split-phase relaying is provided by inverse-time overcurrent relays, generator-differential relaying is recommended because of its higher speed for all except turn faults.

COMBINED SPLIT-PHASE AND OVER-ALL DIFFERENTIAL RELAYING

Figure 11 shows an arrangement that has been used to try to get the benefits of split-phase and over-all differential protection at a saving in current transformers and in relays. However, this arrangement is not as sensitive as the separate conventional split-phase and over-all differential equipments. Sensitivity for turn faults is sacrificed with a percentage-differential relay; with full-load secondary current flowing through the restraining coil, the pickup is considerably higher than with the conventional split-phase equipment, and the equipment will not operate if a single turn is shorted. With the main generator breaker open, the sensitivity for ground faults near the neutral end of the winding that does not have a current transformer may be considerably poorer than with the conventional over-all dif-

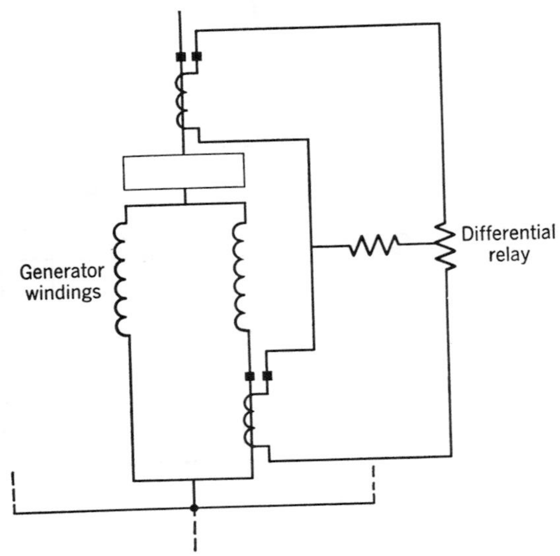

Fig. 11. Combined split-phase and differential relaying (shown for one phase only).

ferential; the current flowing in the winding having the CT is much smaller than one-half of the current flowing in the neutral lead where the over-all differential CT would be. For this reason, the modification shown in Fig. 12 is sometimes used.

SENSITIVE STATOR GROUND-FAULT RELAYING

The protection of unit generator-transformer arrangements is described under the next heading. Here, we are concerned with other than unit arrangements, where the generator's neutral is grounded through such high impedance that conventional percentage-differential relaying equipment is not sensitive enough. The problem here is to get the required sensitivity and at the same time to avoid the possibility of undesired operation because of CT errors with large external-fault currents. Figure 13 shows a solution to the problem: a current-current directional relay is shown whose operating coil is in the neutral of the differential-relay circuit and whose polarizing coil is energized from a CT in the generator neutral. A polarized relay provides greater sensitivity without excessive operating-coil burden; the polarizing CT may have a low enough ratio so that the polarizing

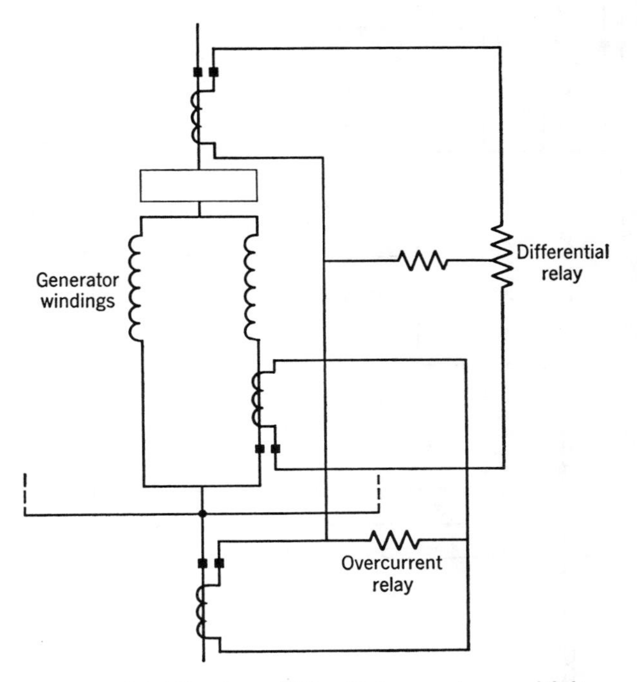

Fig. 12. Modification of Fig. 11 for greater sensitivity.

coil will be "soaked" for the short duration of the fault. Supplementary equipment may sometimes be required to prevent undesired operation because of CT errors during external two-phase-to-ground faults.

STATOR GROUND-FAULT PROTECTION OF UNIT GENERATORS

Figure 14 shows the preferred way[19] to provide ground-fault protection for a generator that is operated as a unit with its power transformer. The generator neutral is grounded through the high-voltage winding of a distribution transformer. A resistor and an overvoltage relay are connected across the low-voltage winding.

It has been found by test that, to avoid the possibility of harmfully high transient overvoltages because of ferroresonance, the resistance of the resistor should be no higher, approximately, than:

$$R = \frac{X_C}{3N^2} \text{ ohms}$$

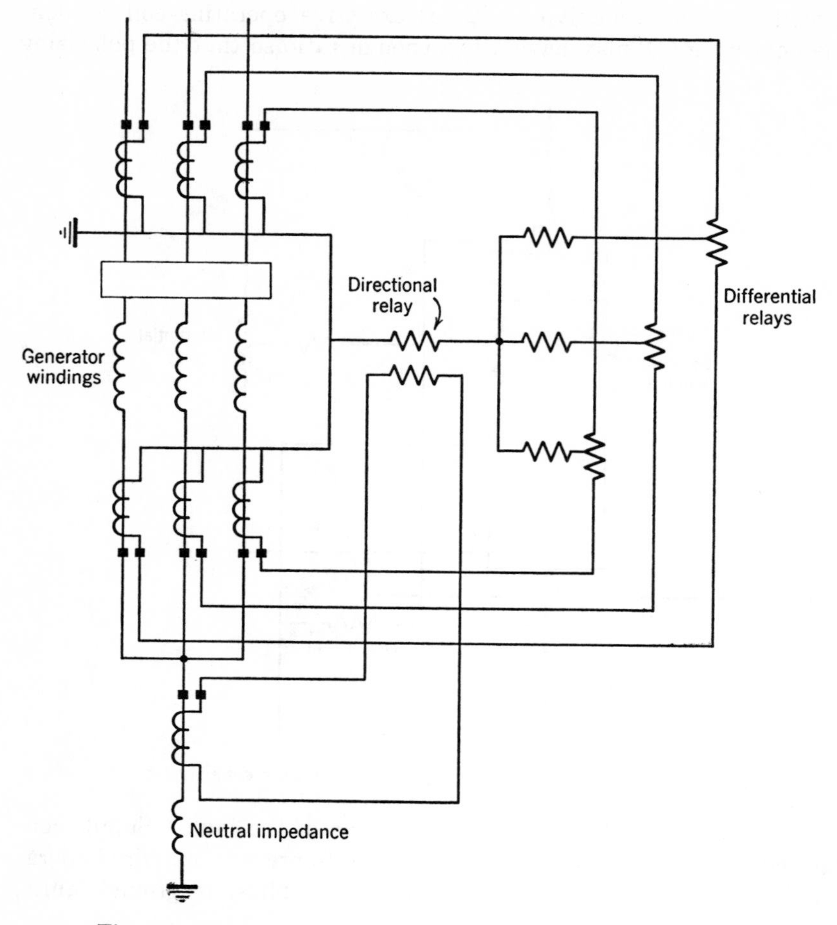

Fig. 13. Sensitive stator ground-fault relaying for generators.

where X_C is the total phase-to-ground-capacity reactance *per phase* of the generator stator windings, the surge protective capacitors or lightning arresters, if used, the leads to the main and station-service power transformers, and the power-transformer windings on the generator side; and N is the open-circuit voltage ratio (or turns ratio) of the high-voltage to the low-voltage windings of the distribution transformer.[20]

The value of R may be less than that given by the foregoing equation. The value of resistance given by the equation will limit the maximum instantaneous value of the transient voltage to ground to about 260% of normal line-to-ground crest value. Further reduction in resistance will not appreciably reduce the magnitude of the transient voltage. The lower the value of R, the more damage will be done by a ground fault, particularly if the relay is not connected to trip the generator breakers. The relaying sensitivity will decrease as R is decreased, because, as can be seen in Fig. 15, more of the available voltage will be consumed in the positive- and negative-phase-sequence impedances and less in the zero-phase-sequence impedance which determines the magnitude of the relay voltage. This decrease in sensitivity is considered by some people to be an advantage because the relay will be less likely to operate for faults on the low-voltage side of the generator potential transformers, as discussed later. In fact, it has been suggested[21] that for this reason, and also to simplify the calculations by making it unnecessary to determine X_C, a resistor be chosen that will limit the fault current to approximately 15 amperes, neglecting the effect of X_C. In other words,

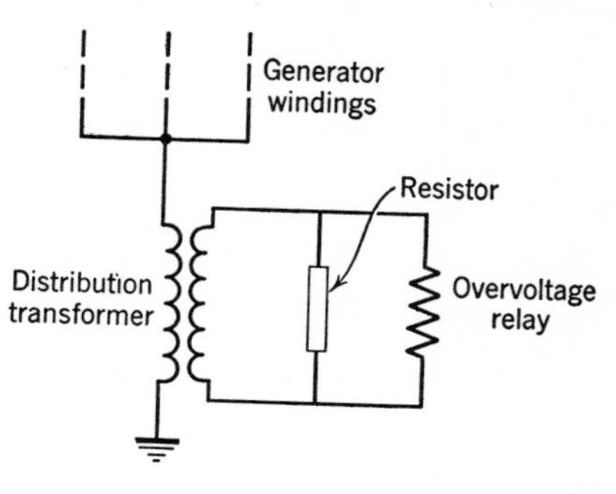

Fig. 14. Stator ground-fault protection of unit generators.

$$R = \frac{10^3 V_G}{15\sqrt{3}\,N^2} \text{ ohms}$$

where V_G is the phase-to-phase-voltage rating of the generator in kilovolts.

It has been suggested that, to avoid large magnetizing-current flow to the distribution transformer when a ground fault occurs, the high-voltage rating of the distribution transformer should be at least 1.5 times the phase-to-neutral-voltage rating of the generator.[20] Its insulation class must fulfill the standard requirements for neutral grounding devices.[22] The low-voltage rating may be 120, 240, or 480 volts, depending on the available or desired voltage rating of the protective relay.

The kva rating to choose for the distribution transformer and for the resistor will depend on whether the user intends to let the overvoltage relay trip the generator main and field breakers or merely sound an alarm. If the relay will merely sound an alarm, the transformer should be continously rated for at least:

$$\text{kva} = \frac{10^3 V_G V_T}{\sqrt{3}\,N^2 R}$$

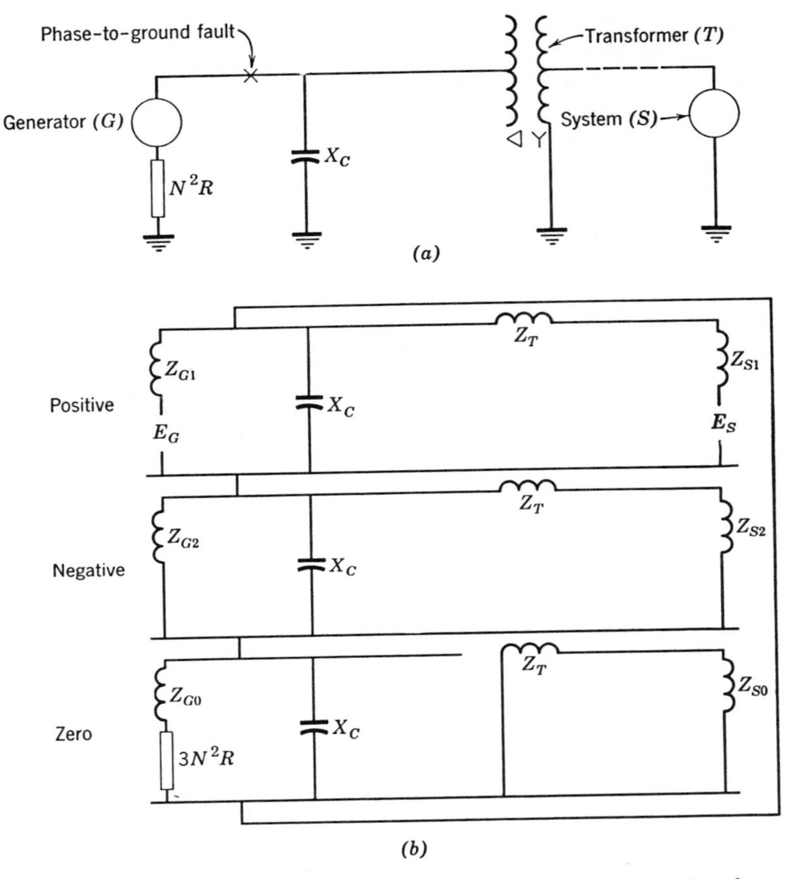

Fig. 15. (*a*) One-line diagram of a system with a unit generator and a phase-to-ground fault. (*b*) Phase-sequence diagram for (*a*).

where V_T is the high-voltage rating of the distribution transformer in kilovolts. Similarly, the continuous rating of the resistor should be at least:

$$\text{kw} = \frac{10^3 V_G{}^2}{3N^2R}$$

If the relay is arranged to trip the generator breakers, short-time transformer and resistor ratings[23] may be used. For example, a 1-minute-rating transformer would have only 21% of the continuous kva rating, and a 10-minute rating would have 40%. However, the lower the transformer rating, the more inductive reactance the transformer will introduce in series with the grounding resistance; for this reason, the 1-minute rating is the lowest considered desirable. The resistor may have either a 10-second or a 1-minute rating, but the 1-minute rating is generally preferred because it is more conservative and not much more expensive. In fact, continuous-rated resistors may even be economical enough.

It is preferred to have the relay trip the generator main and field breakers. Even though the fault current is very low, some welding of the stator laminations may occur if the generator is permitted to continue operating with a ground fault in its winding.[24] Also, in the presence of the fault, the voltage to ground of other parts of the stator windings will rise to $\sqrt{3}$ times normal; should this cause another ground fault to develop, a phase-to-phase fault might result, and additional damage would be done that would have been avoided had the generator breakers been tripped when the first fault occurred. Furthermore, if split-phase protection is not provided and if the ground relay is not permitted to trip, the generator could develop a turn-to-turn fault and there might be considerable iron burning before the fault could spread to another phase and cause the differential relays to operate. In spite of the foregoing, many power companies are willing to risk the possibility of additional damage until they can conveniently remove the faulty generator from service.

A number of power companies simply connect a potential transformer between the generator neutral and ground without any loading resistor. They have operated this way for years, apparently without any difficulty.[19] An overvoltage relay is used as with the distribution-transformer arrangement. The maximum current that can flow in a ground fault is 71% of that with the distribution-transformer-and-resistor combination if the maximum allowable value of R is used, which is not a significant difference. In either case, the arc energy is sufficient to cause damage if immediate tripping is not done.[25] The potential transformer is considerably smaller and cheaper than the distribution-transformer-and-resistor combination, although either one is relatively inexpensive compared with the equipment protected. The principal disadvantage is that the user cannot be certain whether harmfully high transient overvoltages will occur. The only evidence we have is that such overvoltages can occur—at least under laboratory-controlled circumstances. Therefore, until positive evidence to the contrary is presented, the distribution-transformer-and-resistor combination seems to be safer.

If one prefers to let a generator operate with a ground fault in its stator winding, a ground-fault neutralizer will limit the fault current to the smallest value of any of the arrangements, and at the same time will hold the transient voltage to a lower level.[25] However, it is necessary to be sure that surge-protective capacitors, if used, cannot cause harmfully high overvoltages should a defect cause the capacitances to ground to become unbalanced.[26]

Reference 27 describes a modification of the foregoing methods whereby voltage is introduced between the generator neutral and ground for the purpose of obtaining greater sensitivity. Little, if any, application of this principle exists presently in the United States.

The same overvoltage-relay characteristics are required for any of the foregoing generator-neutral-grounding methods. The relay must be sensitive to fundamental-frequency voltages and insensitive to the third-harmonic and multiples of third-harmonic voltages. The relay may require adjustable time delay so as to be selective with other relays for ground faults on the high-voltage side of the main power transformer for which there may be a tendency to operate owing to capacitance coupling between the power-transformer windings, particularly if the high-voltage winding of the power transformer does not have its neutral solidly grounded;[19,20,25] the ground-fault-neutralizer-grounding arrangement has the greatest operating tendency under such circumstances, and the distribution-transformer arrangement has the least. Time delay is desirable also to provide as good selectivity as possible with potential-transformer fuses for faults on the secondary side of potential transformers connected wye-wye. If the relaying equipment is used only to sound an alarm, a combination of relays may be required to get both good sensitivity and a high continuous-voltage rating.

If grounded-neutral wye-wye potential transformers are connected to the generator leads, it may be impossible to get complete selectivity between the relay and the PT (potential-transformer) fuses for certain ground faults on the low-voltage side of the PT's, depending on the fuse ratings and on the relay sensitivity.[21] In other words, the relay may sometimes operate when there is not enough fault current to blow a fuse. Such lack of coordination might be considered an advantage; the relay will protect the PT's from thermal damage for which the fuses could not protect. To make the relay insensitive enough so that it would not operate for low-voltage ground faults would sacrifice too much sensitivity for generator faults. Of course, if the relay has time delay, it will not operate for a momentary short circuit such as might be caused inadvertently during testing. If the relay is used only to sound an alarm, some selectivity may be sacrificed in the interests of sensitivity; the relay will not operate frequently enough to be a nuisance.

SHORT-CIRCUIT PROTECTION OF STATOR WINDINGS BY OVERCURRENT RELAYS

If current transformers are not connected in the neutral ends of wye-connected generator windings, or if only the outgoing leads are brought out, protective devices can be actuated, as in Fig. 16, only by the short-circuit current supplied by the system. Such protection is ineffective when the main circuit breaker is open, or when it is closed if the system has no other generating source, and the following discussion assumes that short-circuit current is available from the system.

If the generator's neutral is not grounded, sensitive and fast ground overcurrent protection can be provided; but, if the neutral is grounded, directional overcurrent relaying should be used for the greatest sensitivity and speed. In either event, directional overcurrent relays should be used for phase-fault protection for the greatest sensitivity and speed.

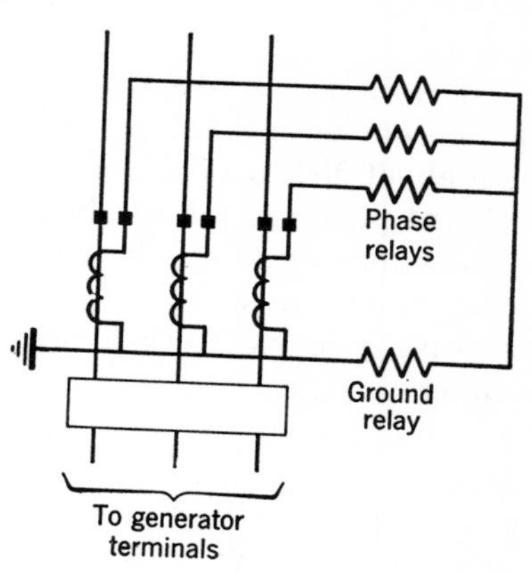

Fig. 16. Generator stator overcurrent relaying.

If non-directional voltage-restrained or -controlled overcurrent relays are used for external-fault back-up protection, they could also serve to protect against generator phase faults.

None of the foregoing forms of relaying will provide nearly as good protection as percentage-differential relaying equipment, and they should not be used except when the cost of bringing out the generator leads and installing current transformers and differential relays cannot be justified.

PROTECTION AGAINST STATOR OPEN CIRCUITS

An open circuit or a high-resistance joint in a stator winding is very difficult to detect before it has caused considerable damage. Split-phase relaying may provide such protection, but only the most sensitive equipment will detect the trouble in its early stages.[17] Negative-phase-sequence-relaying equipment for protection against unbalanced phase currents contains a sensitive alarm unit that will alert an operator to the abnormal condition.

It is not the practice to provide protective-relaying equipment purposely for open circuits. Open circuits are most unlikely in well-constructed machines.

STATOR-OVERHEATING PROTECTION

General stator overheating is caused by overloading or by failure of the cooling system, and it can be detected quite easily. Overheating because of short-circuited laminations is very localized, and it is just a matter of chance whether it can be detected before serious damage is done.

The practice is to embed resistance temperature-detector coils or thermocouples in the slots with the stator windings of generators larger than about 500 to 1500 kva. Enough of these detectors are located at different places in the windings so that an indication can be obtained of the temperature conditions throughout the stator.[28] Several of the detectors that give the highest temperature indication are selected for use with a temperature indicator or recorder, usually having alarm contacts; or the detector giving the highest indication may be arranged to operate a temperature relay to sound an alarm.

Supplementary temperature devices may monitor the cooling system; such equipment would give the earliest alarm in the event of cooling-system failure, but it is generally felt that the stator temperature detectors and alarm devices are sufficient.

Figure 17 shows one form of detector-operated relaying equipment using a Wheatstone-bridge circuit and a directional relay. In another form of equipment, the stator current is used to energize the bridge.

"Replica"-type temperature relays may be used with small generators that do not have temperature detectors. Such a relay is energized either directly by the current flowing in one of the stator windings of the machine or indirectly from current transformers in the stator circuit. The relay is arranged with heating and heat-storage elements so as to heat up and cool down as nearly as possible at the same rate as the machine in response to the same variations in the current. A thermostatic element closes contacts at a selected temperature. It will be evident that such a relay will not operate for failure of the cooling system.

The temperature-detector-operated devices are preferred because they respond more nearly to the actual temperature of the stator. The fact that the actual stator copper temperature is higher than the temperature at the detector[29] should be taken into account in the adjustment of the temperature relay. This difference in temperature may be 25°C or more in hydrogen-cooled machines, being greater at the higher hydrogen pressures. Thus, if the permissible copper temperature is assumed to remain constant with higher loading at higher hydrogen pressure, the temperature-relay setting must be lower.

In unattended stations, temperature relays are arranged to reduce

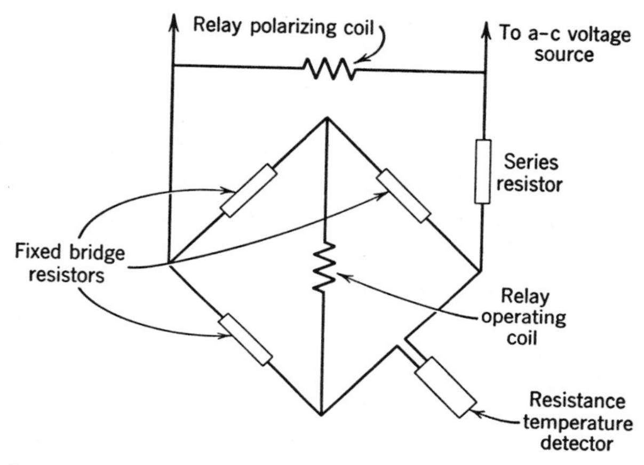

Fig. 17. Stator overheating relaying with resistance temperature detectors.

the load or shut down the unit if it overheats, but in an attended station the relay, if used, merely sounds an alarm. It should not be inferred from the foregoing that a generator may be loaded on the basis of temperature, because such practice is not recommended.

OVERVOLTAGE PROTECTION

Overvoltage protection is recommended for all hydroelectric or gas-turbine generators that are subject to overspeed and consequent overvoltage on loss of load. It is not generally required with steam-turbine generators.

This protection is often provided by the voltage-regulating equipment. If it is not, it should be provided by an a-c overvoltage relay. This relay should have a time-delay unit with pickup at about 110% of rated voltage, and an instantaneous unit with pickup at about 130% to 150% of rated voltage. Both relay units should be compensated against the effect of varying frequency. The relay should be energized from a potential transformer other than the one used for the automatic voltage regulator. Its operation should, preferably, first cause additional resistance to be inserted in the generator or exciter field circuit. Then, if overvoltage persists, the main generator breaker and the generator or exciter field breaker should be tripped.

LOSS-OF-SYNCHRONISM PROTECTION

It is not the usual practice to provide loss-of-synchronism protection at a prime-mover-driven generator. One generator is not likely to lose synchronism with other generators in the same station unless it loses excitation, for which protection is usually provided. Whether a station has one generator or more, if this station loses synchronism with another station, the necessary tripping to separate the generators that are out of step is usually done in the interconnecting transmission system between them; this is discussed at greater length in Chapter 14. However, loss-of-synchronism relaying equipment is available for use at a generating station if desired.

All induction-synchronous frequency converters for interconnecting two systems should have loss-of-synchronism protection on the synchronous-machine side. With synchronous-synchronous sets, such protection may be required on both sides. Operation of the relay should trip the main breaker on the side where the relay is located. The operating characteristics of a relay that can be used for this purpose are shown in Chapter 14.

FIELD GROUND-FAULT PROTECTION

Because field circuits are operated ungrounded, a single ground fault will not cause any damage or affect the operation of a generator in any way. However, the existence of a single ground fault increases the stress to ground at other points in the field winding when voltages are induced in the field by stator transients.[30] Thus, the probability of a second ground occurring is increased. Should a second ground occur, part of the field winding will be by-passed, and the current through the remaining portion may be increased. By-passing part of the field winding will unbalance the air-gap fluxes, and this will unbalance the magnetic forces on opposite sides of the rotor. Depending on what portion of the field is by-passed, this unbalance of forces may be large enough to spring the rotor shaft, and make it eccentric. A calculation of the possible unbalance force for a particular generator gave 40,000 pounds.[31] Cases are on record where the resulting vibration has broken bearing pedestals, allowing the rotor to grind against the stator; such failures caused extensive damage that was costly to repair and that kept the machines out of service for a long time.[31]

The second ground fault may not by-pass enough of the field winding to cause a bad magnetic unbalance, but arcing at the fault may heat the rotor locally and slowly distort it, thereby causing eccentricity and its accompanying vibration to develop slowly in from 30 minutes to 2 hours.

The safest practice is to use protective-relaying equipment to trip the generator's main and field breakers immediately when the first ground fault occurs, and this practice should certainly be followed in all unattended stations. However, many would rather risk the chance of a second ground fault and its possible consequences in an attended station, in order to keep the machine in service until it is more convenient to shut it down;[9] this group would use protective-relaying—or other—equipment, if any—merely to actuate an alarm or an indication when the first ground fault has occurred.

If a generator is to be permitted to operate with a single ground fault in its field, there should at least be provided automatic equipment foı immediately tripping the main and field breakers at an abnormal amplitude of vibration, but at no higher amplitude than necessary to avoid undesired operation on synchronizing or short-circuit transients. Such equipment would minimize the duration of severe vibration should the second ground fault occur at a critical location; obviously, the vibration cannot be stopped instantly because it takes time for the field flux to decay, but this is the best that can be done under the circumstances; the authors of Reference 31 calculated the effect of such prolonged vibration decreasing in amplitude and felt that there was no hazard. However, damage has been known to occur immediately when the second ground occurred and before anything could be done to prevent it. Vibration-detecting equipment should be in service continuously and not be put in service manually after the first ground fault has occurred, because the two ground faults may occur together or in quick succession. In addition, at least an alarm would be desirable, and preferably time-delay automatic tripping of the main and field breakers, at a still lower amplitude of vibration. This lower-set time-delay equipment would minimize the amplitude of vibration caused by rotor distortion because of localized heating. If this lower-set equipment were provided, the high-set vibration equipment could be permitted to shut down the prime mover as well as to trip the main and field breakers. The low-set equipment should preferably not shut down the prime mover; if the vibration is being caused by rotor eccentricity because of local heating, the amplitude might increase to a dangerous amount as the rotor speed decreases, because many generators have a critical speed below normal at which vibration may be materially worse than at normal speed;

instead, it would be preferable to trip the main and field breakers and keep the rotor turning at normal speed for 30 minutes to an hour to cool the rotor and let it straighten itself out.

In spite of the known hazard of extensive damage and a long-time outage, many generators are in service with no automatic protection or even alarm for field grounds, and the majority of the rest have ground-indication equipment only. This can only mean that, during the time that a generator is being operated with one ground in its field, the probability is remote of a second ground occurring and at such a location as to cause immediate damage before an operator can act to correct the condition. The possibility exists, nevertheless, and one should avoid such operation if at all possible.

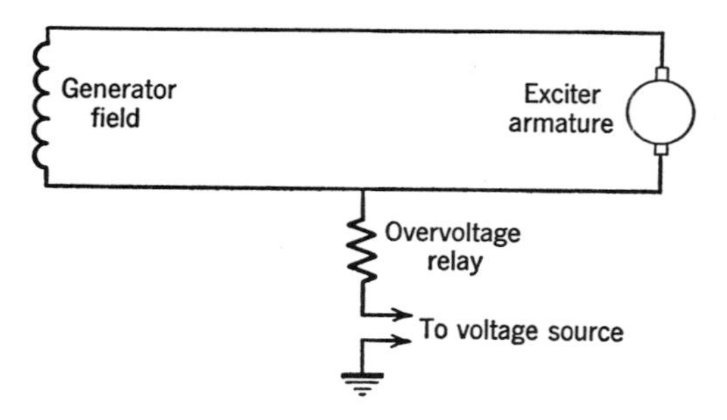

Fig. 18. Generator field ground-fault relaying.

The preferred type of protective-relaying equipment is shown in Fig. 18. Either a-c or d-c voltage may be impressed between the field circuit and ground through an overvoltage relay. A ground anywhere in the field circuit will pick up the relay. If direct current is used, the overvoltage relay can be more sensitive than if alternating current is used; with alternating current, the relay must not pick up on the current that flows normally through the capacitance to ground, and care must be taken to avoid resonance between this capacitance and the relay inductance.

It may be necessary to provide a brush on the rotor shaft that will effectively ground the rotor, especially when a-c voltage is applied.[32] One should not rely on the path to ground through the bearing-oil film for two reasons: (1) the resistance of this path may be high enough so that the relay would not operate if the field became grounded, and (2) even a very small magnitude of current flowing continually through the bearing may pit the surface and destroy the bearing.[33] A brush will probably be required with a steam turbine having steam seals. The brush should be located where it will not by-pass the bearing-pedestal insulation that is provided to prevent the flow of shaft currents. One should consult the turbine manufacturer before deciding that such a brush is not required.

PROTECTION AGAINST ROTOR OVERHEATING BECAUSE OF UNBALANCED THREE-PHASE STATOR CURRENTS

Unbalanced three-phase stator currents cause double-system-frequency currents to be induced in the rotor iron. These currents will quickly cause rotor overheating and serious damage if the generator is permitted to continue operating with such an unbalance.[34,35,36] Unbalanced currents may also cause severe vibration, but the overheating problem is more acute.

Standards have been established for the operation of generators with unbalanced stator currents.[37] The length of time (T) that a generator may be expected to operate with unbalanced stator currents without danger of being damaged can be expressed in the form:

$$\int_o^T i_2{}^2\, dt = K$$

where i_2 is the instantaneous negative-phase-sequence component of the stator current as a function of time; i_2 is expressed in per unit based on the generator rating, and K is a constant. K is 30 for steam-turbine generators, synchronous condensers, and frequency-changer sets; K is 40 for hydraulic-turbine generators, and engine-driven generators. If the integrated value is between that given for K and twice this value, the generator "may suffer varying degrees of damage, and an early inspection of the rotor surface is recommended."[37] If the integrated value is greater than twice that given for K, "serious damage should be expected."

If we let $I_2{}^2$ be the average value of $i_2{}^2$ over the time interval T, we can express the foregoing relation in the handy form $I_2{}^2\ T = 30$ or $I_2{}^2T = 40$, depending on the type of generator. If T is longer than 30 seconds, I_2 may be larger than the foregoing relation would permit, but no general figures can be given that would apply to any machine.[34,36,38]

It has been shown that current-balance relaying equipment operating from the phase currents will operate too quickly for small unbalances and too slowly for large unbalances.[39]

The recommended type of relaying equipment is an inverse-time overcurrent relay operating from the output of a negative-phase-sequence-current filter that is energized from the generator CT's as in Fig. 19.[38,39] The relay's time-current characteristics are of the form $I^2T = K$, so that, with the pickup and time-delay adjustments that are provided, the relay characteristic can be chosen to match closely any machine characteristic. The relay should be connnected to trip the generator's main breaker. Some forms of the relay also include a very sensitive unit to control an alarm for small unbalances.

Extensive studies have shown that, in the majority of cases, the negative-phase-sequence-current relay will properly coordinate with

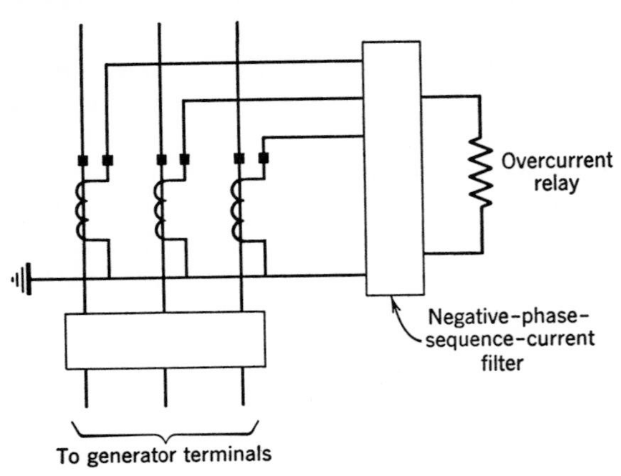

Fig. 19. Negative-phase-sequence-overcurrent relaying for unbalanced stator currents.

other system-relaying equipment.[40,41] Improper coordination is said to be possible where load is supplied at generator voltage and where there are five or more generators in the system. For a unit generator-transformer arrangement, proper coordination is assured.

The fact that the system-relaying equipment will generally operate first might lead to the conclusion that, with modern protective equipment, protection against unbalanced three-phase currents during short circuits is not required.[36,42] This conclusion might be reached also from the fact that there has been no great demand for improvement of the existing forms of protection. The sensitive alarm unit would be helpful to alert an operator in the event of an open circuit under load, for which there may be no other automatic protection. Otherwise, one would apply the negative-phase-sequence-current relay only when the back-up-relaying equipment of the system could not be relied on to remove unbalanced faults quickly enough in the event of primary-relaying failure. However, there are undoubtedly many locations where back-up relaying will not operate for certain faults. Therefore, one should not generalize on this subject but should get the facts for each application. To determine properly whether additional protection is really necessary is a very complicated study. Where additional protection can be afforded, it should be applied.

LOSS-OF-EXCITATION PROTECTION

When a synchronous generator loses excitation, it operates as an induction generator, running above synchronous speed. Round-rotor generators are not suited to such operation because they do not have amortisseur windings that can carry the induced rotor currents. Consequently, a steam-turbine-generator's rotor will overheat rather quickly from the induced currents flowing in the rotor iron, particularly at the ends of the rotor where the currents flow across the slots through the wedges and the retaining ring, if used. The length of time to reach dangerous rotor overheating depends on the rate of slip, and it may be as short as 2 or 3 minutes. Salient-pole generators invariably have amortisseur windings, and, therefore, they are not subject to such overheating.

The stator of any type of synchronous generator may overheat, owing to overcurrent in the stator windings, while the machine is running as an induction generator. The stator current may be as high as 2 to 4 times rated, depending on the slip. Such overheating is not apt to occur as quickly as rotor overheating.

Some systems cannot tolerate the continued operation of a generator without excitation. In fact, if the generator is not disconnected immediately when it loses excitation, widespread instability may very quickly develop, and a major system shutdown may occur. Such systems are those in which quick-acting automatic generator voltage regulators are not employed. When a generator loses excitation, it draws reactive power from the system, amounting to as much as 2 to 4 times the generator's rated load. Before it lost excitation, the generator may have been delivering reactive power to the system. Thus, this large reactive load suddenly thrown on the system, together with the loss of the generator's reactive-power output, may cause widespread voltage reduction, which, in turn, may cause extensive instability unless the other generators can automatically pick up the additional reactive load immediately.

In a system in which severe disturbances can follow loss of excitation in a given generator, automatic quick-acting protective-relaying equipment should be provided to trip the generator's main and field breakers. An operator does not have sufficient time to act under such circumstances. Where system disturbances definitely will not follow loss of excitation, an operator will usually have at least 2 or 3 minutes in which to act in lieu of automatic tripping. Sometimes an emergency excitation source and manual throw-over are provided that may make it unnecessary to remove a generator from service. However, an operator can usually do nothing except remove the generator from service, unless the operator himself has accidentally removed excitation. If a loss-of-excitation condition should not be recognized and a generator should run without excitation for an unknown length of time, it ought to be shut down and carefully examined for damage

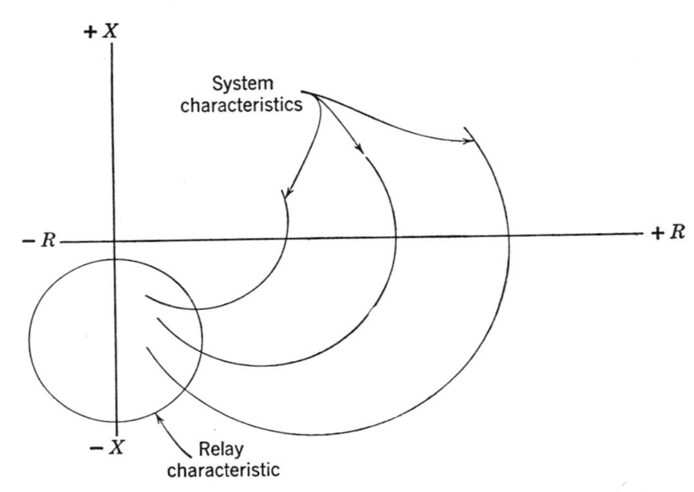

Fig. 20. Loss-of-excitation relay and system characteristics.

before returning it to service. In systems in which severe disturbances may or may not follow loss of excitation in a given generator, the generator must sometimes be tripped when the system does not require it, merely to be sure that the generator will always be tripped when the system *does* require it.

Undercurrent relays connected in the field circuit have been used quite extensively, but the most selective type of loss-of-excitation relay is a directional-distance type operating from the a-c current and voltage at the main generator terminals.[43] Figure 20 shows several loss-of-excitation characteristics and the operating characteristic of one type of loss-of-excitation relay on an *R-X* diagram. No matter what the initial conditions, when excitation is lost, the equivalent generator impedance traces a path from the first quadrant into a region of the fourth quadrant that is entered only when excitation is severely reduced or lost. By encompassing this region within the relay characteristic, the relay will operate when the generator first starts to slip poles and will trip the field breaker and disconnect the generator from the system before either the generator or the system can be harmed. The generator may then be returned to service immediately when the cause of excitation failure is corrected.

PROTECTION AGAINST ROTOR OVERHEATING BECAUSE OF OVEREXCITATION

It is not the general practice to provide protection against overheating because of overexcitation. Such protection would be provided indirectly by the stator-overheating protective equipment or by excitation-limiting features of the voltage-regulator equipment.

PROTECTION AGAINST VIBRATION

Protective-relaying practices and equipment that are described under the headings "Protection against Rotor Overheating because of Unbalanced Three-Phase Stator Currents" and "Field Ground-Fault Protection" prevent or minimize vibration under those circumstances. If the vibration-detecting equipment recommended under the latter heading is used, it will also provide protection if vibration results from a mechanical failure or abnormality. For a steam turbine, it is the general practice to provide vibration recorders that can also be used if desired to control an alarm or to trip. However, it is not the general practice to trip.

PROTECTION AGAINST MOTORING

Motoring protection is for the benefit of the prime mover or the system, and not for the generator. However, it is considered here because, when protective-relaying equipment is used, it is closely associated with the generator.

Steam Turbines. A steam turbine requires protection against overheating when its steam supply is cut off and its generator runs as a motor. Such overheating occurs because insufficient steam is passing through the turbine to carry away the heat that is produced by windage loss. Modern condensing turbines will even overheat at *outputs* of less than approximately 10% of rated load.

The length of time required for a turbine to overheat, when its steam is completely cut off, varies from about 30 seconds to about 30 minutes, depending on the type of turbine. A condensing turbine that operates normally at high vacuum will withstand motoring much longer than a topping turbine that operates normally at high back pressure.

Since the conditions are so variable, no single protective practice is clearly indicated. Instead, the turbine manufacturer's recommendations should be sought in each case. The manufacturer will probably have turbine accessories that will provide an alarm or will shut down the equipment, as required.

For a turbine that will not overheat unless its generator runs as a motor, sensitive power-directional-relaying equipment has been widely used. One type of such relaying equipment is able to operate on power flowing into the generator amounting to about 0.5% of the generator's rated full-load watts. In general, the protective equipment should operate on somewhat less than about 3% of rated power. Sufficient time delay should be provided to prevent undesired operation on transient power reversals such as those occurring during synchronizing or system disturbances.

Hydraulic Turbines. Motoring protection may occasionally be desirable to protect an unattended hydraulic turbine against cavitation of the blades. Cavitation occurs on low water flow that might result, for example, from blocking of the trash gates. Protection is not generally provided for attended units. Protection can be provided by power-directional-relaying equipment capable of operating on motoring current of somewhat less than about 2.5% of the generator's full-load rating.

Diesel Engines. Motoring protection for Diesel engines is generally desirable. The generator will take about 15% of its rated power or more from the system, which may constitute an undesirably high load on the system. Also, there may be danger of fire or explosion from unburned fuel. The engine manufacturer should be consulted if one wishes to omit motoring protection.

Gas Turbines. The power required to motor a gas turbine varies from 10% to 50% of full-load rating, depending on turbine design and whether it is a type that has a load turbine separate from that used to drive the compressor. Protective relays should be applied based primarily on the undesirability of imposing the motoring load on the system. There is usually no turbine requirement for motoring protection.

OVERSPEED PROTECTION

Overspeed protection is recommended for all prime-mover-driven generators. The overspeed element should be responsive to machine speed by mechanical, or equivalent electrical, connection; if electrical, the overspeed element should not be adversely affected by generator voltage.

The overspeed element may be furnished as part of the prime mover, or of its speed governor, or of the generator; it should operate the speed governor, or whatever other shut-down means is provided, to shut down the prime mover. It should also trip the generator circuit breaker; this is to prevent overfrequency operation of loads connected to the system supplied by the generator, and also to prevent possible overfrequency operation of the generator itself from the a-c system. The overspeed device should also trip the auxiliary breaker where auxiliary power is taken from the generator leads. In certain cases, an overfrequency relay may be suitable for providing both of these forms of protection. However, a direct-connected centrifugal switch is preferred.

The overspeed element should usually be adjusted to operate at about 3% to 5% above the full-load rejection speed. Supplementary overspeed protection is required for some forms of gas turbines. Whether such protection is required for any given turbine, and what its adjustment should be, should be specified by the turbine manufacturer.

EXTERNAL-FAULT BACK-UP PROTECTION

Generators should have provision against continuing to supply short-circuit current to a fault in an adjacent system element because of a primary relaying failure. Simple inverse-time overcurrent relaying is satisfactory for single-phase-to-ground faults. For phase faults, a voltage-restrained or voltage-controlled inverse-time-overcurrent relay—or a single-step distance-type relay with definite time delay—is preferred.

Which of the two general types of phase relay to use depends on the types of relays with which the back-up relays must be selective. Thus, if the adjacent circuits have inverse-time-overcurrent relaying, the voltage-restrained or -controlled inverse-time-overcurrent relay should be used. But, if the adjacent circuits have high-speed pilot or distance relaying, then the distance-type relay should be used.

Inverse-time-overcurrent relays for phase-fault-back-up protection are considered decidedly inferior; owing to the decrement in the short-circuit current put out by a generator, the margin between the maximum-load current and the short-circuit current a short time after the fault current has started to flow is too narrow for reliable protection.

Where cross-compounded generators are involved, external-fault-back-up-relaying equipment need be applied to only one unit.

Negative-phase-sequence-overcurrent-relaying equipment to prevent overheating of the generator rotor as a consequence of prolonged unbalanced stator currents is not here considered a form of external-fault-back-up protection. Instead, such relaying is considered a form of primary relaying, and it is treated as such elsewhere. A back-up relay should have characteristics similar to the relays being backed up, and a negative-phase-sequence-overcurrent relay is not the best for this purpose, apart from the fact that such a relay would not operate for three-phase faults.

When a unit generator-transformer arrangement is involved, the external-fault-back-up relay is generally energized by current and voltage sources on the low-voltage side of the power transformer. Then the connections should be such that the distance-type units measure distance properly for high-voltage faults.

BEARING-OVERHEATING PROTECTION

Bearing overheating can be detected by a relay actuated by a thermometer-type bulb inserted in a hole in the bearing, or by a resistance-temperature-detector relay, such as that described for stator-overheating protection, with the detector embedded in the bearing. Or, where lubricating oil is circulated through the bearing under pressure, the temperature of the oil may be monitored if the system has provision for giving an alarm if the oil stops flowing.

Such protection is provided for all unattended generators where the size or importance of the generator warrants it. Such protection for attended generators is generally only to sound an alarm.

OTHER MISCELLANEOUS FORMS OF PROTECTION

The references given later under the heading "Protection of the Prime Mover" describe other protective features provided for generators and their associated equipment. These forms of protection are generally mechanical and are not generally classified with protective-relaying equipment.

GENERATOR POTENTIAL-TRANSFORMER FUSING AND FUSE BLOWING

Unless special provision is made, the blowing of a potential-transformer fuse may cause certain relays to trip the generator breakers. Such relays are those types employing voltage restraint, such as voltage-controlled or distance-type relays used for loss-of-excitation or external-fault-back-up protection. It is not necessarily a complete loss of voltage that causes such undesired tripping; with a three-phase voltage supply consisting of two or three potential transformers, the blowing of a fuse may change the magnitude and phase relations of certain secondary voltages through the mechanism of the potentiometer effect of other devices connected to the PT's. Such an effect can cause a relay to operate undesirably when complete loss of voltage would not cause undesired operation.

The proper solution to this problem is not the complete removal of all fuses. The preferred practice is to fuse both primary and secondary circuits.[2] However, the secondary fuses may be omitted from the circuits of relays or other devices where correct operation is so essential that it is "preferable to incur hazards associated with the possible destruction of the PT by a sustained secondary short circuit rather than to risk interruption of the voltage supply to such devices as the result of a momentary short circuit."[2] Advantage is usually taken of this clause not to fuse the secondary, and the record with this practice has been very good. Primary fuses should not be omitted, but they must be chosen so that they will not blow on magnetizing-current inrush or other transients.[44]

When secondary fusing is used because of the better protection that it gives the potential transformers, the exposure of critical devices to the effects of accidental fuse blowing can be minimized by fusing their circuits separately, or by fusing all circuits except those of the critical devices.

When separate secondary fusing is not enough assurance against the consequences of fuse blowing, a voltage-balance relay may be used that compares the magnitudes of the voltages of the voltage source under consideration with the voltages of another source that are always approximately equal to the voltages of the first source unless a fuse blows. Such a relay can be arranged to prevent undesired operation of critical relays and to actuate an alarm when a fuse blows. Not only preventing undesired relay operation but also knowing immediately that a fuse has blown are important. With wye-wye potential transformers, a set of auxiliary PT's connected wye-broken-delta, with a voltage relay energized by the voltage across the open corner of the delta, can be used to open the trip circuit when one or more fuses blow.

STATION AUXILIARY PROTECTION

Power-plant auxiliaries are treated somewhat differently from similar equipment used elsewhere. It is generally felt that they deserve higher-quality protective equipment. At the same time, however, certain so-called "essential" auxiliaries are kept in service under manual supervision during overload conditions that would ordinarily call for tripping.[45,46] Reference 46 stresses the importance of keeping auxiliary motors running during system disturbances, and describes techniques for accomplishing this. Reference 47 is a collection of several papers on the effect of reduced voltage and frequency on power-plant capabilities. The protection of station auxiliaries will be treated in more detail where the protection of motors, transformers, and busses is described.

PROTECTION OF THE PRIME MOVER

Except for the protection against motoring and overspeed, the protection of the prime mover and its associated mechanical equipment is not treated in this book. References to some excellent papers on this subject, and also on the subject of fire protection, are given in the Bibliography.[8,48]

Motor Protection

This section deals with the protection of attended synchronous motors, induction motors, synchronous condensers, and the motors of frequency converters. Motors in unattended stations must be protected against all harmful abnormal conditions.[1] The protection of very small motors is not specifically described, although the same basic principles apply; this subject is treated in detail in the National Electrical Code.[3] The practices described here for large motors are at least equal to those covered by the Code, and are generally more comprehensive. However, it is recommended that the Code be consulted whenever it applies. The protection of fire-pump motors is not included here, because it is completely described elsewhere.[4]

SHORT-CIRCUIT PROTECTION OF STATOR WINDINGS

Overcurrent protection is the basic type that is used for short-circuit protection of stator windings. The equipment for this type of protection ranges from fuses for motor voltages of 600 volts and lower, through direct-acting overcurrent tripping elements on circuit breakers, to separate overcurrent relays and circuit breakers for voltages of 2200 volts and higher.

Protection should be provided against a fault in any ungrounded conductor between the interrupting device and the motor, including its stator windings. Where fuses or direct-acting tripping devices are used, there must be one protective element in each ungrounded conductor. Where relays and current transformers are used with so-called "a-c tripping" from the output of the current transformers, a CT and relay are required for each ungrounded conductor. However, if battery or capacitor tripping is provided, three current transformers with two phase relays and one ground relay will suffice for a three-phase circuit whether or not the source neutral is grounded.

Motors Other than Essential Service. For all except "essential-service" motors, it is the practice to provide both inverse-time and instantaneous phase and ground overcurrent relays for automatic tripping. The inverse-time phase relays are generally adjusted to pick up at somewhat less than about 4 times rated motor current, but to have enough time delay so as not to operate during the motor-starting period. The instantaneous phase relays are adjusted to pick up a little above the locked-rotor current. The inverse-time ground relays are adjusted to pick up at no more than about 20% of rated current or about 10% of the maximum available ground-fault current, whichever is smaller. The instantaneous ground-relay pickup should be from about 2.5 to 10 times rated current; this relay may be omitted if the maximum available ground-fault current is less than about 4 times rated current, or if the pickup has to be more than about 10 times rated current to avoid undesired tripping during motor starting or external faults. If a CT, like a bushing CT, is used with all three phase conductors of the motor circuit going through the opening in the core, a very sensitive instantaneous overcurrent relay can be used that will operate for ground faults within about 10% of the winding from the neutral end.

Percentage-differential relaying is provided for large motors. It is the practice of manufacturers[2] to recommend such protection for motors of the following ratings: (*a*) 2200 volts to 4999 volts, inclusive, 1500 hp and higher; (*b*) 5000 volts and higher, 501 hp and higher. The advantage of percentage-differential relaying is that it will provide faster and more sensitive protection than overcurrent relaying, but at the same time it will not operate on starting or other transient overcurrents.

References to excellent articles on the subject of industrial-motor protection are given in the Bibliography.[49]

Essential-Service Motors. For essential-service motors, the inverse-time phase overcurrent relays are usually omitted, leaving the instantaneous phase relays, and the inverse-time and instantaneous ground relays, or the differential relays if applicable. The reason for the omission is to trip the motor breaker automatically only for short circuits and not to trip for any other reason. This is because the tripping of such a motor may force a partial or complete shutdown of a generator or other service with which the motor is associated, and hence any unnecessary tripping must be avoided. As will be seen when we consider stator overheating protection, supplementary protection against phase overcurrents less than locked-rotor values is provided.

STATOR-OVERHEATING PROTECTION

All motors need protection against overheating resulting from overload, stalled rotor, or unbalanced stator currents. For complete protection, three-phase motors should have an overload element in each phase; this is because an open circuit in the supply to the power transformer feeding a motor will cause twice as much current to flow in one phase of the motor as in either of the other two phases, as shown in Fig. 21. Consequently, to be sure that there will be an overload ele-

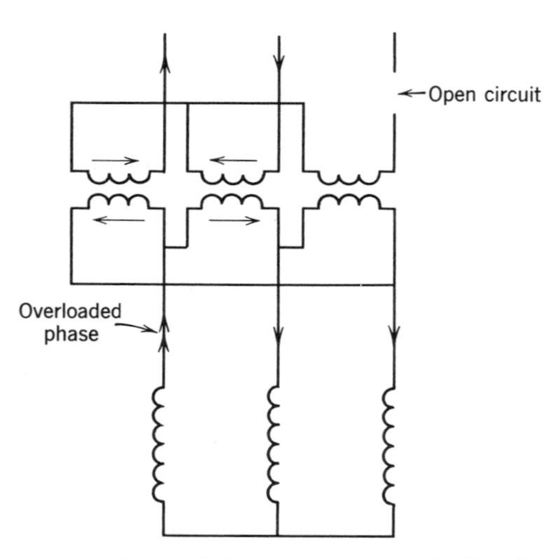

Fig. 21. Illustrating the need for overcurrent protection in each phase.

ment in the most heavily loaded phase no matter which power-transformer phase is open-circuited, one should provide overload elements in all three phases. In spite of the desirability of overload elements in all three phases, motors rated about 1500 hp and below are generally provided with elements in only two phases, on the assumption that the open-phase condition will be detected and corrected before any motor can overheat.

Single-phase motors require an overload element in only one of the two conductors.

Motors Other than Essential Service. Except for some essential-service motors, whose protection will be discussed later, it is the practice for motors rated less than about 1500 hp to provide either replica-type thermal-overload relays or long-time inverse-time-overcurrent relays or direct-acting tripping devices to disconnect a motor from its source of supply in the event of overload. Which type of relay to use is largely a matter of personal preference. Other things being equal, the replica type will generally provide the best protection because, as shown in Fig. 22, its time-current characteristic more nearly matches the heating characteristic of a motor over the full range of overcurrent; also, it may take into account the heating effect

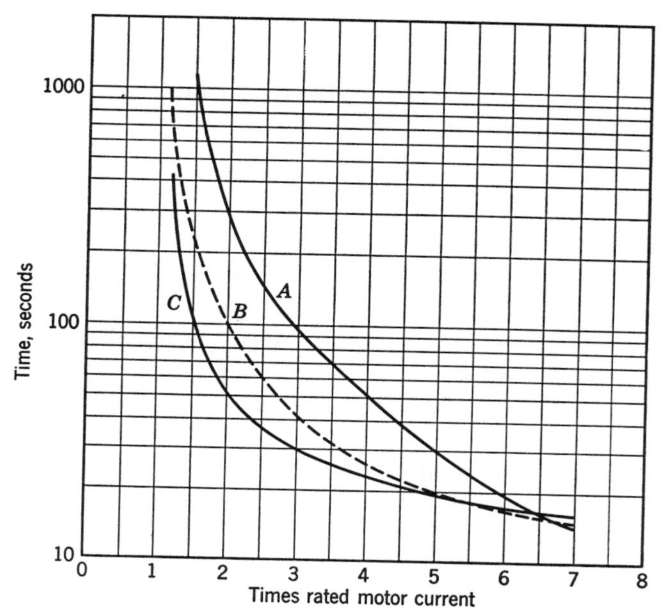

Fig. 22. Typical motor-heating and protective-relay characteristics. *A*, motor; *B*, replica relay; *C*, inverse-time relay.

of the load on the motor before the overload condition occurred. The inverse-time-overcurrent relay will tend to "overprotect" at low currents and to "underprotect" at high currents, as shown in Fig. 22. However, the overcurrent relay is very easy to adjust and test, and it is self-reset. For continuous-rated motors without service factor or short-time overload ratings, the protective relays or devices should be adjusted to trip at not more than about 115% of rated motor current. For motors with 115% service factor, tripping should occur at not more than about 125% of rated motor current. For motors with special short-time overload ratings, or with other service factors, the motor characteristic will determine the required tripping characteristic, but the tripping current should not exceed about 140% of rated motor current. The manufacturer's recommendations should be obtained in each case.

The overload relays will also provide protection in the event of phase-to-phase short circuits, and in practice one set of such relays serves for both purposes wherever possible. A survey of the practice of a number of power companies[45] showed that a single set of long-time inverse-time-overcurrent relays, adjusted to pick up at 125% to 150% of rated motor current, is used for combined short-circuit and overload protection of non-essential auxiliary motors; they are supplemented by instantaneous overcurrent relays adjusted as already described. Such inverse-time overload relays must withstand short-circuit currents without damage for as long as it takes to trip the breaker. Also the minimum requirements as to the number of relays or devices for either function must be fulfilled.

Motors rated higher than about 1500 hp are generally provided with resistance temperature detectors embedded in the stator slots between the windings. If such temperature detectors are provided, a single relay operating from these detectors is used instead of the replica-type or inverse-time-overcurrent relays. Also, current-balance relays capable of operating on about 25% or less unbalance between the phase currents should be supplied. If the motor does not have resistance temperature detectors, but is provided with current-balance relays, a single replica-type thermal overload relay may be substituted for the resistance-temperature-detector relay.

Specially cooled or ventilated motors may require other types of protective equipment than those recommended here. For such motors, the manufacturer's recommendations should be obtained.

Reference 50 gives more useful information on the subject of industrial-motor protection.

Essential-Service Motors. The protection recommended for some essential-service motors is based on minimizing the possibility of unnecessarily tripping the motor, even though such practice may sometimes endanger the motor. In other words, long-time inverse-time-overcurrent relays are provided for all motor ratings, but they merely control an alarm and leave tripping in the control of an operator. Then, for motors that can suffer locked rotor, supplementary instantaneous overcurrent relays, adjusted to pick up at about 200% to 300% of rated motor current are used, and their contacts are connected in series with the contacts of the inverse-time-overcurrent relays to trip the motor breaker automatically. The instantaneous relays should be of the high-reset type to be sure that they will reset when the current returns to normal after the starting inrush has subsided. The protection provided by this type of equipment is illustrated in Fig. 23.

For essential-service motors for which automatic tripping is desired in addition to the alarm for overloads between about 115% of rated current and the pickup of the instantaneous overcurrent relays,

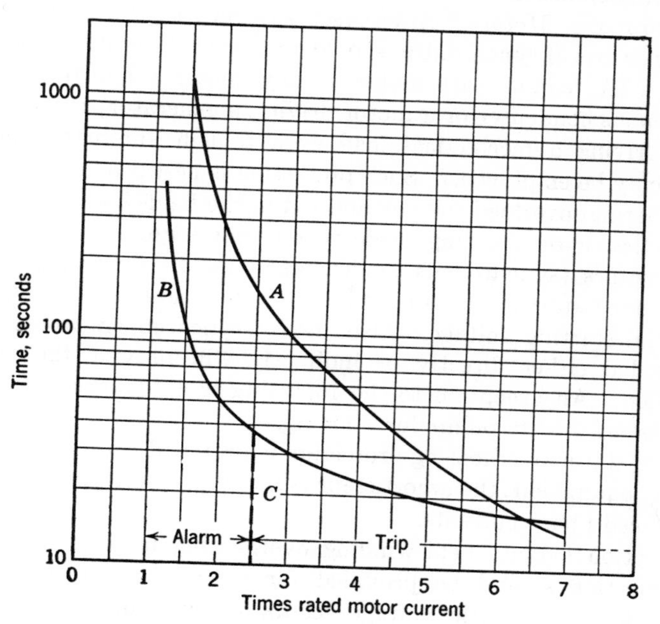

Fig. 23. Protection characteristic for essential-service motors. *A*, motor; *B*, inverse-time relay; *C*, instantaneous relay.

thermal relays of either the replica type or the resistance-temperature-detector type should be used, depending on the size of the motor. Such relays permit operation for overloads as far as possible beyond the point where the alarm will be sounded, but without damaging the motor to the extent that it must be repaired before it can be used again.

ROTOR-OVERHEATING PROTECTION

Squirrel-Cage Induction Motors. The replica-type or the inverse-time-overcurrent relays, recommended for protection against stator overheating, will generally protect the rotor except where high-inertia load is involved; such applications should be referred to the manufacturer for recommendations. Where resistance-temperature-detector relaying is used, a single replica-type or inverse-time-overcurrent relay should be added for rotor protection during starting.

Wound-Rotor Induction Motors. General recommendations for this type of motor cannot be given except that the rotor may not be protected by the stator-overheating protective equipment that has been described. Each application should be referred to the manufacturer for recommendations.

Synchronous Motors. Amortisseur-overheating protection during starting or loss of synchronism should be provided for all "loaded-start" motors. (A loaded-start motor is any motor other than either a synchronous condenser or a motor driving a generator; it includes any motor driving a mechanical load even though automatic unloading means may be employed.) Such protection is best provided by a time-delay thermal overload relay connected in the field-discharge circuit.[51]

Amortisseur-overheating protection is not required for "unloaded-start" motors (synchronous condensers or motors driving generators). An unloaded-start motor is not likely to fail to start on the application of normal starting voltage. Also, loss-of-synchronism protection that is provided either directly or indirectly will provide the necessary protection. An exception to the foregoing is a condenser or a motor that has an oil-lift pump for starting.

Where stator-overheating protection is provided by current-balance-relaying equipment, the amortisseur is indirectly protected also against unbalanced phase currents.

Protection against field-winding overheating because of prolonged overexcitation should be provided for synchronous motors or condensers with automatic voltage regulators without automatic field-current-limiting features. A thermal overload relay with time delay or a relay that responds to an increase in the field-winding resistance with increasing temperature may be used. In an attended station, the relay would merely control an alarm.

LOSS-OF-SYNCHRONISM PROTECTION

All loaded-start synchronous motors should have protection against loss of synchronism, generally arranged to remove the load and the excitation temporarily and to reapply them when permissible. Otherwise, the motor is disconnected from its source.

For unloaded-start motors except the synchronous motor of a frequency converter, the combination of undervoltage protection, loss-of-excitation protection, and the d-c generator overcurrent protection that is generally furnished will provide satisfactory loss-of-synchronism protection. Should additional protection be required, it can be provided by an inverse-time-overcurrent relay energized by the current in the running connection and arranged to trip the main breaker. Usually, automatic resynchronizing is not required.

All frequency converters interconnecting two systems should have loss-of-synchronism protection on the synchronous-machine side. With synchronous-synchronous sets, protection may be required on both sides. The protective-relaying equipment should be arranged to trip the main breaker on its side.

UNDERVOLTAGE PROTECTION

All a-c motors except essential-service motors should have protection against undervoltage on at least one phase during both starting and running. For polyphase motors larger than about 1500 hp, polyphase undervoltage protection is generally provided.

Wherever possible, the protective equipment should have inverse-time-delay characteristics.

"Undervoltage release," which provides only temporary shutdown on voltage failure and which permits automatic restart when voltage is re-established, should not be used with such equipment as machine tools, etc., where such automatic restart might be hazardous to personnel or detrimental to process or equipment.

LOSS-OF-EXCITATION PROTECTION

All unloaded-start synchronous motors that do not have loss-of-synchronism protection as described elsewhere, and that do not have automatic voltage regulators, should have loss-of-excitation protection in the form of a low-set, time-delay-reset undercurrent relay whose coil is in series with the field winding.

If a motor has loss-of-synchronism protection, amortisseur-overheating protection, and stator-overheating protection, these equipments indirectly provide loss-of-excitation protection.

FIELD GROUND-FAULT PROTECTION

The same equipment as that described for generators may be used if the size or importance of the motor warrants it.

Bibliography

1. "Automatic Station Control, Supervisory and Telemetering Equipments," *Publ. C37.2,* American Standards Association, Inc., 70 East 45th St., New York 17, N. Y.

2. "Power Switchgear Assemblies," *Publ. SG5,* National Electrical Manufacturers Assoc., 155 East 44th St., New York 17, N. Y.

3. *National Electrical Code,* National Electrical Board of Fire Underwriters, 85 John St., New York 38, N. Y.; 222 West Adams St., Chicago 6, Ill.; 465 California St., San Francisco 4, Calif.

4. "Standards for the Installation and Operation of Centrifugal Fire Pumps," *Publ. 20,* National Fire Protection Assoc., 60 Batterymarch St., Boston 10, Mass.

5. *Silent Sentinels,* Westinghouse Electric Corp., Meter Division, Newark, N. J.

6. *The Art of Protective Relaying* and *An Introduction to Protective Relays,* General Electric Co., Schenectady 5, N. Y.

7. "Bibliography of Relay Literature," by AIEE Committee:

	AIEE Trans.		
Years of Publication	Vol.	Year	Pages
1927–1939	60	1941	1435–1447
1940–1943	63	1944	705– 709
1944–1946	67, Pt. I	1948	24– 27
1947–1949	70, Pt. I	1951	247– 250
1950–1952	74, Pt. III	1955	45– 48

8. "Protection of Turbine Generators and Boilers by Automatic Tripping," by H. A. Bauman, J. M. Driscoll, P. T. Onderdonk, and R. L. Webb, *AIEE Trans., 72,* Part III (1953), pp. 1248–1255. Discussions, pp. 1255–1260.

"Human Error Causes Half Power Plant Outages," *Elec. World,* Nov. 15, 1954, p. 112.

9. "Relay Protection of A-C Generators," by AIEE Committee, *AIEE Trans., 70* (1951), pp. 275–281. Discussions, pp. 281–282.

10. "Power Systems Engineering Faces a New Challenge," by E. H. Snyder, *EEI Bulletin,* June, 1954, pp. 193–196.

11. "Principles and Practices of Relaying in the United States," by E. L. Harder and W. E. Marter, *AIEE Trans., 67* (1948), pp. 1075–1081.

12. "Matching Ratio Errors Improves Differential Relays," by E. C. Wentz, *Elec. World, 99,* No. 3 (January 16, 1932), pp. 154–155.

13. "A High-Speed Differential Relay for Generator Protection," by W. K. Sonnemann, *AIEE Trans., 59* (1940), pp. 608–612. Discussions, pp. 1250–1252.

14. "A New Generator Differential Relay," by A. J. McConnell, *AIEE Trans., 62* (1943), pp. 11–13. Discussions, p. 381.

15. "Characteristics of Split-Phase Currents As a Source of Generator Protection," by H. R. Sills and J. L. McKeever, *AIEE Trans., 72,* Part III (1953), pp. 1005–1013. Discussions, pp. 1013–1016.

16. "Generator Protected by a 'Floating' CT," by E. G. Ratz, *Elec. World, 124* (August 18, 1945), pp. 115–116.

17. "Shipshaw Relay Protection," by J. T. Madill and F. H. Duffy, *AIEE Trans., 68,* Part I (1949), pp. 50–58.

18. "Protect Generator from Internal Faults," by Giuseppe Calabrese, *Elec. World, 109* (May 21, 1938), p. 76 (1704). Correction, *110,* No. 7 (Aug. 12, 1938), p. 435.

19. "Application Guide for the Grounding of Synchronous Generator Systems," by AIEE Committee, *AIEE Trans., 72,* Part III (1953), pp. 517–526. Discussions, pp. 526–530. (Also included in *AIEE No. 954,* Oct., 1954.)

20. *Electrical Transmission and Distribution Reference Book,* Westinghouse Electric Corp., East Pittsburgh, Pa.

21. Discussions of Reference 19, pp. 526–530.

22. "Neutral Grounding Devices," *Publ. 32,* The American Institute of Electrical Engineers, 33 West 39th St., New York 18, N. Y.

23. "Guide for Loading Oil-Immersed Distribution and Power Transformers," *Publ. C57.32,* American Standards Assoc., Inc., 70 East 45th St., New York 17, N. Y.

24. "Iron Burning Caused by Low Current Arcs," by R. Pohl, *AEG Mitteilungen,* Berlin, Germany, Jan., 1930.

25. "Ground-Fault Neutralizer Grounding of Unit-Connected Generators," by H. R. Tomlinson, *AIEE Trans., 72,* Part III (1953), pp. 953–960.

26. Discussions of Reference 25, pp. 960–961.

27. "Sensitive Ground Relaying of A-C Generators," by E. T. B. Gross, *AIEE Trans., 71,* Part III (1952), pp. 539–541. Discussions, pp. 541–544.

28. "Rotating Electrical Machinery," *Publ. C50–1943,* American Standards Assoc., 70 East 45th St., New York 17, N. Y.

29. "Temperature Drop to Resistance Temperature Detector in Stator Windings of Turbine Generators," by R. P. Jerrard, *AIEE Trans., 73,* Part III (1954), pp. 665–669. Discussions, pp. 669–670.

"Controls for Operation of Steam Turbine-Generator Units," by O. N. Bryant, C. C. Sterrett, and D. M. Sauter, *AIEE Trans., 73,* Part III (1954), pp. 79–88.

"Turbine Generator Stator Winding Temperatures at Various Hydrogen Pressures," by J. R. M. Alger, C. E. Kilbourne, and D. S. Snell, *AIEE Trans., 74,* Part III (1955), pp. 232–250. Discussions, pp. 250–251.

30. Discussions of Reference 31, pp. 1355–1357.

31. "Vibration Protection for Rotating Machinery," by R. L. Webb and C. S. Murray, *AIEE Trans., 63* (1944), pp. 534–537.

32. "Detection of Grounds in Generator Field Windings," by J. E. Barkle, C. C. Sterrett, and L. L. Fountain, *AIEE Trans., 74,* Part III (1955), pp. 467–470. Discussions, pp. 470–472.

33. "Bearings Pitted by Ground Alarm Surges," by T. M. Blakeslee, *Elec. World, 119* (Jan. 9, 1943), pp. 57 (81)–58 (82).

34. "Turbine Generator Rotor Heating during Single Phase Short Circuits," by M. D. Ross and E. I. King, *AIEE Trans., 72,* Part III (1953), pp. 40–45.

35. "Effects of Negative Sequence Currents on Turbine-Generator Rotors," by E. I. Pollard, *AIEE Trans., 72,* Part III (1953), pp. 404–406. Discussions, p. 406.

36. "Short-Circuit Capabilities of Synchronous Machines for Unbalanced Faults," by P. L. Alger, R. F. Franklin, C. E. Kilbourne, and J. B. McClure, *AIEE Trans., 72,* Part III (1953), pp. 394–403. Discussions, pp. 403–404.

37. "Rotating Electrical Machinery," *Publ. C50,* American Standards Assoc.. Inc., 70 East 45th St., New York 17, N. Y.

38. "Protection of Generators against Unbalanced Currents," by J. E. Barkle and W. E. Glassburn, *AIEE Trans., 72,* Part III (1953), pp. 282–285. Discussions, pp. 285–286.

"Backup Protection for Generators," by E. T. B. Gross and L. B. LeVesconte, *AIEE Trans., 72,* Part III (1953), pp. 585–589. Discussions, pp. 589–592.

39. "A Negative-Phase-Sequence-Overcurrent Relay for Generator Protection," by W. C. Morris and L. E. Goff, *AIEE Trans., 72,* Part III (1953), pp. 615–618. Discussions, pp. 618–621.

40. "Application of Relays for Unbalanced Faults on Generators," by J. E. Barkle and Frank Von Roeschlaub, *AIEE Trans., 72,* Part III (1953), pp. 277–281. Discussions, pp. 281–282.

41. "Generator Negative-Sequence Currents for Line-to-Line Faults," by R. F. Lawrence and R. W. Ferguson, *AIEE Trans., 72,* Part III (1953), pp. 9–16.

42. Discussion by Mr. R. L. Webb of "Turbine Generator Controls, Protections, and Accessories," by G. W. Cunningham and M. A. Eggenberger, *AIEE Trans., 73,* Part III (1954), pp. 455–464. Discussions, pp. 464–465.

43. "A New Loss-of-Excitation Relay for Synchronous Generators," by C. R. Mason, *AIEE Trans., 68* (1949), pp. 1240–1245.

"Loss-of-Field Protection for Synchronous Machines," by R. L. Tremaine and J. L. Blackburn, *AIEE Trans., 73,* Part III (1954), pp. 765–772. Discussions, pp. 772–777.

44. "How to Fuse Potential Transformer Primary Circuits," by C. L. Schuck, *Gen. Elec. Rev., 44* (July 1941), p. 385.

45. "Protection of Powerhouse Auxiliaries," by AIEE Committee, *AIEE Trans., 65* (1946), pp. 746–750. Discussions, pp. 1115–1116.

46. "Electric Drives for Steam-Electric Generating-Station Auxiliaries," by Wm. R. Brownlee and J. A. Elzi, *AIEE Trans., 64* (1945), pp. 741–745.

47. "Symposium on the Effect of Low Frequency and Low Voltage on Thermal Plant Capability and Load Relief during Power System Emergencies," *AIEE Trans., 73,* Part III (1954), pp. 1628–1668.

48. "Controls for Operation of Steam Turbine-Generator Units," by O. N. Bryant, C. C. Sterrett, and D. M. Sauter, *AIEE Trans., 73,* Part III (1954), pp. 79–88.

"Turbine and Boiler Protection and Interlocking on the American Gas and Electric Co. System," by H. C. Barnes and C. P. Lugrin, *AIEE Trans., 73,* Part III (1954), pp. 884–893. Discussions, pp. 893–894.

"Turbine Generator Controls, Protections, and Accessories," by G. W. Cunningham and M. A. Eggenberger, *AIEE Trans., 73,* Part III (1954), pp. 455–464. Discussions, pp. 464–465.

"Fire Protection in Electric Stations," by H. A. Bauman and W. E. Rossnagel, *AIEE Trans., 70* (1951), pp. 1040–1045.

49. "Protection of Branch Circuits for Auxiliary Motors," by William Deans, *Iron Steel Engr.,* June, 1939, pp. 17–25. Discussions, pp. 25–30.

"Selecting A-C Overcurrent Protective Device Settings for Industrial Plants," by F. P. Brightman, *AIEE Trans., 71,* Part II (1952), pp. 203–211.

"More about Setting Industrial Relays," by F. P. Brightman, *AIEE Trans., 73,* Part III-A (1954), pp. 397–405. Discussions, pp. 405–406.

"Relay Protective Methods for Steel Mill Service," by H. A. Travers, *Iron Steel Engr.,* Nov., 1939, pp. 36–43.

Magnetic Control of Industrial Motors, by G. W. Heumann, John Wiley & Sons, New York, 1954.

50. "Selection of Thermal Overload Motor Protection," by C. W. Kuhn, *Elec. World,* April 1, 1944, pp. 61–63 (*121,* pp. 1225–1227).

"Determining the Correct Type of Thermal Protection for Electric Motors Used in the Pipe Line Industry," by Bradford Kleinman, *Petroleum Engr.,* October, 1953, pp. D-14 to D-18.

"Selecting Thermal Relays to Protect Three-Phase Motors," by H. Wilson, *Power, 77,* No. 2, February, 1933, pp. 78–80.

51. "Protecting Squirrel-Cage Winding of Synchronous Motors," by M. N. Halberg, *Elec. World,* November 11, 1933, p. 627.

11 TRANSFORMER PROTECTION

This chapter describes the protection practices for transformers of the following types whose three-phase bank rating is 501 kva and higher:

Power transformers
Power autotransformers
Regulating transformers
Step voltage regulators
Grounding transformers
Electric arc-furnace transformers
Power-rectifier transformers

Contrasted with generators, in which many abnormal circumstances may arise, transformers may suffer only from winding short circuits, open circuits, or overheating. In practice relay protection is not provided against open circuits because they are not harmful in themselves. Nor in general practice, even for unattended transformers, is overheating or overload protection provided; there may be thermal accessories to sound an alarm or to control banks of fans, but, with only a few exceptions, automatic tripping of the transformer breakers is not generally practiced. An exception is when the transformer supplies a definite predictable load. External-fault back-up protection may be considered by some a form of overload protection, but the pickup of such relaying equipment is usually too high to provide effective transformer protection except for prolonged short circuits. There remains, then, only the protection against short circuits in the transformers or their connections, and external-fault back-up protection. Moreover, the practices are the same whether the transformers are attended or not.

Power Transformers and Power Autotransformers

THE CHOICE OF PERCENTAGE-DIFFERENTIAL RELAYING FOR SHORT-CIRCUIT PROTECTION

It is the practice of manufacturers to recommend percentage-differential relaying for short-circuit protection of all power-transformer banks whose three-phase rating is 1000 kva and higher.[1] A survey of a large number of representative power companies showed that a minority favored differential relaying for as low as 1000-kva banks, but that they were practically unanimous in approving differential relaying for banks rated 5000 kva and higher.[2] To apply these recommendations to power autotransformers, the foregoing ratings should be taken as the "equivalent physical size" of autotransformer banks, where the equivalent physical size equals the rated capacity times $[1 - (V_L/V_H)]$, and where V_L and V_H are the voltage ratings on the low-voltage and high-voltage sides, respectively.

The report of an earlier survey[3] included a recommendation that circuit breakers be installed in the connections to all windings when banks larger than 5000 kva are connected in parallel. The more recent report is not very clear on this subject, but nothing has transpired that would change the earlier recommendation. The protection of parallel banks without separate breakers and the protection of a single bank in which a transmission line terminates without a high-voltage breaker will be considered later.

The differential relay should operate a hand-reset auxiliary that will trip all transformer breakers. The hand-reset feature is to minimize the likelihood of a transformer breaker being reclosed inadvertently, thereby subjecting the transformer to further damage unnecessarily.

Where transmission lines with high-speed distance relaying terminate on the same bus as a transformer bank, the bank should have high-speed relaying. Not only is this required for the same reason that the lines require it, but also it permits the second-zone time of the distance relays "looking" toward the bus to be set lower and still be selective.

CURRENT-TRANSFORMER CONNECTIONS FOR DIFFERENTIAL RELAYS

A simple rule of thumb is that the CT's on any wye winding of a power transformer should be connected in delta, and the CT's on any delta winding should be connected in wye. This rule may be broken, but it rarely is; for the moment let us assume that it is inviolate. Later, we shall learn the basis for this rule. The remaining problem is how to make the required interconnections between the CT's and the differential relay.

Two basic requirements that the differential-relay connections must satisfy are: (1) the differential relay must not operate for load or external faults; and (2) the relay must operate for severe enough internal faults.

If one does not know what the proper connections are, the procedure is first to make the connections that will satisfy the requirement of not tripping for external faults. Then, one can test the connections for their ability to provide tripping for internal faults.

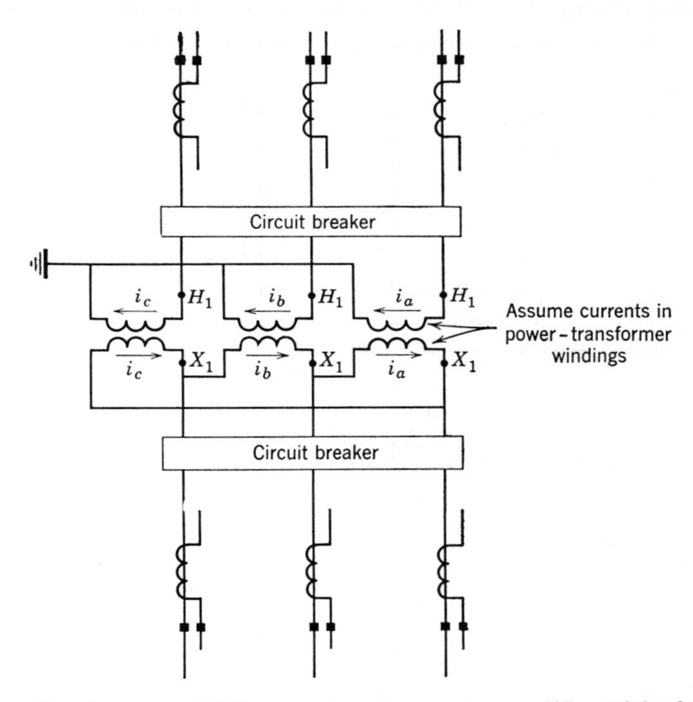

Fig. 1. Development of CT connections for transformer differential relaying, first step.

As an example, let us take the wye-delta power transformer of Fig. 1. The first step is arbitrarily to assume currents flowing in the power-transformer windings in whichever directions one wishes, but to observe the requirements imposed by the polarity marks that the currents flow in opposite directions in the windings on the same core, as shown in Fig. 1. We shall also assume that all the windings have the same number of turns so that the current magnitudes are equal, neglecting the very small exciting-current component. (Once the proper connections have been determined, the actual turn ratios can very easily be taken into account.)

On the basis of the foregoing, Fig. 2 shows the currents that flow in the power-transformer leads and the CT primaries for the general external-fault case for which the relay must not trip. We are assuming that no current flows into the ground from the neutral of the wye winding; in other words, we are assuming that the three-phase currents add vectorially to zero.

The next step is to connect one of the sets of CT's in delta or in wye, according to the rule of thumb already discussed; it does not matter how the connction is made, i.e., whether one way or reversed.

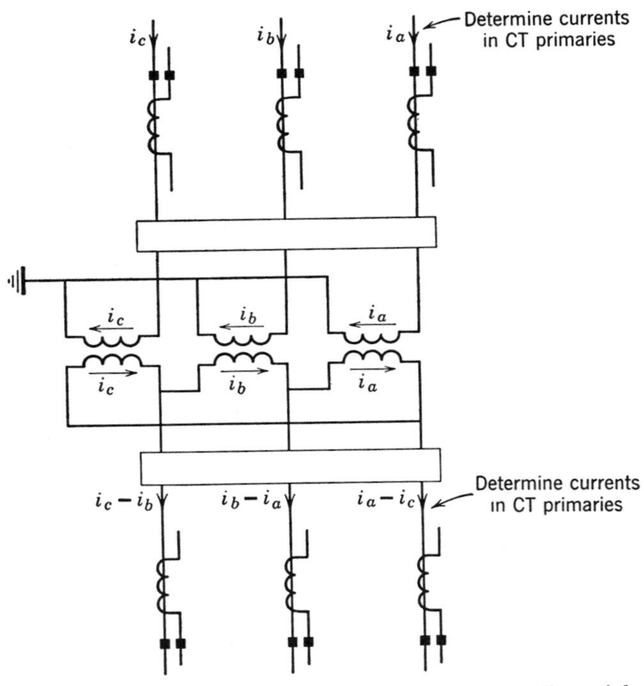

Fig. 2. Development of CT connections for transformer differential relaying, second step.

Then, the other set of CT's must be connected also according to the rule, but, since the connections of the first set of CT's have been chosen, it does matter how the second set is connected; this connection must be made so that the secondary currents will circulate between the CT's as required for the external-fault case. A completed connection diagram that meets the requirements is shown in Fig. 3. The connections would still be correct if the connections of both sets of CT's were reversed.

Proof that the relay will tend to operate for internal faults will not be given here, but the reader can easily satisfy himself by drawing current-flow diagrams for assumed faults. It will be found that protection is provided for turn-to-turn faults as well as for faults between phases or to ground if the fault current is high enough.

We shall now examine the rule of thumb that tells us whether to connect the CT's in wye or in delta. Actually, for the assumption

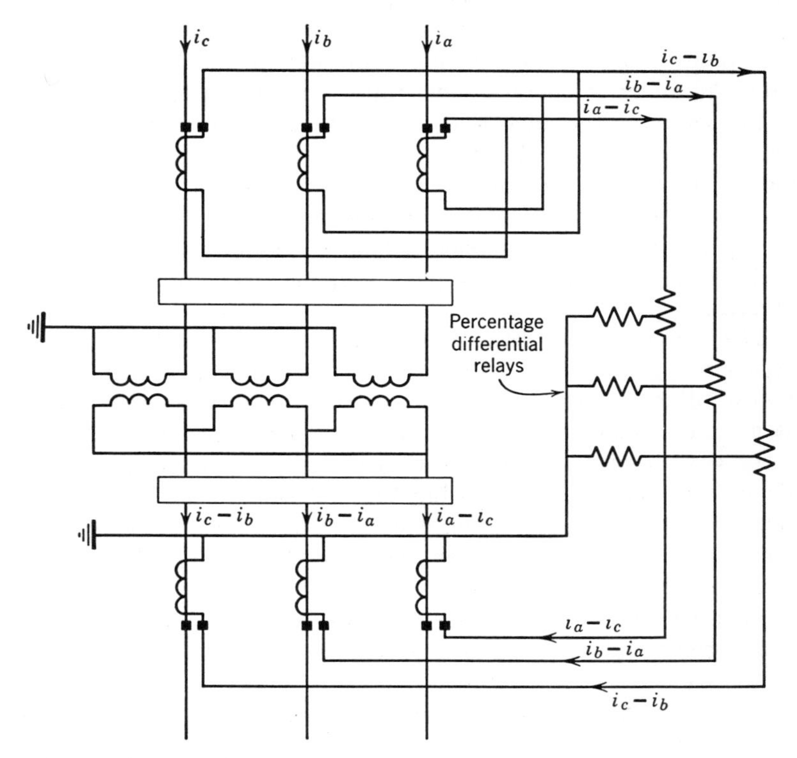

Fig. 3. Completed connections for percentage-differential relaying for two-winding transformer.

made in arriving at Fig. 2, namely, that the three-phase currents add vectorially to zero, we could have used wye-connected CT's on the wye side and delta-connected CT's on the delta side. In other words, for all external-fault conditions except ground faults on the wye side of the bank, it would not matter which pair of CT combinations was used. Or, if the neutral of the power transformer was not grounded, it would not matter. The significant point is that, when ground current can flow in the wye windings for an external fault, we must use the delta connection (or resort to a "zero-phase-sequence-current shunt" that will be discussed later). The delta CT connection circulates the zero-phase-sequence components of the currents inside the delta and thereby keeps them out of the external connections to the relay. This is necessary because there are no zero-phase-sequence components of current on the delta side of the power transformer for a ground fault on the wye side; therefore, there is no possibility of the zero-phase-sequence currents simply circulating between the sets of CT's and, if the CT's on the wye side were not delta connected, the zero-phase-sequence components would flow in the operating coils and cause the relay to operate undesirably for external ground faults.

Incidentally, the fact that the delta CT connection keeps zero-phase-sequence currents out of the external secondary circuit does not mean that the differential relay cannot operate for single-phase-to-ground faults in the power transformer; the relay will not receive zero-phase-sequence components, but it will receive—and operate on—the positive- and negative-phase-sequence components of the fault current.

The foregoing instructions for making the CT and relay interconnections apply equally well for power transformers with more than two windings per phase; it is only necessary to consider two windings at a time as though they were the only windings. For example, for three-winding transformers consider first the windings *H* and *X*. Then, consider *H* and *Y*, using the CT connections already chosen for the *H* winding, and determine the connections of the *Y* CT's. If this is done properly, the connections for the *X* and *Y* windings will automatically be compatible.

Figure 4 shows schematic connections for protecting the main power transformer and the station-service power transformer where a generator and its power transformer operate as a unit. To simplify the picture, only a one-line diagram is shown with the CT and power-transformer connections merely indicated. It will be noted that one restraint coil is supplied by current from the station-service-bus side of the breaker on the low-voltage side of the station-service power transformer in parallel with the CT in the neutral end of the generator winding; this is to obtain the advantage of overlapping adjacent protective zones around a circuit breaker, as explained in Chapter 1. A separate differential relay is used to protect the station-service power transformer because the relay protecting the main power transformer is not sensitive enough to provide this protection; with a steam-turbine generator, the station-service bank is no larger than about 10% of the size of the main bank, and, consequently, the CT's used for the main bank have ratios that are about 10 times as large as would be desired for the most sensitive protection of the station-service transformer. With a hydroelectric-turbine generator, the station-service transformer is more nearly 1% of the size of the main transformer; consequently, the impedance of the station-service transformer is so high that a fault on its low-voltage side cannot operate the relay protecting the main transformer even if the CT's are omitted from the low-voltage side of the station-service transformer; therefore,

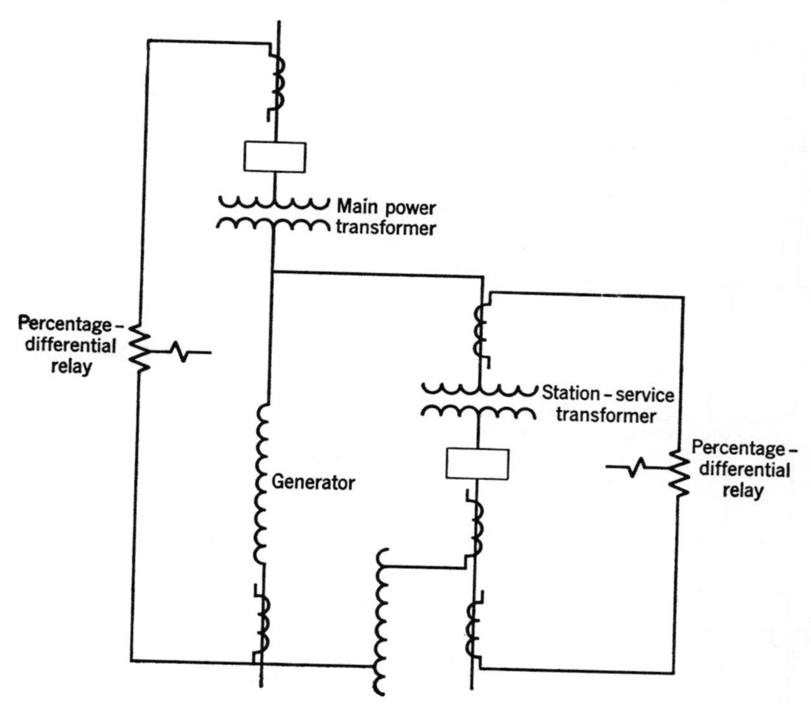

Fig. 4. Schematic connections for main and station-service-transformer protection.

for hydroelectric generators it is the practice to omit these CT's and to retain separate differential protection for the station-service bank. In order to minimize the consequential damage should a station-service-transformer fault occur, separate high-speed percentage-differential relaying should be used on the station-service transformer as for the main power transformer.

Figure 5 shows the usual way to protect a Scott-connected bank. This arrangement would not protect against a ground fault on phase b', but, since this is on the low-voltage side where a ground-current source is unlikely, such a possibility is of little significance. A more practical objection to Fig. 5, but still of secondary significance, is that, for certain turn-to-turn or phase-to-phase faults, only one relay unit can operate. This is contrasted with the general practice of providing three relay units to protect three-phase banks where, for any phase-to-phase fault, two relay units can operate, thereby giving double assurance that at least one unit will cause tripping. However, since

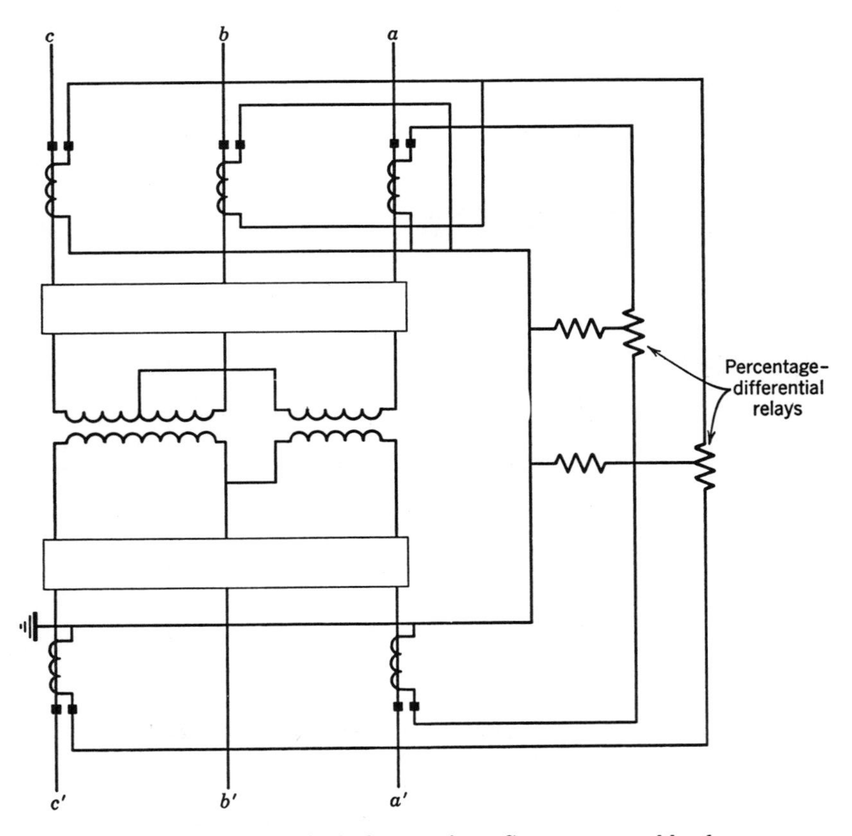

Fig. 5. Usual method of protecting a Scott-connected bank.

Scott-connected banks are used only at or near the load, it is questionable if the added cost of slightly more reliable protection can be justified. An alternative that does not have the technical disadvantages of Fig. 5 is shown in Fig. 6. Reference to other forms of Scott-connected bank and their differential protection is given in the Bibliography.[4]

Differentially connected CT's should be grounded at only one point. If more than one set of wye-connected CT's is involved, the neutrals should be interconnected with insulated wire and grounded at only one point. If grounds are made at two or more different points, even to a low-resistance ground bus, fault currents flowing in the ground or ground bus may produce large differences of potential between

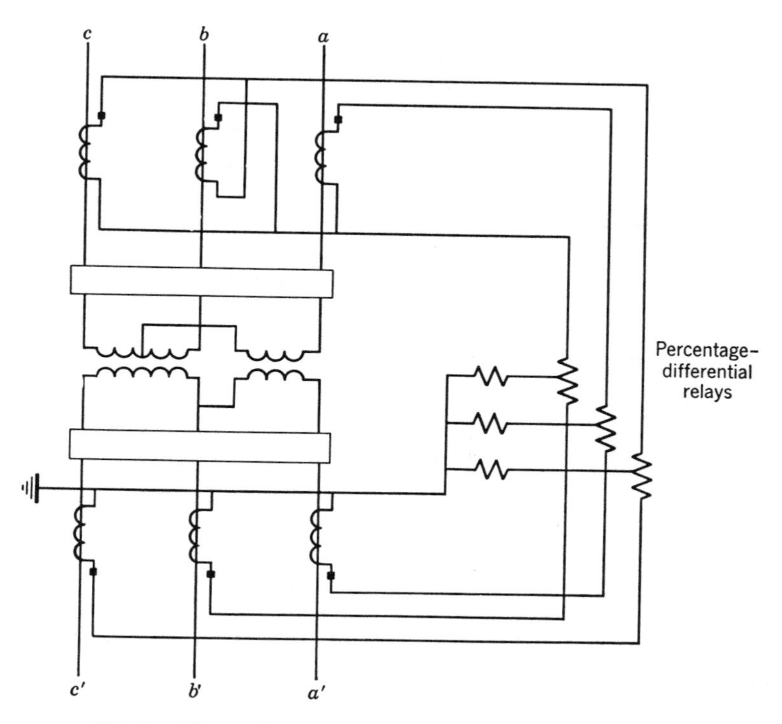

Fig. 6. Alternative protection of a Scott-connected bank.

the CT grounds, and thereby cause current to flow in the differential circuit. Such a flow of current might cause undesired tripping by the differential relays or damage to the circuit conductors.

THE ZERO-PHASE-SEQUENCE-CURRENT SHUNT

The zero-phase-sequence-current shunt was described in Chapter 7. Such a shunt is useful where it is necessary to keep the zero-phase-sequence components of current out of the external secondary circuits of wye-connected CT's. Such a shunt would permit one to connect the CT's in wye on the wye side of a power transformer and in delta on the delta side. Advantage is seldom taken of this possibility because there is usually no hardship in using the conventional connections, and in fact the conventional connections are usually preferred. The shunt is occasionally useful for the application of Fig. 7, where a grounding transformer on the delta side of a wye-delta power transformer is to be included within the zone of protection of the main

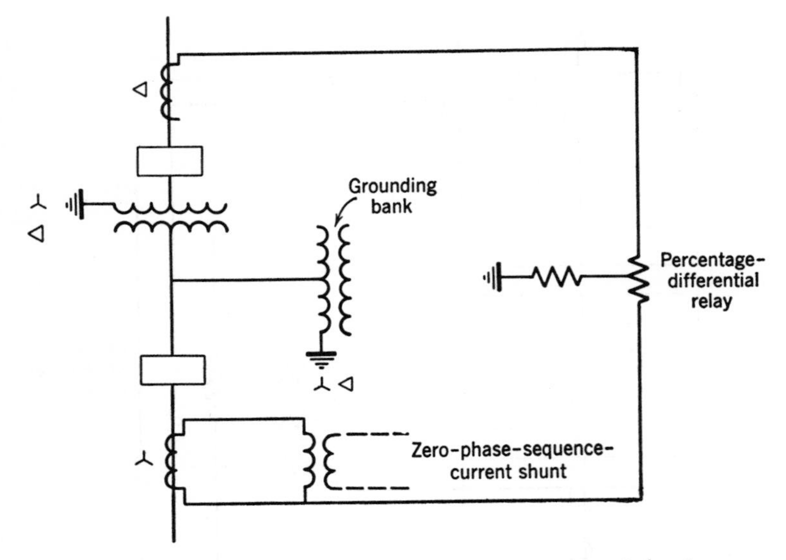

Fig. 7. Application of a zero-phase-sequence-current shunt.

bank. It is emphasized that, as indicated in Fig. 7, the neutral of the relay connection should not be connected to the neutral of the CT's or else the effectiveness of the shunt will be decreased. Also, the CT's chosen for the shunt should not saturate for the voltages that can be impressed on them when large phase currents flow.

CURRENT-TRANSFORMER RATIOS FOR DIFFERENTIAL RELAYS

Most differential relays for power-transformer protection have taps, or are used with auxiliary autotransformers having taps, to compensate for the CT ratios not being exactly as desired. Where there is a choice of CT ratio, as with relaying-type bushing CT's, the best practice is to choose the highest CT ratio that will give a secondary current as nearly as possible equal to the lowest-rated relay tap. The purpose of this is to minimize the effect of the connecting circuit between the CT's and the relay (for the same reason that we use high voltage to minimize transmission-line losses). For whatever relay tap is used, the current supplied to the relay under maximum load conditions should be as nearly as possible equal to the continuous rating for that tap; this assures that the relay will be operating at its maximum sensitivity when faults occur. If the current supplied is only half the tap rating, the relay will be only half as sensitive, etc.

When choosing CT ratios for power transformers having more than two windings per phase, one should assume that each winding can carry the total rated phase load. The proper matching of the CT ratios and relay or autotransformer taps depends on the current-transformation ratios between the various power-transformer windings and not on their full-load-current ratings. This is because the relations between the currents that will flow in the windings during external faults will not depend on their rated-current values but on the current-transformation ratios.

CURRENT-TRANSFORMER ACCURACY REQUIREMENTS FOR DIFFERENTIAL RELAYS

It is generally necessary to make certain CT accuracy calculations when applying power-transformer differential relays. These calculations require a knowledge of the CT characteristics either in the form of ratio-correction-factor curves or secondary-excitation and impedance data.

Two types of calculations are generally required. First, it is necessary to know approximately what CT errors to expect for external faults. Percentage-differential relays for power-transformer protection generally have adjustable percent slopes. This subject will be treated in more detail later, but the knowledge of what the CT errors will be is one factor that determines the choice of the percent slope. The other type of calculation is to avoid the possibility of locking-in for internal faults, as was described in Chapter 10 for generator differential protection; such a calculation is particularly necessary with the "harmonic-current-restraint" relay, a type that will be described later. For detailed application procedures, the manufacturers' bulletins should be followed.

The example given in Chapter 10 of a method for calculating steady-state CT errors in a generator differential-relay circuit is also applicable to the power-transformer relay, with minor exceptions. The fact that some CT's may be in delta introduces a slight complication, but the circuit calculation is still simple.

A study based on certain equipment of the manufacturer with whom the author is associated showed the minimum requirements for bushing CT's to be as in the accompanying table. The fact that relaying-type bushing CT's may be operated on their lowest turns-ratio tap makes

Number of Secondary Turns	ASA Accuracy Rating (Full Winding)
120	10L200
240	10L400
400	10L400
600	10L400
800	10L800
1000	10L800
1200	10L800

it necessary that the rating of the full winding be higher than if the full winding were used.

CHOICE OF PERCENT SLOPE FOR DIFFERENTIAL RELAYS

Percentage-differential relays are generally available with different percent slopes; they may have adjustment so that a single relay can have any one of several slopes. The purpose of the percent-slope characteristic is to prevent undesired relay operation because of "unbalances" between CT's during external faults arising from an accumulation of unbalances for the following reasons: (1) tap-changing in the power transformer; (2) mismatch between CT currents and relay tap ratings; and (3) the difference between the errors of the CT's on either side of the power transformer. Many power transformers have taps that will give $\pm X\%$ change in transformation ratio. It is the practice to choose CT ratios and relay or autotransformer taps to balance the currents at the midpoint of the tap-changing range; on that basis, the most unbalance that can occur from this cause is $X\%$. The maximum unavoidable mismatch between CT currents and relay tap ratings is one-half of the difference between two relay tap ratings, expressed in percent. The percent difference between CT errors must be determined for the external fault that produces the greatest error; the best that we can do is to calculate this on a steady-state basis. We should assume that all three unbalances are in the same direction to get the total maximum possible unbalance. Then add at least 5% to this value, and the new total is the minimum percent slope that should be used.

PROTECTING A THREE-WINDING TRANSFORMER WITH A TWO-WINDING PERCENTAGE-DIFFERENTIAL RELAY

Unless there is a source of generation back of only one side of a power transformer, a two-winding percentage-differential relay should not be used to protect a three-winding transformer. Figure 8 shows that, when a two-winding relay is used, the CT secondaries on two sides of the power transformer must be paralleled. If there is a source of generation back of one of these sides, the conditions shown by the arrows of Fig. 8 could exist. For an external fault on the other side there may be sufficient unbalance between the CT currents, either

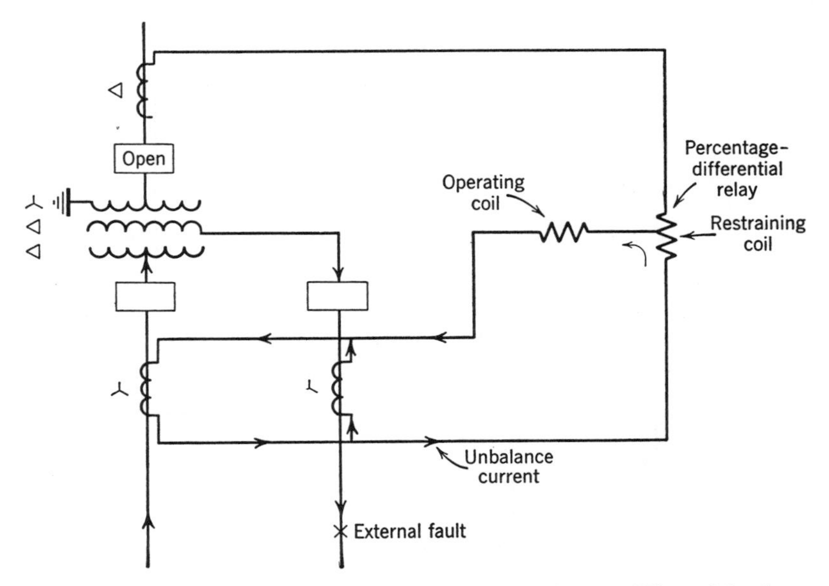

Fig. 8. A misapplication of a two-winding transformer differential relay.

because of mismatch or errors or both, to cause the differential relay to operate undesirably. The relay would not have the benefit of through-current restraint, which is the basis for using the percentage-differential principle. Instead, only the unbalance current would flow in all of the operating coil and in half of the restraining coil; in effect, this constitutes a 200% unbalance, and it is only necessary that the unbalance current be above the relay's minimum pickup for the relay to operate.

Of course, if the two sides where CT's are paralleled in Fig. 8 supply load only and do not connect to a source of generation, a two-winding relay may be used with impunity.

Figure 9 shows that, if a three-winding relay is used, there will always be through-current restraint to restrain the relay against undesired operation.

A further advantage of a three-winding relay with a three-winding transformer is that, where relay types are involved having taps for matching the CT secondary currents, it is often unnecessary to use any auxiliary CT's. Thus, a three-winding relay may even be used with advantage where a two-winding relay might suffice. There is no disadvantage, other than a slight increase in cost, in using a three-winding relay on a two-winding transformer; no harm is done if one of the restraint circuits is left unconnected.

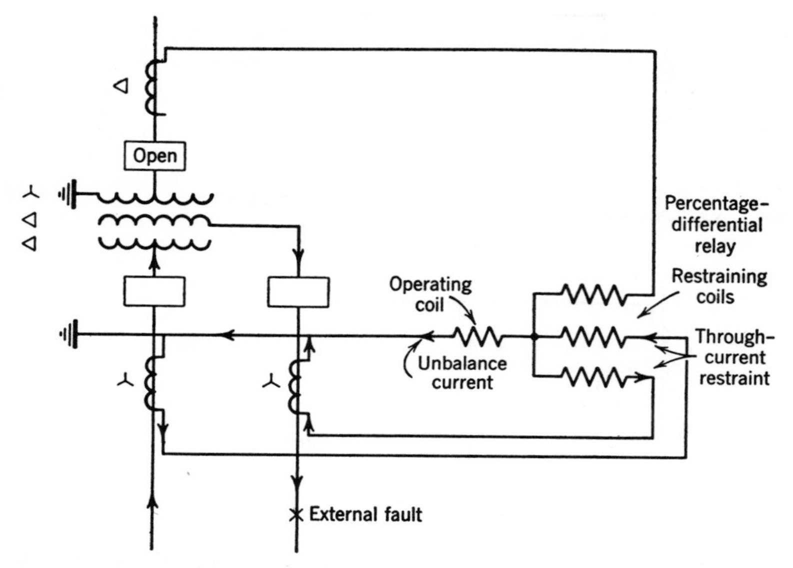

Fig. 9. Illustrating the advantage of a three-winding relay with a three-winding transformer.

EFFECT OF MAGNETIZING-CURRENT INRUSH ON DIFFERENTIAL RELAYS

The way in which CT's are connected and the way in which CT ratios and relay taps are chosen for differential relaying neglect the power-transformer exciting-current component. Actually, this component causes current to flow in the relay's operating coil, but it is so small under normal load conditions that the relay has no tendency to operate. However, any condition that calls for an instantaneous change in flux linkages in a power transformer will cause abnormally large magnetizing currents to flow, and these will produce an operating tendency in a differential relay.[5,6,7]

The largest inrush and the greatest relay-operating tendency occur when a transformer bank has been completely de-energized and then a circuit breaker is closed, thereby applying voltage to the windings on one side with the windings on the other side still disconnected from load or source. Reference 5 gives data as to the magnitudes and durations of such inrush currents. Considerably smaller but still possibly troublesome inrushes occur when a transformer with connected load is energized[7] or when a short circuit occurs or is disconnected.[8]

Another troublesome inrush problem will be discussed later under the heading "Protection of Parallel Transformer Banks."

The occasional tripping because of inrush when a transformer is energized is objectionable because it delays putting the transformer into service. One does not know but that the transformer may have a fault in it. Consequently, the safest thing to do is to make the necessary tests and inspection to locate the trouble, if any, and this takes considerable time.

Percentage-differential relays operating with time delay of about 0.2 second or more will often "ride over" the inrush period without operating. Where high-speed relays are required, it is generally necessary to use relay equipment that is especially designed to avoid undesired tripping on the inrush current.

Three methods that are used for preventing operation on inrush current will now be described.

Desensitizing. One type of desensitizing equipment consists of an undervoltage relay with "*b*" contacts and having time-delay pickup and reset; these contacts are connected in series with a low-resistance resistor that shunts the operating coil of the differential relay in each phase. This is shown schematically in Fig. 10 for the differential relay of one phase. The undervoltage relay is energized from a potential transformer connected to the power-transformer leads between the power transformer and its low-voltage breaker. When the power transformer is de-energized, the undervoltage relay resets, and its contacts complete the shunt circuit across the operating coil of the differential relay. The undervoltage relay will not pick up and open its contacts until a short time after the power transformer has been energized, thereby desensitizing the differential relay during the magnetizing-current-inrush period. During normal operation of the power transformer, the desensitizing circuit is open, thereby not interfering with the differential-relay sensitivity should a fault occur in the power transformer. Should a transformer fault occur that would reset the undervoltage relay, its time delay would prevent desensitizing the differential relay until after it had had more than sufficient time to operate if it was going to do so.

One disadvantage of such a desensitizing method is that it might delay tripping should a short circuit occur during the magnetizing-inrush period while the differential relay is desensitized. If the fault were severe enough to lower the voltage sufficiently so that the desensitizing relay could not pick up, tripping would depend on the current being high enough to operate the differential relay in its desensitized state. This is a rather serious disadvantage in view of the fact that one of the most likely times for a fault to occur is when the bank is being energized. The other disadvantage is that this equipment cannot desensitize the differential relay against the possibility of

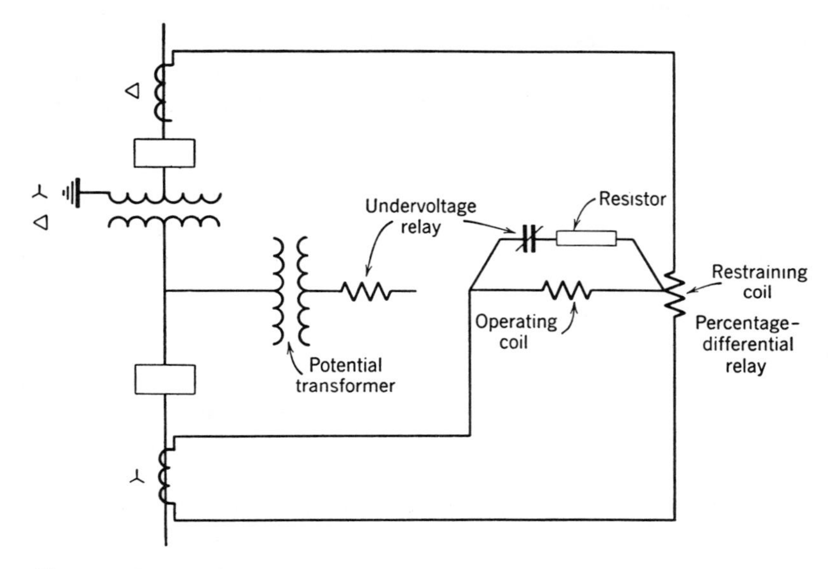

Fig. 10. Desensitizing equipment to prevent differential-relay tripping on magnetizing inrush.

undesired operation during the magnetizing inrush after the clearing of an external fault. This is not so serious a disadvantage because desensitizing of the type described here is used only with relays having about 0.2-second time delay, and there is practically no problem of tripping on voltage recovery with such relays.

Tripping Suppressor.[9] An improvement over the desensitizing principle is called the "tripping suppressor." Three high-speed voltage relays, connected to be actuated by either phase-to-phase or phase-to-neutral voltage, control tripping by the percentage-differential relays. If all three voltage relays pick up during the inrush period, thereby indicating either a sound transformer or one with very low fault current, a timer is energized that closes its "*a*" contact in the tripping circuit of the differential relays after enough time delay so that tripping on inrush alone would not occur. But, for any fault that will operate a differential relay and also reduce the voltage enough so that at least one voltage relay will not pick up, tripping occurs immediately. In other words, tripping is delayed only for very-low-current faults that affect the voltage only slightly.

Any external fault that lowers the voltage enough to cause a significant inrush when the fault is cleared from the system will reset one or more of the voltage relays, thereby resetting the timer and opening the trip circuit long enough to assure that the differential relays will have reset if they had any tendency to operate.

The tripping suppressor is usable with either high-speed or slower differential relays, but its widest application is with high-speed relays. In fact, high-speed relays that are not inherently selective between inrush and fault currents require tripping suppressors.

Harmonic-Current Restraint.[10] The principle of "harmonic-current restraint" makes a differential relay self-desensitizing during the magnetizing-current-inrush period, but the relay is not desensitized if a short circuit should occur in the transformer during the magnetizing-inrush period. This relay is able to distinguish the difference between magnetizing-inrush current and short-circuit current by the difference in wave shape. Magnetizing-inrush current is characterized by large harmonic components that are not noticeably present in short-circuit current. A harmonic analysis of a typical magnetizing-inrush-current wave was as shown in the accompanying table.

Harmonic Component	Amplitude in Percent of Fundamental
2nd	63.0
3rd	26.8
4th	5.1
5th	4.1
6th	3.7
7th	2.4

Figure 11 shows how the relay is arranged to take advantage of the harmonic content of the current wave to be selective between faults and magnetizing inrush.

Figure 11 shows that the restraining coil will receive from the through-current transformer the rectified sum of the fundamental and harmonic components. The operating coil will receive from the differential-current transformer only the fundamental component of the differential current, the harmonics being separated, rectified, and fed back into the restraining coil.

The direct-current component, present in both magnetizing-inrush and offset fault current, is largely blocked by the differential-current and the through-current transformers, and produces only a slight momentary restraining effect.

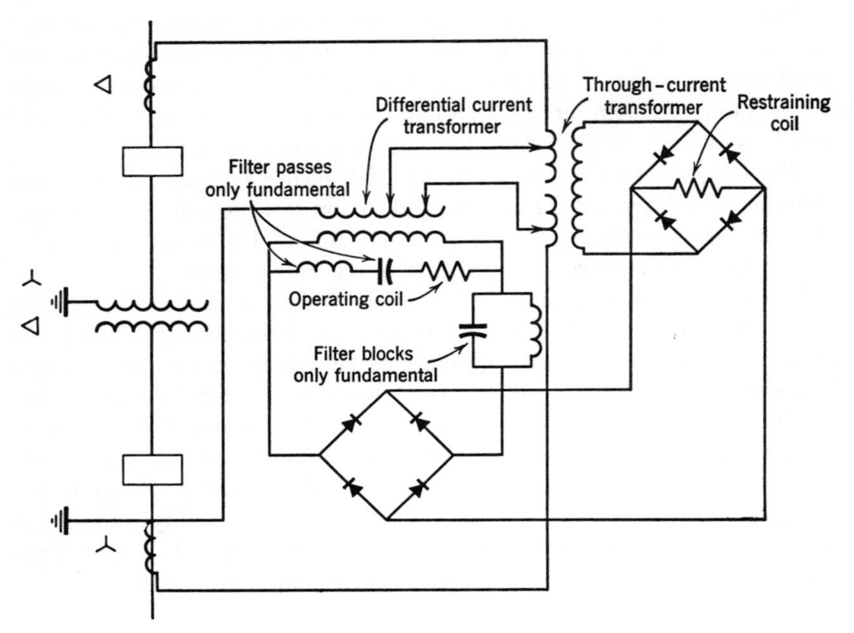

Fig. 11. Harmonic-current-restraint percentage-differential relay.

PROTECTION OF PARALLEL TRANSFORMER BANKS

From the standpoint of protective relaying, the operation of two transformer banks in parallel without individual breakers is to be avoided. In order to obtain protection equivalent to that when individual breakers are used, the connections of Fig. 12 would be required. To protect two equally rated banks as a unit, using only CT's on the source sides of the common breakers and a single relay is only half as sensitive as protecting each bank from its own CT's; this is because the CT ratios must be twice as high as if individual CT's were used for each bank, both banks being assumed to have the same rating, and as a result the secondary current for a given fault will be only half as high. If one bank is smaller than the other, its protection will be less than half as sensitive. With more than two banks, the protection is still poorer.

When parallel transformer banks having individual breakers are located some distance away from any generating station, a possibly troublesome magnetizing-current-inrush problem may arise.[11] If one bank is already energized and a second bank is then energized, magnetizing-current inrush will occur—not only to the bank being energized but also to the bank that is already energized. Moreover,

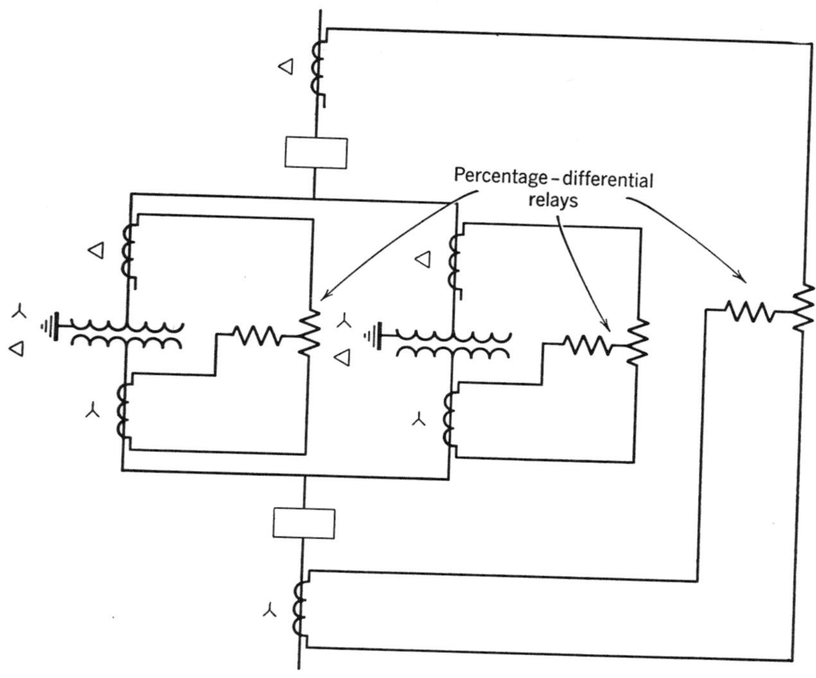

Fig. 12. The protection of parallel transformer banks with common breakers.

the inrush current to both banks will decay at a much slower rate than when a single bank is energized with no other banks in parallel. The magnitude of the inrush to the bank already connected will not be as high as that to the bank being switched, but it can easily exceed twice the full-load-current rating of the bank; the presence of load on the bank will slightly reduce its inrush and increase its rate of decay.

Briefly, the cause of the foregoing is as follows: The d-c component of the inrush current to the bank being energized flows through the resistance of transmission-line circuits between the transformer banks and the source of generation, thereby producing a d-c voltage-drop component in the voltage applied to the banks. This d-c component of voltage causes a build-up of d-c magnetizing current in the already-connected bank, the rate of which is the same as the rate at which the d-c component of magnetizing current is decreasing in the bank just energized. When the magnitudes of the d-c components in both banks become equal, there is no d-c component in the transmission-line circuit feeding the banks, but there is a d-c component circulating in the loop circuit between the banks. The time constant of this trapped d-c circulating current, depending only on the constants of the loop circuit, is much longer than the time constant of the d-c component

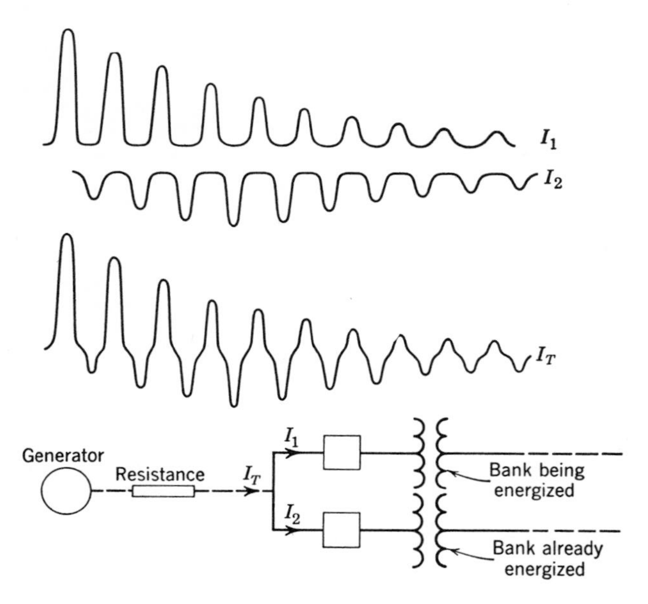

Fig. 13. Prolonged inrush currents with parallel transformers.

in the transmission-line circuit feeding the banks. Figure 13 shows the circuits involved and the magnetizing-current components in each circuit.

The significance of the foregoing is two-fold. First, desensitizing means already described for preventing differential-relay operation on magnetizing-current inrush are not effective in the bank that is already energized. Only time delay in the operation of the differential relay will be effective in preventing undesired tripping. However, if the banks are protected by separate relays having tripping suppression or harmonic restraint, no undesired tripping will occur. Second, if the banks are protected as a unit, even the harmonic-current restraint type may cause undesired tripping because, as shown in Fig. 13, the total-current wave very shortly becomes symmetrical and does not contain the necessary even harmonics required for restraint.

SHORT-CIRCUIT PROTECTION WITH OVERCURRENT RELAYS

Overcurrent relaying is used for fault protection of transformers having circuit breakers only when the cost of differential relaying cannot be justified. Overcurrent relaying cannot begin to compare with differential relaying in sensitivity.

Three CT's, one in each phase, and at least two overcurrent phase relays and one overcurrent ground relay should be provided on each side of the transformer bank that is connected through a circuit breaker to a source of short-circuit current. The overcurrent relays should have an inverse-time element whose pickup can be adjusted to somewhat above maximum rated load current, say about 150% of maximum, and with sufficient time delay so as to be selective with the relaying equipment of adjacent system elements during external faults. The relays should also have an instantaneous element whose pickup can be made slightly higher than either the maximum short-circuit current for an external fault or the magnetizing-current inrush.

When the transformer bank is connected to more than one source of short-circuit current, it may be necessary for at least some of the overcurrent relays to be directional in order to obtain good protection as well as selectivity for external faults.

The overcurrent relays for short-circuit protection of transformers provide also the external-fault back-up protection discussed elsewhere.

GAS-ACCUMULATOR AND PRESSURE RELAYS

A combination gas-accumulator and pressure relay, called the "Buchholz" relay after its inventor, has been in successful service for over 30 years in Europe and for 10 years in Canada.[12] This relay is applicable only to a so-called "conservator-type" transformer in which the transformer tank is completely filled with oil, and a pipe connects the transformer tank to an auxiliary tank, or "conservator," which acts as an expansion chamber. In the piping between the main tank and the conservator are the two elements of the relay. One element is a gas-collecting chamber in which gas evolved from the slow breakdown of insulation in the presence of a small electric arc is collected; when a certain amount of gas has been collected a contact closes, usually to sound an alarm. The collected gas may be drawn into a gas analyzer to determine what kind of insulation is being broken down and thereby to learn whether lamination, core-bolt, or major insulation is being deteriorated. This gas analyzer is not a part of the Buchholz relay. The other element contains a vane that is operated by the rush of oil through the piping when a severe fault occurs, to close contacts that trip the transformer breakers.

The gas-accumulator element of the Buchholz type of relay has not had extensive use in the United States, partly because the value of such protection "has been underestimated,"[2] and partly because conservator-type transformers are not being built here in any quantity. From Canada, where such relays are widely used, come very favorable reports of the protection that they provide on conservator-type transformers.[13,15]

However, pressure relays, applicable to gas-cushioned transformers, are being used to an increasing extent in the United States. A relay operating in response to rate-of-rise of pressure has been introduced that uses the pressure in the gas cushion.[14] Such relays are valuable supplements to differential or other forms of relaying, particularly for transformers with complicated circuits that are not well suited to differential relaying, such as certain regulating and rectifier transformers; they will be considered later.

Many of those familiar with the Buchholz relay feel that the gas-accumulator element is more valuable than the pressure element. The gas-accumulator element gives early warning of incipient faults, permitting the transformer to be taken out of service and repaired before extensive damage is done. How valuable this feature is depends on how large a proportion of the total number of faults is of the incipient type, such as failures of core-bolt or lamination insulation, and high-resistance or defective joints in windings. Also, the gas-accumulator feature is valuable only if there is also in service a thoroughly reliable protective equipment that will quickly disconnect the transformer when a short circuit occurs.

From the foregoing it will be evident that gas-accumulator and pressure relays are valuable principally as supplements to other forms of protection. In the first place, a transformer must be of the type that lends itself to this type of protection. Then, protection is provided only for faults inside the transformer tank; differential or other types of relaying must be provided for protection in the event of external bushing flashovers or faults in the connections between a transformer and its circuit breakers. Where sensitive and reliable gas-accumulator and pressure relays are applicable, the other relaying equipment need not be nearly as sensitive, and therefore the problem of preventing undesired operation on magnetizing-current inrush is greatly simplified. In fact, it has been suggested that, where gas and pressure relaying is used, it is good practice to "try again" if a differential or other relay operates when a transformer bank is energized, so long as the gas or pressure elements do not indicate any internal fault.[15]

GROUNDING PROTECTIVE RELAY

On grounded-neutral systems, protection can be provided by insulating a transformer tank from ground except for a connection to ground through a CT whose secondary energizes an overcurrent relay. Such an arrangement will give sensitive protection for arc-overs to the tank or to the core, but it will not respond to turn faults or to faults in the leads to the transformer.

REMOTE TRIPPING

When a transmission line terminates in a single transformer bank, the practice is frequently to omit the high-voltage breaker and thereby avoid considerable expense. Such practice is made possible by what is called "transferred tripping" or, preferably, "remote tripping."[16]

Remote tripping is the tripping of the circuit breaker at the other end of the transmission line for faults in the power transformer. The protective relays at that other end of the line are not sensitive enough to detect turn faults inside the transformer bank. Consequently, the transformer bank's own differential-relaying equipment trips the bank's low-voltage breaker and initiates tripping of the breaker at the other end of the line in one of two basic ways.

One way to cause the distant relays to operate and trip their breaker is to throw a short circuit on the line at the high-voltage terminals of the power transformer.[16,17] This is done by arranging the transformer-differential relays to trip the latch of a spring-closed air-break-type disconnecting switch that grounds one or three phases of the line. A three-phase switch is used if there is automatic reclosing at the other end of the line; this is to protect the transformer against further damage by preventing the reapplication of voltage to the transformer. If automatic reclosing is not used, and if the station is attended, a single-phase switch is sufficient.

The principal disadvantage of the grounding-disconnect method of remote tripping is that it is relatively slow. To the closing time of the switch must be added the operating time of the relaying equipment at the other end and the tripping time of the breaker there; this total time may amount to about a half second or more, which is long for transformer protection. Of course, if a three-phase grounding switch is used, the transformer is de-energized as soon as the switch closes. Another disadvantage is that, where automatic reclosing is used, the system is subjected to the shock of one or more reclosings on a short circuit. It may be necessary to delay reclosing to be sure that the grounding switch is closed first when high-voltage transformer-bushing flashovers occur. That these disadvantages are not always too serious is shown by the fact that about half of the installations in this country use this method.

The other way to trip the distant breaker is with a pilot.[16,18] Any of the types of pilot (wire, carrier-current, or microwave) may be used, depending on the circumstances. In any event, the equipment must be free of the possibility of undesired tripping because of extraneous causes; this is achieved by transmitting a tripping signal that is not apt to be duplicated otherwise. One of the most successful methods is the so-called "frequency-shift" system;[18] not only is this system most reliable but it is also high speed, requiring only about 3 cycles to energize the trip coil of the distant breaker after the transformer-differential relay has closed its tripping contacts. By using two frequency-shift channels, the equipment can be tested without removing it from service.

An inherent advantage of remote tripping over a pilot is that the received tripping signal can also block automatic reclosing. It may be necessary, however, to delay reclosing a few cycles to be sure that reclosing is blocked when high-voltage transformer-bushing flashovers occur.

EXTERNAL-FAULT BACK-UP PROTECTION

A differentially protected transformer bank should have inverse-time-overcurrent relays, preferably energized from CT's other than those associated with the differential relays, to trip fault-side breakers when external faults persist for too long a time. An exception is the transformer bank of a unit generator-transformer arrangement where the generator's external-fault back-up relays provide all the necessary back-up protection. The back-up relays should preferably be operated from CT's located as in Fig. 14; this makes it unnecessary to adjust the relays so as not to operate on magnetizing-current inrush and hence permits greater sensitivity and speed if desired. When the transformer is connected to more than one source of short-circuit current, back-up relays in all the circuits are required, and at least some may need to be directional, as indicated in Fig. 15, for good protection and selectivity. Each set of back-up relays should trip only its associated breaker, also as indicated in Fig. 15.

When a transformer has overcurrent relaying for short-circuit protection because the cost of differential relaying cannot be justified, the same overcurrent relays are used for back-up protection. It is realized that combining the two functions may work to the disadvantage of one or both, but this is the price that one must pay to minimize the investment.

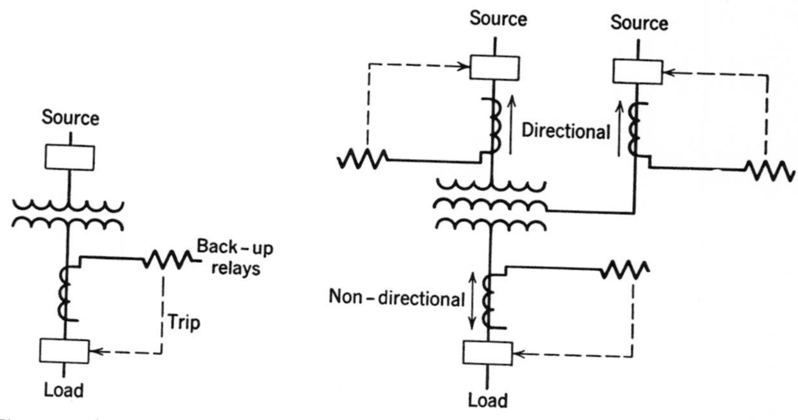

Fig. 14. Back-up relaying for transformer connected to one source.

Fig. 15. Back-up relaying with two sources.

Regulating Transformers

Regulating transformers may be of the "in-phase" type or the "phase-shifting" type. The in-phase type provides means for increasing or decreasing the circuit voltage at its location under load without changing the phase angle. The phase-shifting type changes the phase angle —and usually also the voltage magnitude—under load.

A regulating transformer may be used alone in a circuit or in conjunction with a power transformer. Or the regulating-transformer function may be built into a power transformer.

PROTECTION OF IN-PHASE TYPE

Figure 16 shows schematically the relay equipment that is recommended for protection against internal short circuits. Percentage-differential relaying, like that for generators, should be used to protect the series winding and its connections to its breakers. If the regulating transformer is close enough to a power transformer in the same circuit, the differential-protection zone of the power transformer may be extended to include the regulating transformer. The percent slope of the differential relay should be high enough to accommodate the full range of voltage change, as already mentioned for tap-changing power transformers.

The exciting windings need separate protective equipment because the equipment protecting the series winding is not sensitive enough

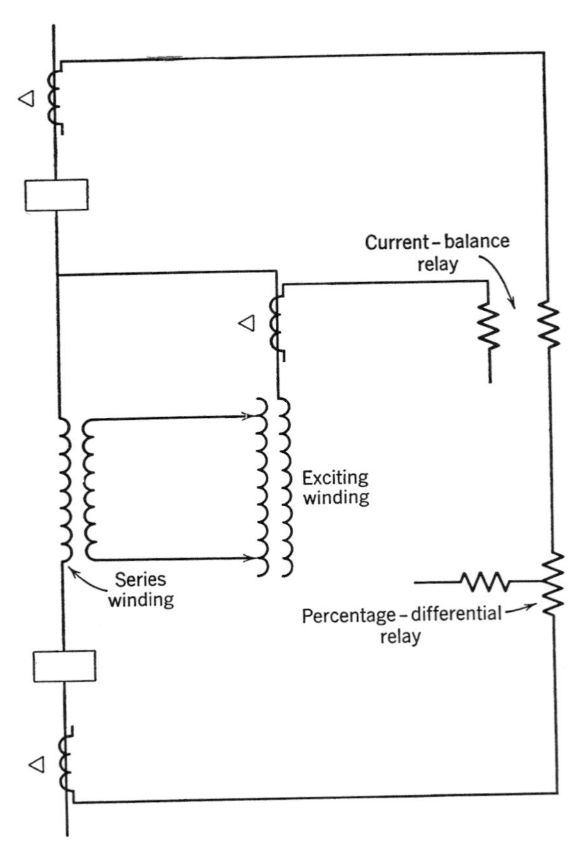

Fig. 16. Protection of an in-phase regulating transformer.

for the exciting windings. This is because the full-load-current rating of the exciting winding is much less than that of the series winding and the short-circuit current is proportionally less; for example, in a regulating transformer that changes the circuit voltage by $\pm 10\%$, the full-load-current rating of the exciting winding will be 10% of that of the series winding. The situation is the same as that already described for protecting two different-sized power transformers with one differential relay. In practice a current-balance relay protects the exciting winding, as shown in Fig. 16. So long as there is no fault in the exciting windings, the exciting current of a ±10% transformer will never exceed 10% of the rated series-winding current; the current-balance relay will operate whenever the ratio of exciting-winding current to series-winding current is about 25% higher than the maximum normal ratio under conditions of maximum buck or boost.

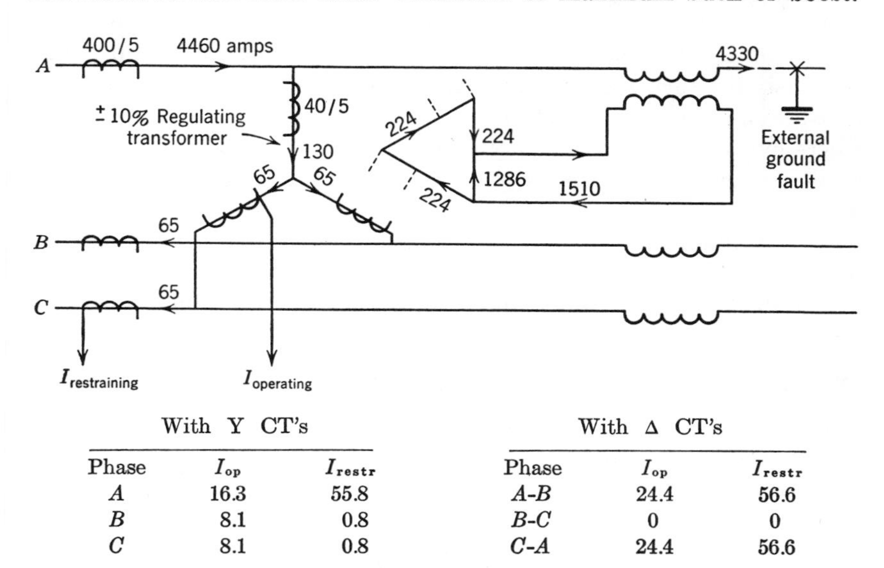

With Y CT's		
Phase	I_{op}	I_{restr}
A	16.3	55.8
B	8.1	0.8
C	8.1	0.8

With Δ CT's		
Phase	I_{op}	I_{restr}
A-B	24.4	56.6
B-C	0	0
C-A	24.4	56.6

Fig. 17. Illustrating why delta-connected CT's are required for regulating-transformer protection.

A very important precaution is that the CT's supplying the current-balance relay must always be delta connected. This is so whether the neutral of the exciting windings is grounded or not. Figure 17 shows the results of a study of an actual application where an external phase-to-ground fault would cause the current-balance relays of phases *B* and *C* to operate incorrectly if the CT's were wye connected.

Wherever possible, it is recommended that gas-accumulator and pressure relaying supplement the other protective equipment. Or, if the regulating-transformer tank can be insulated from ground, a grounding protective relay would be recommended because of the more sensitive protection that it would provide.

PROTECTION OF PHASE-SHIFTING TYPE

Wherever possible, the phase-shifting type of regulating transformer is protected in the same manner as the in-phase type. However, with conventional percentage-differential relaying, a 10° phase shift is about all that can be tolerated; such a phase shift requires that the differential relays have about a 40% slope and that relays in two phases operate before tripping is permitted, in order not to trip undesirably for external faults.

When phase shifts of more than about 10° are involved, special forms of relaying equipment are necessary. Certain modifications to conventional differential relaying may sometimes be possible, but the basis for such modifications is too complicated to consider here. Gas-accumulator and pressure relaying take on more importance where over-all differential relaying is not completely adequate. Complete percentage-differential protection can often be provided for wye windings if CT's are made available at both ends of each winding,[19] or differential protection against ground faults only can be provided if CT's at the neutral ends are lacking. Overcurrent relaying can protect against ground faults in a delta winding connected to a grounded-neutral source.

EXTERNAL-FAULT BACK-UP PROTECTION

The external-fault back-up relays of the power transformer or circuit associated with the regulating transformer will provide the necessary back-up protection.

Step Voltage Regulators

If circuit breakers are provided, pressure relaying should be used for regulators whose equivalent physical size is about 1000 kva or more.

Grounding Transformers

Two types of grounding transformer are in general use: (1) the wye-delta transformer, and (2) the zig-zag transformer. The neutral of either type may be grounded directly or through current-limiting impedance. It is assumed here that neither load nor a source of generation is connected to the delta winding of the wye-delta transformer, and that the zig-zag transformer does not have another winding connected to load or generation; should either type have such connections, it would be treated as an ordinary power transformer.

Figure 18 shows the recommended way to protect either type of bank. For external ground faults, only zero-phase-sequence currents flow through the primaries of the delta-connected CT's. Therefore, current will flow only in the external-fault back-up overcurrent relay, and its time delay should be long enough to be selective with other relays that should operate for external faults. The other three relays will provide protection for short circuits on the grounding-transformer side of the CT's. These relays may be sensitive and quite fast because, except for magnetizing current and small currents that may flow through the relays because of CT errors, current will flow only when short circuits requiring tripping occur. The pickup of the overcurrent relays should be 25% to 50% of the grounding-trans-

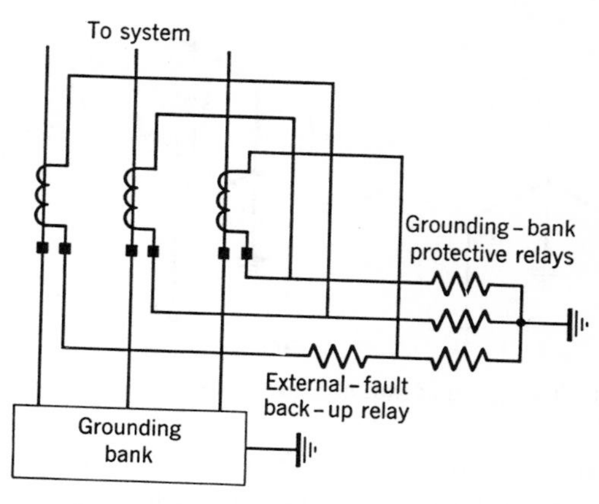

Fig. 18. Grounding-bank protection.

former's continuous-current rating, and the primary-current rating of the CT's should be about the continuous-current rating of the power transformer.

An interesting fact in connection with either type of grounding bank is that, under certain conditions, it is impossible to have certain types of fault in the bank without the short-circuit current's being limited by some magnetizing impedance. For example, certain types of fault can occur without the limiting effect of magnetizing impedance only if there is another grounding bank to provide a zero-phase-sequence-circulating-current path for the currents in the faulted bank; this other grounding bank may or may not have a delta winding connected to a source of generation. Or the fault must occur between certain points of the windings, and the presence of another grounding bank may or may not be necessary. Examples of the foregoing facts are shown in Figs. 19(a), 19(b), and 19(c) for a zig-zag bank. Remember that, unless fault current can flow in windings on the same core in such a way that the ampere-turns cancel, the current will be limited by some magnetizing impedance. However, if enough of a winding is shorted out, considerable overvoltage impressed on the remaining portion would cause large magnetizing currents to flow because of saturation. Figure 19(*a*) is an example of a type of short circuit where the current is limited by some magnetizing impedance of a winding. Figure 19(*b*) shows a type of short circuit that can occur without requiring the presence of another grounding bank; here, the fault is assumed to occur between the middle points of the two windings involved, and the relative

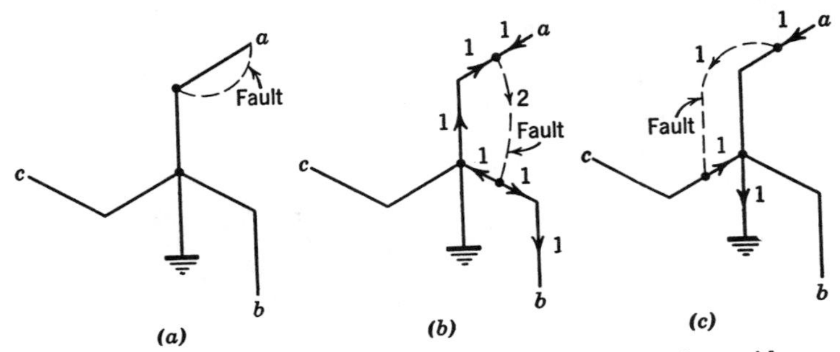

Fig. 19. Examples of faults in zig-zag banks. High-voltage side.

magnitudes and directions of the currents are shown. Figure 19(*c*) shows a type of fault that requires the presence of a grounding bank with or without a delta connected to a source of generation; here again, the fault is between the middle points of the two windings involved. A good exercise for the reader is to trace the flow of current back through the other grounding bank, and also to apply other types of short circuit, to see if there is any way in which current can flow to cancel the ampere-turns on each core involved. Figures 19(*a*), 19(*b*), and 19(*c*) are not the only examples of the three different conditions.

Because faults can occur that will not cause high currents to flow, gas-accumulator relaying, if applicable, would provide valuable supplementary protection.

Electric Arc-Furnace Transformers

Electric arc-furnace power transformers are not protected with percentage-differential relays because of the complications that would be introduced by the very frequent tap changing on the power transformer. Every time a furnace-transformer tap was changed, the low-voltage CT ratio or a tap on the relay would have to be changed.

Also, the connections of the furnace-transformer primary windings are usually changed from delta to wye and back again, which would require changing the CT connections.

Protection against short circuits inside the power transformer should be provided by inverse-time phase (and ground if required) overcurrent relays operating from the current on the high-voltage side of the power transformer. The phase relays should have torque-control coils and should be adjusted to pick up at currents only slightly in excess of the transformer's rated full-load current; they should have time delay only long enough to prevent operation on transformer magnetizing-current inrush. High-speed overcurrent relays on the low-voltage side of the transformer, adjusted to pick up at current slightly above rated full load but slightly below the current that will pick up the high-voltage phase relays, should be arranged to control the operation of the high-voltage phase relays through their torque-control coils so as to permit the high-voltage relays to operate only when the low-voltage relays do not operate. In this way, the high-voltage relays may normally be sensitive and fast so as to provide as good protection to the transformer as it is possible to provide with overcurrent relays, while at the same time avoiding undesired operation on external faults, the most common of which are short circuits in the furnace.

For primary protection against short circuits between the "back-up" breaker and the power transformer, and for back-up protection against faults in the transformer or beyond it, inverse-time phase (and ground if required) overcurrent relays should be provided. These relays should obtain their current from the source side of the back-up breaker. This so-called "back-up" breaker is the breaker that is provided to interrupt short-circuit currents in the transformer or on the high-voltage side, and it may serve several transformers.

Both of the foregoing groups of relays should trip the back-up breaker.

Power-Rectifier Transformers

Inverse-time-overcurrent relays are recommended for internal short-circuit protection. The inverse-time elements should have time-delay adjustment with just sufficient delay to be selective with the d-c protective equipment for external d-c short circuits or overloads. The instantaneous elements should be adjustable so as barely not to operate for low-voltage faults or magnetizing-current inrush, including an allowance for overtravel.

A temperature relay operating in conjunction with a resistance-temperature detector should be provided to sound an alarm or trip the transformer breaker as desired.

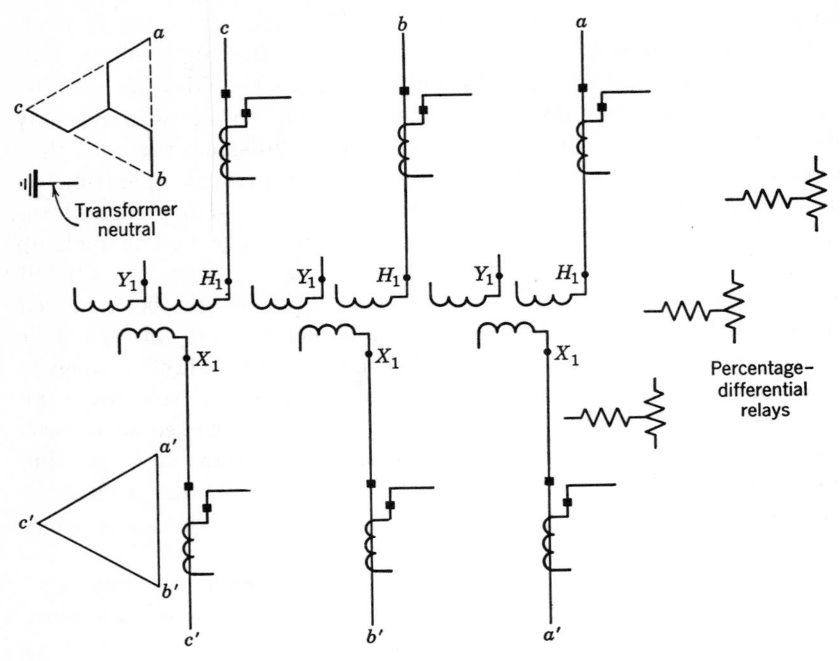

Fig. 20. Illustration for Problem 1.

Problems

1. Given three single-phase power transformers having windings as shown in Fig. 20. Complete the connections of the power transformers so as to obtain a zig-zag connection on the high-voltage side and a delta connection on the low-voltage side, using the partial connections shown, the voltage diagrams to be as shown. Connect the CT's to the percentage-differential relays so as to obtain protection of the transformer bank for internal faults but so that undesired tripping will not occur for external faults. Assume a 1/1 turn ratio between each pair of power-transformer windings, and assume that any desired ratio is available for the CT's. Add the CT-secondary ground connection.

2. Given a wye-delta power transformer protected as shown in Fig. 21. An external three-phase fault occurs, and fault currents flow through the transformer with the magnitudes as shown. Will the differential relay operate to trip?

3. Repeat Problem 2 except with a three-phase fault between the high-voltage breaker and the transformer. Assume that the system supplies 4000 amperes three-phase to the fault, the current supplied by the power transformer being the same as in Problem 2.

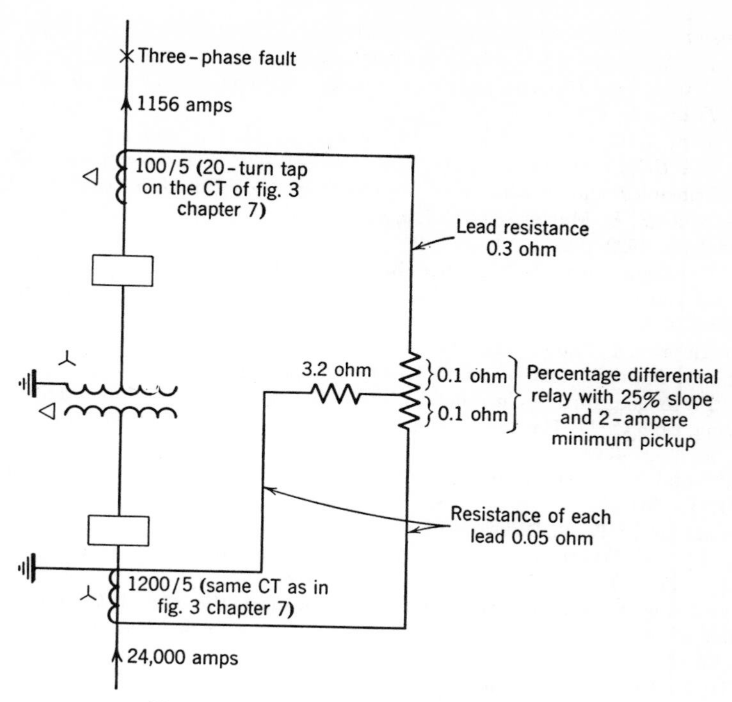

Fig. 21. Illustration for Problems 2 and 3.

Bibliography

1. "Power Switchgear Assemblies," *Publ. SG5,* National Electrical Manufacturers Assoc., 155 East 44th St., New York 17, N. Y.

2. "Relay Protection of Power Transformers," by AIEE Committee, *AIEE Trans., 66* (1947), pp. 911–915. Discussions, pp. 915–917.

3. "Recommended Practices for the Protection of Electrical Apparatus," by AIEE Committee, *AIEE Trans., 52* (1933), pp. 607–613.

4. "A Practical Discussion of Problems in Transformer Differential Protection," by P. W. Shill, *AIEE Trans., 61* (1942), pp. 854–858. Discussions, pp. 1067–1069.

"Scott 2-3-Phase Banks Differential Relays," by V. P. Brodsky, *Elec. World,* May 8, 1937, pp. 80–82 (*107,* pp. 1590–1592).

5. "Report on Transformer Magnetizing Current and Its Effect on Relaying and Air Switch Operation," by AIEE Committee, *AIEE Trans., 70,* Part II (1951), pp. 1733–1739. Discussions, pp. 1739–1740.

6. "The Inrush of Magnetizing Current in Single-Phase Transformers," by L. A. Finzi and W. H. Mutschler, Jr., *AIEE Trans., 70,* Part II (1951), pp. 1436–1438.

"Transformer Magnetizing Inrush Currents and Influence on System Operation," by L. F. Blume, G. Camilli, S. B. Farnham, and H. A. Petersen, *AIEE Trans., 63* (1944), pp. 366–374. Discussions, p. 423.

Transformer Engineering, by L. F. Blume, A. Boyajian, G. Camilli, T. S. Lennox, S. Minneci, and V. M. Montsinger, John Wiley & Sons, New York, 1951.

7. "Transformer Current and Power Inrushes under Load," by E. B. Kurtz, *AIEE Trans., 56* (1937), pp. 989–994.

8. "Some Utility Ground-Relay Problems," by H. C. Barnes and A. J. McConnell, *AIEE Trans., 74,* Part III (1955), pp. 417–428. Discussions, pp. 428–433.

9. "Principles and Practices of Relaying in the United States," by E. L. Harder and W. E. Marter, *AIEE Trans., 67,* Part II (1948), pp. 1005–1022. Discussions, pp. 1022–1023.

10. "Harmonic-Current-Restrained Relays for Differential Protection," by L. F. Kennedy and C. D. Hayward, *AIEE Trans., 57* (1938), pp. 262–266. Discussions, pp. 266–271.

"An Improved Transformer Differential Relay," by C. A. Matthews, *AIEE Trans., 73,* Part III (1954), pp. 645–649. Discussions, pp. 649–650.

11. "Prolonged Inrush Currents with Parallel Transformers Affect Differential Relaying," by C. D. Hayward, *AIEE Trans., 60* (1941), pp. 1096–1101. Discussions, pp. 1305–1308.

12. "Simplicity in Transformer Protection," by E. T. B. Gross, *Elec. Eng., 66* (1947), pp. 564–569.

Discussion by E. T. B. Gross of Reference 14.

13. "Typical Transformer Faults and Gas Detector Relay Protection," by J. T. Madill, *AIEE Trans., 66* (1947), pp. 1052–1060.

"Gas Detector Relays," by A. L. Hough, *Reports for the 57th Annual Convention of the Canadian Electrical Assoc., Engineering Section,* 1947, pp. 56–59.

14. "A Sudden Gas Pressure Relay for Transformer Protection," by R. L. Bean and H. L. Cole, *AIEE Trans., 72,* Part III (1953), pp. 480–483. Discussions, p. 483.

15. Discussions of Reference 2.

16. "Remote Tripping Schemes," by AIEE Committee, *AIEE Trans., 72,* Part III (1953), pp. 142–150. Discussions, pp. 150–151.

17. "Grounding Switch Protects Transformer Installed near Center of 50-Mile Line," *Elec. World,* March 4, 1944, p. 58.

"Short-Circuit Switch in Lieu of Breaker," by J. F. Sinnot, *Elec. World, 121,* April 29, 1944, pp. 50, 51.

"Protection of Stations without High-Voltage Switching," by AIEE Committee, *AIEE Trans., 68* (1949), pp. 226–231. Discussions, pp. 231–232.

18. "A New Carrier-Current Frequency-Shift System for Use with Differential Protection of Transformer Banks," by R. W. Beckwith, *AIEE Trans., 70,* Part I (1951), pp. 832–835. Discussions, p. 835.

19. "Relay Protection for a Large Regulating Transformer," by W. E. Marter, *Elec. J., 36,* No. 3, (March, 1939), pp. 86–88.

12 BUS PROTECTION

Although bus protection for new installations is now a simple application problem, this has not always been true, as attested by the many different ways in which protection has been provided. This problem always has been—and still is in many existing installations—primarily a current-transformer problem. A bus has no peculiar fault characteristics, and it would lend itself readily to current-differential protection if its CT's were suitable.

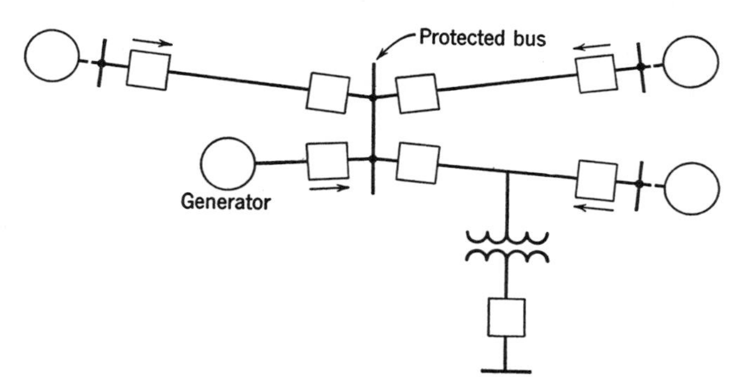

Fig. 1. Bus protection by back-up relays.

Before considering the more modern equipments for bus protection, let us first examine the various forms of protection that have been used and that are still in service; new uses for some of these may still be found infrequently today.

PROTECTION BY BACK-UP RELAYS

The earliest form of bus protection was that provided by the relays of circuits over which current was supplied to a bus, at locations such as shown by the arrows on Fig. 1. In other words, the bus was included within the back-up zone of these relays. This method was relatively slow speed, and loads tapped from the lines would be interrupted unnecessarily, but it was otherwise effective. Some preferred this method to one in which the inadvertent operation of a single relay would trip all the connections to the bus.

THE FAULT BUS[1]

The fault-bus method consists of insulating the bus-supporting structure and its switchgear from ground; interconnecting all the framework, circuit-breaker tanks, etc.; and providing a single ground connection through a CT that energizes an overcurrent relay, as illustrated schematically in Fig. 2. The overcurrent relay controls a multicontact auxiliary relay that trips the breakers of all circuits

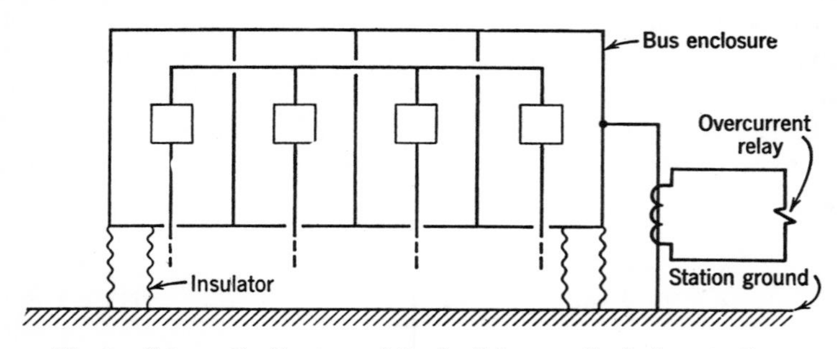

Fig. 2. Schematic diagram of the fault-bus method of protection.

connected to the bus. The maximum effectiveness is obtained by this method when the switchgear is of the isolated-phase construction, in which event all faults will involve ground. However, it is possible to design other types of switchgear with special provisions for making ground faults the most probable. Of course, if interphase faults not involving ground can occur, and if CT's and conventional differential relaying have to be used for protection against such faults, the fault-bus method would probably not be justified.

This method is most applicable to new installations, particularly of the metal-clad type, where provision can be made for effective insulation from ground. It has been more favored for indoor than for outdoor installations. Certain existing installations may not be adaptable to fault-bus protection, owing to the possibility of other paths for short-circuit current to flow to ground through concrete-reinforcing rods or structural steel. It is necessary to insulate cable sheaths from the switchgear enclosure or else cable ground-fault currents may find their way to ground through the fault-bus CT and improperly trip all the switchgear breakers. An external flashover of an entrance bushing will also improperly trip all breakers, unless the bushing support is insulated from the rest of the structure and independently grounded.

If a sectionalized bus structure is involved, the housing of each section must be insulated from adjoining sections, and separate fault-bus relaying must be employed for each section. The fault-bus method does not provide overlapping of protective zones around circuit breakers; and, consequently, supplementary relaying is required to protect the regions between bus sections.

Some applications have used a fault-detecting overcurrent relay, energized from a CT in the neutral of the station-grounding transformer or generator, with its contact in series with that of the fault-bus relay, to guard against improper tripping in the event that the fault-bus relay should be operated accidentally; without some such provision, the accidental grounding of a portable electric tool against the switchgear housing might pass enough current through the fault-bus ground to operate the relay and trip all the breakers, unless the pickup of the relay was higher than the current that it could receive under such a circumstance.

Consideration should be given to the possibility of people being able to make contact between the switchgear housing and ground, to avoid the possibility of contact with high voltage should a ground fault occur; although the ground connection would have very low impedance, high current flowing through it could produce dangerously high voltage. This requirement also makes it desirable to ground all relay, meter, and control circuits to the switchgear housing rather than by means of a separate connection to the station ground.

DIRECTIONAL-COMPARISON RELAYING

The principle of "directional comparison" used for transmission-line relaying has been adapted to bus protection in order to avoid the problem of matching current-transformer ratings and characteristics.[2] Basically, the contacts of directional relays in all source circuits and the contacts of overcurrent relays in purely load circuits are interconnected in such a way that, if fault current flows toward the bus, the equipment will operate to trip all bus breakers unless sufficient current is flowing away from the bus in any circuit.

This principle has been used only with ground relays, on the basis that most bus faults start as ground faults, or at least that they very quickly involve ground. This greatly reduces the cost of the equipment over that if phase relays were also used; even so, it is still more costly than most relaying equipments for bus protection. Of course, if it saves investment in new current transformers and their installation cost, it would be economically attractive.

The principal disadvantage of this type of equipment is the greater maintenance required and the greater probability of failure to operate because of the large number of contacts in series in the trip circuit. Another disadvantage is that connections from the current transformers in all the circuits must be run all the way to the relay panel. Of course, if only ground relays were used, only two connections to the CT's of each circuit would be required. If phase relays were used, they would depend on bus voltage for polarization, and, therefore, they might not operate for a metallic short circuit that reduced the voltage practically to zero.

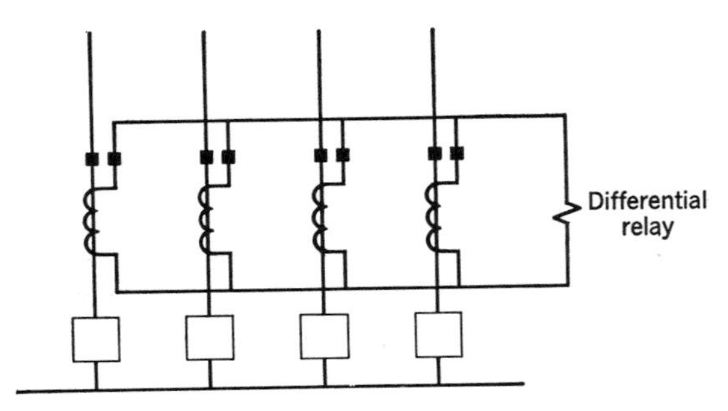

Fig. 3. Bus protection by current-differential relaying.

CURRENT-DIFFERENTIAL RELAYING WITH OVERCURRENT RELAYS

The principle of current-differential relaying has been described. Figure 3 shows its application to a bus with four circuits. All the CT's have the same nominal ratio and are interconnected in such a way that, for load current or for current flowing to an external fault beyond the CT's of any circuits, no current should flow through the relay coil, assuming that the CT's have no ratio or phase-angle errors. However, the CT's in the faulty circuit may be so badly saturated by the total fault current that they will have very large errors; the other CT's in circuits carrying only a part of the total current may not saturate so much and, hence, may be quite accurate. As a consequence, the differential relay may get a very large current, and, unless the relay has a high enough pickup or a long enough time delay or both, it will operate undesirably and cause all bus breakers to be tripped.

The greatest and most troublesome cause of current-transformer saturation is the transient d-c component of the short-circuit current.

It is easy to determine if the CT's in the faulty circuit will be badly saturated by a fault-current wave having a d-c component, by using the following approximate formula:

$$B_{\text{max}} = \frac{(\sqrt{2})(10^8)R_2 I_1 T N_1}{A N_2^2}$$

where B_{max} = maximum flux density in CT core in lines per square inch.
R_2 = resistance of CT-secondary winding and leads up to, but not including, the relay circuit, in ohms.
I_1 = rms magnitude of symmetrical component of primary fault current in amperes.
T = time constant of primary d-c component in seconds.
A = cross-sectional area of CT core in square inches.
N_1 = number of primary turns (=1 for bushing CT's).
N_2 = number of secondary turns.

For values of B_{max} greater than about 100,000 lines per square inch, there will be saturation in modern CT's, and for values greater than about 125,000 lines there will be appreciable saturation, the degree being worse the higher the value of B_{max}. For instantaneous relays, B_{max} should be no more than about 40,000 lines because the residual flux can be as high as about 60,000 lines. For CT's that are 10 or 15 years old, about 77,500 lines represents saturation flux density.

Considering the effect of the d-c time constant, it will become evident that the nearer a bus is to the terminals of a generator, the greater will be the CT saturation. Typical d-c time constants for different circuit elements are:

Generators	0.3 second
Transformers	0.04 second
Lines	0.01 second

It makes a tremendous difference, therefore, whether the fault-current magnitude is limited mostly by line impedance or by generator impedance.

If the d-c component will not badly saturate the CT's, it is a relatively simple matter to calculate the error characteristics of the CT's by the methods already described, and to find out how much current will flow in the differential relay. Knowing this magnitude of current and the magnitude for which the differential relay must operate for bus faults, one can choose the pickup and time settings that will give the best protection to the bus and still provide selectivity for external faults.

But, if the d-c saturation is severe, and it usually is, the problem is much more difficult, particularly if instantaneous relaying is desired. Two methods of analysis have been presented. One is a method for first calculating the differential current and then determining the response of an overcurrent relay to this current.[3] The other is a method whereby the results of a comprehensive study can be applied directly to a given installation for the purpose of estimating the response of an overcurrent relay.[4] Because of the approximations and the uncertainties involved (and probably also because of the

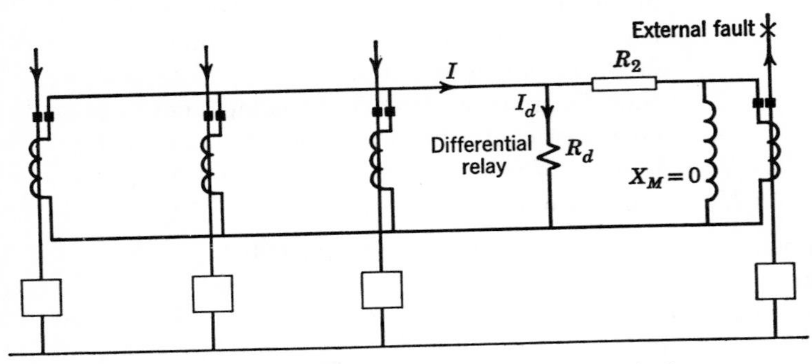

Fig. 4. Distribution of current for an external fault.

labor involved), neither of these two methods is used very extensively, but, together with other investigations, they provide certain very useful guiding principles.

Perhaps the most useful information revealed by these and other[5,6] studies is the effect of resistance in the differential branch on the magnitude of current that can flow in that branch. Figure 4 is a one-line diagram of the CT's and differential relay for a bus with four feeders, showing an external short circuit on one of the feeders. Figure 4 shows the equivalent circuit of the CT in the feeder having the short circuit. If that CT is assumed to be so completely saturated that its magnetizing reactance is zero, neglecting the air-core mutual reactance, the secondary current (I) from all the other CT's will divide between the differential branch and the saturated CT secondary, and the rms value of the differential current (I_d) will be no higher than:

$$I_d = I\left(\frac{R_2}{R_d + R_2}\right)$$

where R_2 includes the secondary-winding resistance of the CT in the faulty circuit. The effect of relay-coil saturation must be taken into account in using this relation. Of course, the differential current will quickly decrease as the fault-current wave becomes symmetrical. However, studies have shown that, depending on the circuit resistances, the rms value of the differential current (I_d) can momentarily approach the fault-current magnitude (I) expressed in secondary terms. Where this is possible, instantaneous overcurrent relays are not applicable unless sufficient resistance can be added to the differential branch. The amount of such additional resistance should not be enough to cause too high voltages when very high currents flow to a bus fault. Nor should the resistance be so high that the CT's could not supply at least about 1.5 times pickup current under minimum bus-fault-current conditions. If we assume the CT's in the faulty circuit to be so badly saturated that their magnetizing reactance is zero, and that all the other CT's maintain their nominal ratio, the division of current between the differential relays and the secondaries of the saturated CT's, and the effects of adding resistance to the differential branch, may be calculated assuming symmetrical sinusoidal currents; the results will be conservative in that the differential relays will not have as great an operating tendency as the calculations would indicate.

The foregoing furnishes a practical rule for obtaining the best possible results with any current-differential-relaying application. This rule is to make the junction point of the CT's at a central location with respect to the CT's and to use as large-diameter wire as practical for the interconnections. The fact that the CT secondary windings have appreciable resistance makes it impractical to try to go too far toward reducing the lead resistance. Resistance in the leads from the junction point to the differential relays is beneficial to a certain extent, as already mentioned.

Another rule that is generally followed is to choose CT ratings so that the maximum magnitude of external-fault current is less than about 20 times the CT rating.[7] Some allow this multiple to go to 30 or more, and others[8] try to keep it below 10. In an existing installation with multiratio bushing CT's, use the highest turns ratio.

To prevent differential-relay operation should a CT open-circuit, the relay pickup is often made no less than about twice the load current of the most heavily loaded circuit;[8] if the magnitude of ground-fault current is sufficiently limited by neutral-grounding impedance, lower pickup may be required and additional fault-detecting means may be required to prevent operation should a CT open-circuit, such as by the use of an overcurrent relay energized from a CT in the grounded-neutral source, and with its "*a*" contacts in series with the trip circuit.

Some instantaneous overcurrent relays are used for current-differential relaying, but inverse-time induction-type overcurrent relays are the most common; the induction principle makes these relays less responsive to the d-c and harmonic components of the differential current resulting from CT errors because of saturation. Time delay is most helpful to delay differential-relay operation long enough for the transient differential current due to CT errors to subside below the relay's pickup; from 0.2 second to 0.5 second is sufficient for most applications. The fact that a relay will overtravel after the current has dropped below the pickup value should be taken into account.

Where not all the CT's are of the same ratio, sometimes the practice is to provide current-differential relaying only for ground faults. To accomplish this, auxiliary CT's are connected in the neutral of the CT's of each circuit, and the ratios of these auxiliary CT's are chosen to compensate for the differences in ratio of the main CT's. The ratio accuracy of such auxiliary CT's should be investigated to determine their suitability. Of course, auxiliary CT's might be used to permit phase-fault relaying, but it would be much more expensive. In general, auxiliary CT's should be avoided whenever possible.

PARTIAL-DIFFERENTIAL RELAYING

Partial-differential relaying is a modification of current-differential relaying whereby only the CT's in generating-source (either local or distant) circuits are paralleled, as illustrated in Fig. 5. This is not done because of any advantage to be gained by omitting the other CT's in the purely load circuits, but either because there are no CT's to be used or because those that are available are not suitable for complete current-differential relaying.

Two types of partial-differential relaying have been used, one type employing overcurrent relays and the other employing distance relays. The protection provided by the overcurrent type is much like that provided by back-up relays in the individual source circuits. The overcurrent type must have enough time delay to be selective with the relays of the load circuits for external faults in these circuits. Also, it must have a pickup higher than the total maximum-load current of all source circuits. The only advantages of partial-differential relaying with overcurrent relays are (1) that local protective equipment is provided for bus protection, and (2) that back-up protection is provided for the load circuits.

A second type of partial-differential-relaying equipment uses distance relays.[9,10] This type is applicable where all the load circuits have current-limiting reactors, as illustrated in Fig. 6. So long as two or more of the load circuits are not paralleled a short distance from the bus, the reactors introduce enough reactance into the cir-

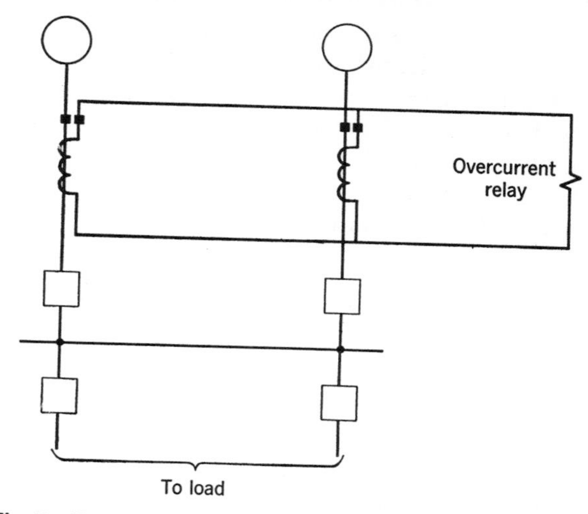

Fig. 5. Partial-differential relaying with overcurrent relays.

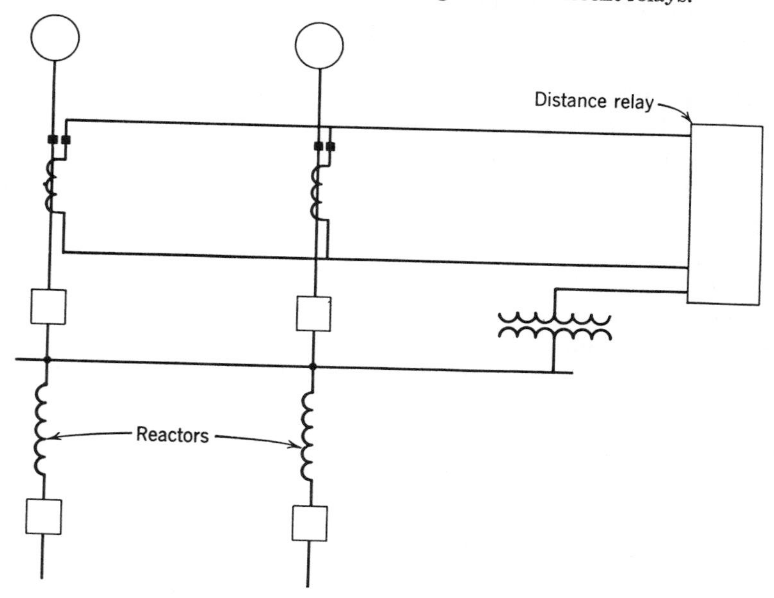

Fig. 6. Partial-differential relaying with distance relays.

cuits so that the distance relays can select between faults on the bus side and faults on the load side of the reactors. In some actual applications, only ground distance relays have been used on the basis that all bus faults will involve ground sooner or later. Because a fault in one of the source circuits that badly saturates its CT's will tend to cause the distance relays to operate undesirably, such a possibility must be carefully investigated. Otherwise, this type of relaying can be both fast and sensitive.

One application has been described in which distance relays were used for station-service bus protection and there were no reactors in the load circuits.[10] Instead, selectivity with the load-circuit relays was obtained by adding a short time delay to the operating time of the distance relays.

CURRENT-DIFFERENTIAL RELAYING WITH PERCENTAGE-DIFFERENTIAL RELAYS

As in differential relaying for generators and transformers, the principle of percentage-differential relaying is a great improvement over overcurrent relays in a differential CT circuit. The problem of providing enough restraining circuits has been largely solved by so-called "multirestraint" relays.[5] By judicious grouping of circuits and by the use of two relays per phase where necessary, sufficient restraining circuits can generally be provided. Further improvement in selectivity is provided by the "variable-percentage" characteristic,[5] like that described in connection with generator protection; with this characteristic, one should make sure that very high internal-fault currents will not cause sufficient restraint to prevent tripping.

This type of relaying equipment is available with operating times of the order of 3 to 6 cycles (60-cycle basis). It is not suitable where high-speed operation is required.

As in current-differential relaying with overcurrent relays, the problem of calculating the CT errors is very difficult. The use of percentage restraint and the variable-percentage characteristic make the relay quite insensitive to the effects of CT error. Nevertheless, it is recommended that each application be referred to the manufacturer together with all the necessary data.

A disadvantage of this type of equipment is that all CT secondary leads must be run to the relay panel.

VOLTAGE-DIFFERENTIAL RELAYING WITH "LINEAR COUPLERS"

The problem of CT saturation is eliminated at its source by air-core CT's called "linear couplers."[11] These CT's are like bushing CT's but they have no iron in their core, and the number of secondary turns is much greater. The secondary-excitation characteristic of these CT's is a straight line having a slope of about 5 volts per 1000 ampere-turns. Contrasted with conventional CT's, linear couplers may be operated without damage with their secondaries open-circuited. In fact, very little current can be drawn from the secondary, because so much of the primary magnetomotive force is consumed in magnetizing the core.

The foregoing explains why the linear couplers are connected in a voltage-differential circuit, as shown schematically in Fig. 7. For normal load or external-fault conditions, the sum of the voltages in-

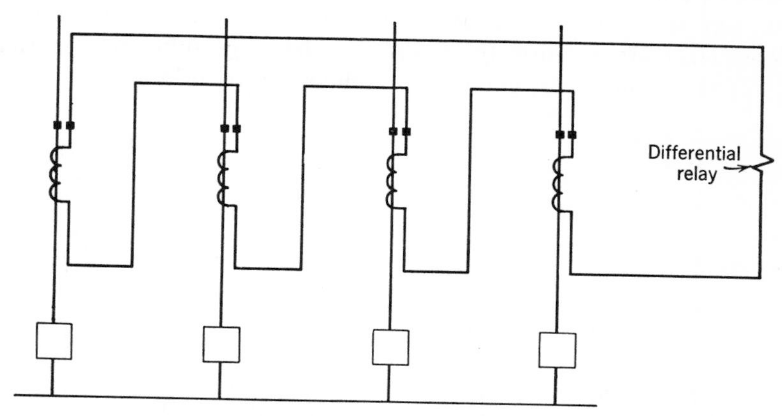

Fig. 7. Bus protection with voltage-differential relaying.

duced in the secondaries is zero, except for the very small effects of manufacturing tolerances, and there is practically no tendency for current to flow in the differential relay.

When a bus fault occurs, the voltages of the CT's in all the source circuits add to cause current to flow through all the secondaries and the coil of the differential relay. The differential relay, necessarily requiring very little energy to operate, will provide high-speed protection for a relatively small net voltage in the differential circuit.

The application of the linear-coupler equipment is most simple, requiring only a comparison of the possible magnitude of the differential voltage during external faults, because of differences in the characteristics of individual linear couplers, with the magnitude of the voltage when bus faults occur under conditions for which the fault-current magnitude is the lowest. Except when ground-fault current is severely limited by neutral impedance, there is usually no selectivity problem. When such a problem exists, it is solved by the use of additional more-sensitive relaying equipment, including a supervising relay that permits the more-sensitive equipment to operate only for a single-phase-to-ground fault.[12]

CURRENT-DIFFERENTIAL RELAYING WITH OVERVOLTAGE RELAYS

A type of high-speed relaying equipment employing current-differential relaying with overvoltage relays also eliminates the problem of current-transformer saturation, but in a different manner from that described using linear couplers. With this equipment, conventional bushing CT's (or other CT's with low-impedance secondaries) are used, and they are differentially connected exactly as for current-differential relaying already described; the only difference is that overvoltage rather than overcurrent relays are used.[13]

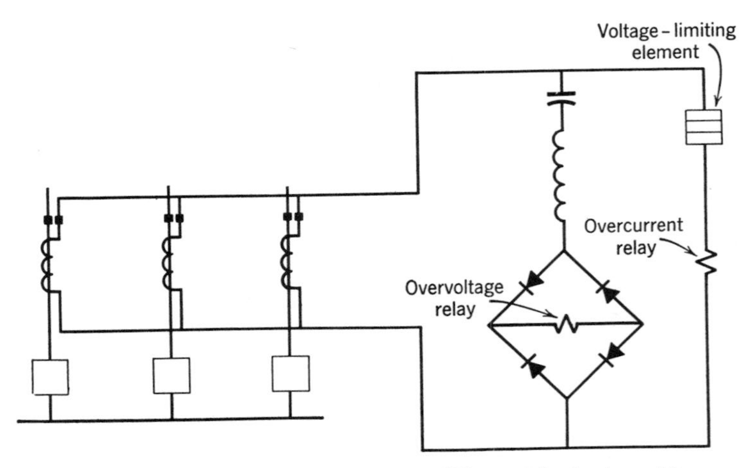

Fig. 8. Bus protection using current-differential relaying with overvoltage relays.

In effect, this equipment carries to the limit the beneficial principle already described of adding resistance to the differential branch of the circuit. However, in this equipment, the impedance of the overvoltage-relay's coil is made to appear to the circuit as resistance by virtue of a full-wave rectifier, as illustrated in Fig. 8. Hence, the efficiency of the equipment is not lowered as it would be if a series resistor were used.

The capacitance and inductance, shown in series with the rectifier circuit, are in series resonance at fundamental system frequency; the purpose of this is to make the relay responsive to only the fundamental component of the CT secondary current so as to improve the relay's selectivity. It has the disadvantage, however, of slowing the voltage-relay's response slightly, but this is not serious in view of the high-speed operation of an overcurrent-relay element now to be described.

Because the effective resistance of the voltage-relay's coil circuit is so high, being approximately 3000 ohms, a voltage-limiting element must be connected in parallel with the rectifier branch, or else excessively high CT secondary voltages would be produced when bus faults occurred. As shown in Fig. 8, an overcurrent-relay unit in series with the voltage limiter provides high-speed operation for bus faults involving high-magnitude currents. Since the overcurrent unit is relied on only for high-magnitude currents, its pickup can easily be made high enough to avoid operation for external faults.

The procedure for determining the necessary adjustments and the resulting sensitivity to low-current bus faults is very simple and straightforward, requiring only a knowledge of the CT secondary excitation characteristics and their secondary impedance.

For the best possible results, all CT's should have the same rating, and should be a type, like a bushing CT with a distributed secondary winding, that has little or no secondary leakage reactance.

COMBINED POWER-TRANSFORMER AND BUS PROTECTION

Figure 9 shows a frequently encountered situation in which a circuit breaker is omitted between a transformer bank and a low-voltage bus. If the low-voltage bus supplies purely load circuits without any back-feed possible from generating sources, the CT's in all the load circuits may be paralleled and the transformer-differential relay's zone of protection may be extended to include the bus.

Figure 10 shows two parallel high-voltage lines feeding a power-transformer bus with no circuit breaker between the transformer and the bus. As shown in the figure, a three-winding type of percentage-differential relay will provide good protection for the bus and the transformer.

In Fig. 11, the two high-voltage lines are from different stations and may constitute an interconnection between parts of a system. Consequently, much higher load currents may flow through these circuits than the rated load current of the power transformer. Therefore, the CT ratios in the high-voltage circuits may have to be much higher than one would desire for the most sensitive protection of the power transformer. And therefore, the protective scheme of Fig. 10, though generally applicable, is not as sensitive to transformer faults as the arrangement of Fig. 11. Bushing CT's can generally be added to most power transformers, but it is considerably less expensive and less

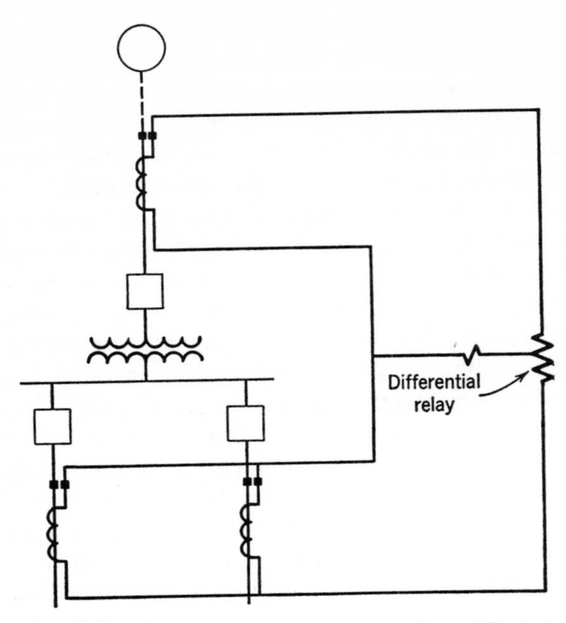

Fig. 9. Combined transformer and bus protection with a two-winding percentage-differential relay.

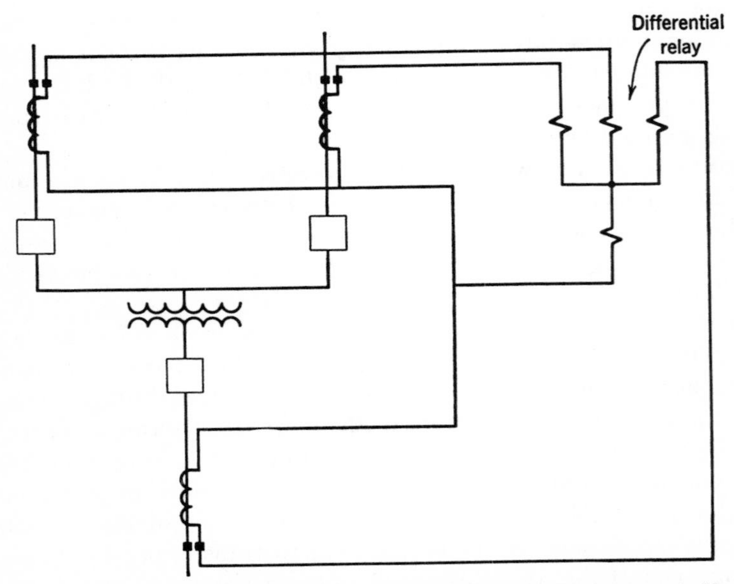

Fig. 10. Combined transformer and bus protection with a three-winding percentage-differential relay.

troublesome if the power transformers are purchased with the two sets of CT's already installed. It is almost axiomatic that, whenever circuit breakers are to be omitted on the high-voltage side of power transformers, two sets of bushing CT's should be provided on the

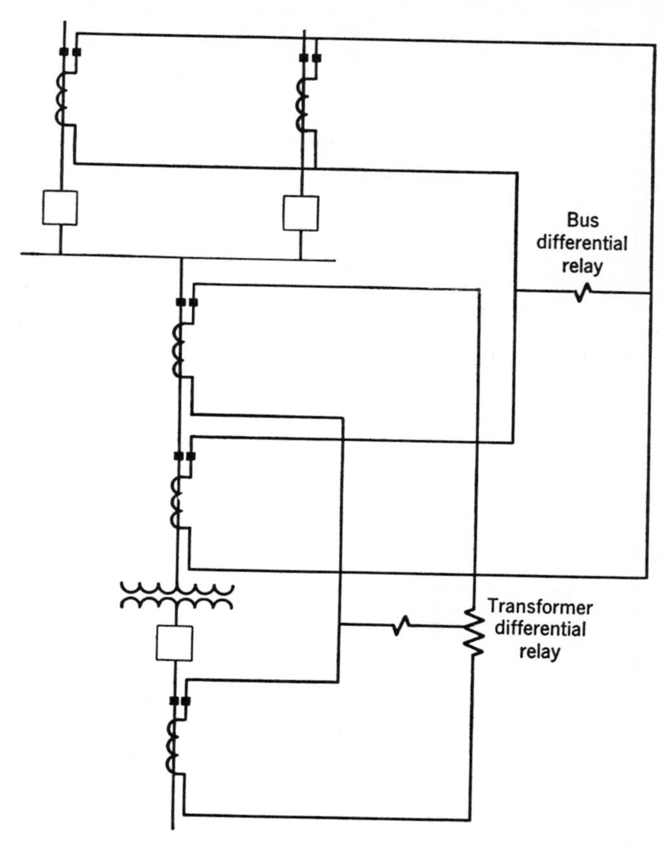

Fig. 11. Preferred alternative to Fig. 10 when high-voltage lines are from different stations.

transformer high-voltage bushings. The arrangement of Fig. 11 can be extended to accommodate more high-voltage lines or more power transformers, although, as stated in Chapter 11, it is not considered good practice to omit high-voltage breakers when two or more power-transformer banks rated 5000 kva or higher are paralleled.

Figure 12 shows a protective arrangement that has been used for the combination of two power transformers as shown. It could not be extended to accommodate any more transformers or high-voltage circuits, and it does not provide as sensitive protection as the arrangement of Fig. 11, but it saves one set of bushing CT's and a set of bus-differential relays.

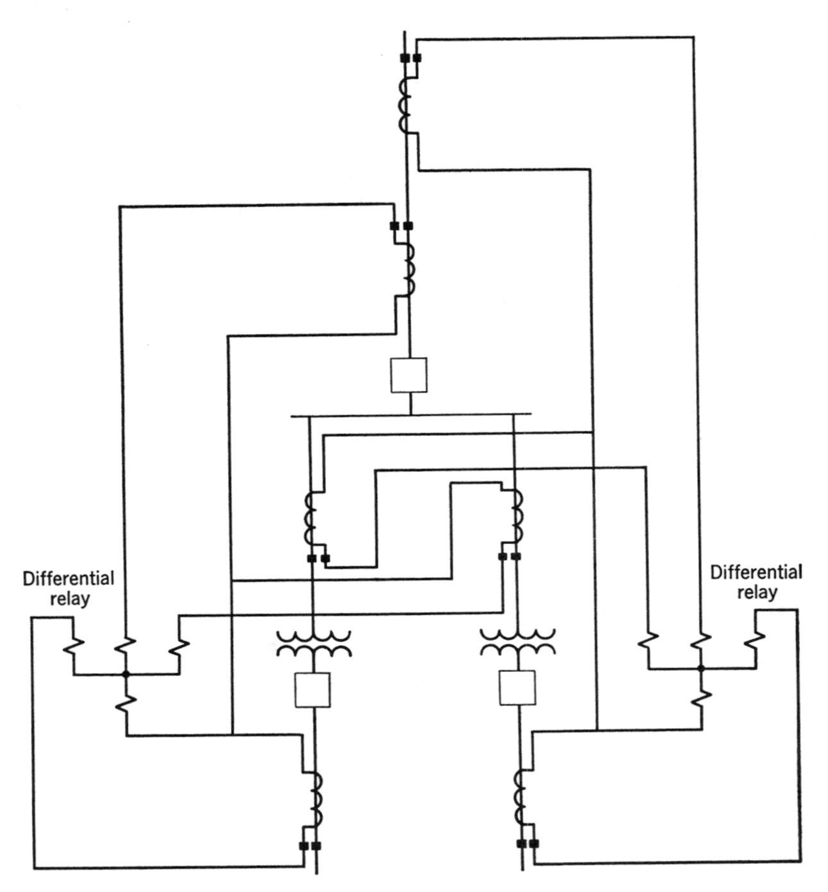

Fig. 12. Complete protection of two transformers and a bus with two three-winding percentage-differential relays.

RING-BUS PROTECTION

No separate relaying equipment is provided for a ring bus. Instead, the relaying equipments of the circuits connected to the bus include the bus within their zones of protection, as illustrated in Fig. 13. The relaying equipment of each circuit is indicated by a box lettered to correspond to the protected circuit, and is energized by the parallel-connected CT's in the branches that feed the circuit.

A separate voltage supply is required for the protective relays of each circuit. Also, the CT ratios must be suitable for the largest magnitude of load current that might flow around the ring, which might be too high for the desired protection of a given circuit.

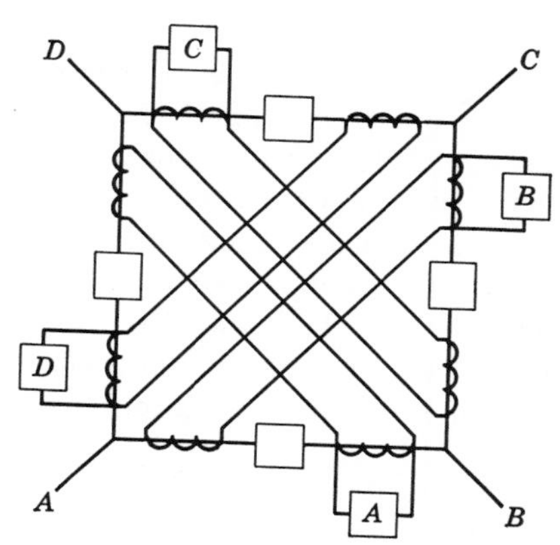

Fig. 13. Protection of a ring bus.

THE VALUE OF BUS SECTIONALIZING

Although the design of busses does not fall in the category of bus relaying, it is well to keep in mind that bus sectionalizing helps to minimize interference with service when a bus fault occurs. For some busses, sectionalizing is an essential feature of design if stability is to be maintained after a bus fault. With bus sectionalizing, each bus section can be protected separately, and the likelihood of a fault in one section interfering with the service of another section is thereby minimized.

BACK-UP PROTECTION FOR BUS FAULTS

If one or more bus breakers fail to trip in the event of a bus fault, back-up protection is provided by the relaying equipments at the far ends of the circuits that continue to feed current directly to the fault.

Occasionally, relaying equipment is provided at a bus location for back-up protection of adjoining circuits. This is done only when it is impossible to provide the desired back-up protection in the conventional manner described in Chapter 1. This matter is treated further under the subject of line protection.

GROUNDING THE SECONDARIES OF DIFFERENTIALLY CONNECTED CT'S

The intent here is to emphasize the fact that differentially connected CT's should be grounded *at only one point.*[14] In spite of the fact that various publications have warned against separately grounding the CT's of each circuit, the survey reported in Reference 7 found separate grounding to be practiced by many. Metering practices have been so strongly ingrained that people have replaced grounds that were purposely omitted from differentially connected CT's.

The correct practice is to interconnect the neutrals of all differentially connected CT's with the same kind of *insulated* wire as that used for the other CT interconnections. Then, the neutral interconnection is grounded at *one point only*. Since the grounding is for the protection of personnel, the best place to make the ground is at the differential-relay panel where connection is made to the neutral of wye-connected relay coils.

The reason why the ground should be made at only one point is to avoid improper relay operation and damage to the CT interconnections. If grounds are made at two or more locations, circulating currents may be caused to flow in the differential circuit because of differences of potential between the grounding points, owing to the flow of large ground currents during ground faults.

AUTOMATIC RECLOSING OF BUS BREAKERS

A few installations of outdoor automatic substations, whose busses are not enclosed, employ automatic reclosing of the bus breakers. In at least one installation, a single circuit connected to a generating source is first reclosed and, if it stays in, the remaining circuits are then reclosed—all automatically. Somewhat the same philosophy applies to outdoor open busses as to transmission lines, namely, that many faults will be non-persisting if quickly cleared, and, hence, that automatic reclosing will usually be successful. However, substations are generally better protected against lightning than lines, and their exposure to lightning is far less. Hence, one can expect that a larger percentage of bus faults will be persisting.

PRACTICES WITH REGARD TO CIRCUIT-BREAKER BY-PASSING

Most users of bus-differential protection take the bus protective-relaying equipment completely out of service, either automatically or manually, and do not substitute any other equipment for temporary protection, when circuit breakers are to be by-passed for maintenance purposes or when any other abnormal set-up is to be made.[7] Of course, the bus still has time-delay protection because the back-up equipment in the circuits connected to the bus should function for bus faults.

Others use a wide diversity of temporary forms of bus relaying.[7]

ONCE-A-SHIFT TESTING OF DIFFERENTIAL-RELAYING EQUIPMENT

Most power companies make maintenance checks once every 6 to 12 months or longer, but a few follow the practice of testing their bus-differential-relaying CT secondary circuits every shift for open circuits or short circuits. This can be done with permanently installed testing equipment arranged to measure current in the relay operating coils, and to superimpose on the circuit a testing current or voltage, as required. Where fault detectors are employed to supervise tripping, the equipment is sometimes arranged to sound an alarm if a CT circuit should open and cause a differential relay to operate.

Such testing is felt by some to be particularly desirable where the differential relay's pickup is high enough so that the relays will not operate on the current that they will get during normal load conditions if an open circuit or a short circuit should occur in a CT secondary circuit. Such trouble would not be discovered until after the relays had operated undesirably for an external fault or had failed to operate for a bus fault.

Of course, the only difference between bus-differential CT circuits and the CT circuits of relaying equipment protecting other system elements is that for the bus there are more CT's involved and that a larger part of the power system is affected. It is not that the bus equipment is inferior to the other, but that the consequences of failure are more serious.

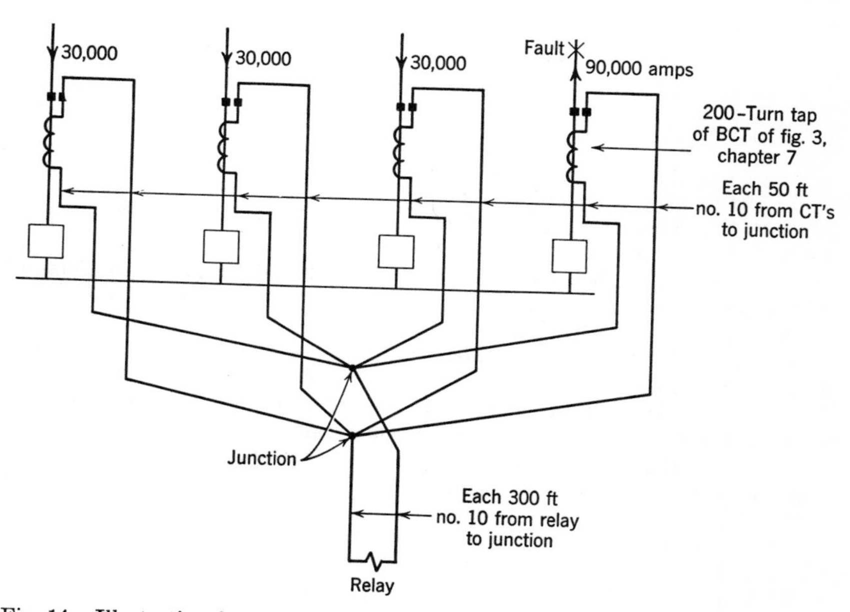

Fig. 14. Illustration for problem. Assume resistance of no. 10 wire is 1.018 ohms per 1000 ft at 25° C.

Problem

Given a bus-differential-relaying equipment as shown in Fig. 14, using 4/16-ampere inverse-time-overcurrent relays whose time-current characteristics are shown in Fig. 3 of Chapter 3. The total short-circuit current to a bus fault under minimum generating conditions will be 5000 amperes; and it has been decided to adjust the relays to pick up at 2400 amperes (12-ampere tap). For an external three-phase fault involving the current magnitudes and directions as shown, the breaker in the faulty circuit will interrupt the flow of short-circuit current in 0.3 second by virtue of other protective relaying equipment not shown.

In view of the large magnitude of external-fault current, it is suspected that the differential relays may tend to operate, and, if so, they are to be adjusted to have a 0.5-second time delay so as to be selective with the faulty-feeder relaying equipment. What time-lever setting should be chosen for this purpose?

Assume that each CT in the faulty feeder saturates completely so as to have no secondary-current output, and that the total output current of the other CT's divides between the saturated CT's and the relays. Take into account the fact that the other CT's will have ratio error, and for this purpose assume that their primary currents are symmetrical. Also, take into account the fact that the relay coils saturate and, hence, that their impedance decreases as shown in the acompanying table when current of large magnitude flows in each coil.

Multiple of Pickup Current (4-amp. tap)	Coil Impedance, ohms
1	0.3
3	0.15
10	0.08
20	0.07

Bibliography

1. "Damage Local and Less with Fault-Bus Relaying," by I. C. Eppley, *Elec. World, 120* (August 21, 1943), pp. 633, 634.

"Fault-Bus Differential Scheme Protects Cahokia Bus," by G. W. Gerell, *Elec. World, 117* (April 18, 1942), pp. 1335, 1336.

"The Fault Ground Bus, Its Use and Design in Brunot Island Switch House of the Duquesne Light Co.," by R. M. Stanley and F. C. Hornibrook, *AIEE Trans., 49* (1930), pp. 201–211. Discussions, pp. 211–212.

"Ground Fault Relaying with Metal-Enclosed Busses," by S. M. Dean and W. F. Wetmore, *Elec. World, 95,* No. 21 (May 24, 1930), pp. 1024–1028.

2. "Bus Protection Independent of Current Transformer Characteristics," by G. Steeb, *AIEE Trans., 60* (1941), pp. 859–862. Discussions, pp. 1365–1368.

"Directional Relays Provide Directional-Type Protection on Large Industrial Plant Power System," by M. M. Gilbert and R. N. Bell, *AIEE Conference Paper CP55–143.*

3. "Current Transformers and Relays for High-Speed Differential Protection with Particular Reference to Off-Set Transient Currents," by E. C. Wentz and W. K. Sonnemann, *AIEE Trans., 59* (1940), pp. 481–488. Discussions, pp. 1144–1148.

4. "Transient Characteristics of Current Transformers during Faults," by C. Concordia, C. N. Weygandt, and H. S. Shott, *AIEE Trans., 61* (1942), pp. 280–285. Discussions, p. 469.

5. "Considerations in Applying Ratio Differential Relays for Bus Protection," by R. M. Smith, W. K. Sonnemann, and G. B. Dodds, *AIEE Trans., 58* (1939), pp. 243–249. Discussions, pp. 249–252.

6. "Transient Characteristics of Current Transformers during Faults—Part II," by F. S. Rothe and C. Concordia, *AIEE Trans., 66* (1947), pp. 731–734.

7. "Bus Protection," by AIEE Committee, *AIEE Trans., 58* (1939), pp. 206–211. Discussions, pp. 211–212.

8. "Bus Protection Gives Service Reliability," by P. Sporn, *Elec. World, 106,* No. 19 (May 9, 1936), pp. 55, 56, 59 (1357, 1358, 1401).

9. "Relay Protects Bus with Feeder Reactors," by W. K. Sonnemann, *Elec. World, 115* (January 25, 1941), p. 342.

"Distance-Relay Protection for Sectionalized Bus," by C. A. Molsberry, *Elec. World, 124* (August 18, 1945), pp. 96–98.

10. "Reactance Relays Discriminate between Load-Transfer Currents and Fault Currents on 2,300-Volt Station Service Generator Bus," by G. B. Dodds and W. E. Marter, *AIEE Trans., 71,* Part III (1952), pp. 1124–1128. Discussions, p. 1128.

11. "Linear Couplers for Bus Protection," by E. L. Harder, E. H. Klemmer, W. K. Sonnemann, and E. C. Wentz, *AIEE Trans., 61* (1942), pp. 241–248. Discussions, p. 463.

"Linear Couplers, Field Tests and Experience at York and Middletown, Pennsylvania," by E. L. Harder, E. H. Klemmer, and R. E. Neidig, *AIEE Trans., 65* (1946), pp. 107–113.

12. "A New Single-Phase-to-Ground Fault-Detecting Relay," by W. K. Sonnemann, *AIEE Trans., 61* (1942), pp. 677–680. Discussions, pp. 995–996.

13. "Instantaneous Bus-Differential Protection Using Bushing Current Transformers," by H. T. Seeley and Frank von Roeschlaub, *AIEE Trans., 67* (1948), pp. 1709–1718. Discussions, p. 1719.

14. "Application Guide for Grounding of Instrument Transformer Secondary Circuits and Cases," *Publ. 52,* American Institute of Electrical Engineers, 33 West 39th St., New York 18, N. Y.

13 LINE PROTECTION WITH OVERCURRENT RELAYS

Lines are protected by overcurrent-, distance-, or pilot-relaying equipment, depending on the requirements. Overcurrent relaying is the simplest and cheapest, the most difficult to apply, and the quickest to need readjustment or even replacement as a system changes. It is generally used for phase- and ground-fault protection on station-service and distribution circuits in electric utility and in industrial systems, and on some subtransmission lines where the cost of distance relaying cannot be justified. It is used for primary ground-fault protection on most transmission lines where distance relays are used for phase faults, and for ground back-up protection on most lines having pilot relaying for primary protection. However, distance relaying for ground-fault primary and back-up protection of transmission lines is slowly replacing overcurrent relaying. Overcurrent relaying is used extensively also at power-transformer locations for external-fault back-up protection, but here, also, there is a trend toward replacing overcurrent with distance relays.

It is generally the practice to use a set of two or three overcurrent relays for protection against interphase faults and a separate overcurrent relay for single-phase-to-ground faults. Separate ground relays are generally favored because they can be adjusted to provide faster and more sensitive protection for single-phase-to-ground faults than the phase relays can provide. However, the phase relays alone are sometimes relied on for protection against all types of faults. On the other hand, the phase relays must sometimes be made to be inoperative on the zero-phase-sequence component of ground-fault current. These subjects will be treated in more detail later.

Overcurrent relaying is well suited to distribution-system protection for several reasons. Not only is overcurrent relaying basically simple and inexpensive but also these advantages are realized in the greatest degree in many distribution circuits. Very often, the relays do not need to be directional, and then no a-c voltage source is required. Also, two phase relays and one ground relay are permissible. And finally, tripping reactor or capacitor tripping (described elsewhere) may be used.

In electric-utility distribution-circuit protection, the greatest advantage can be taken of the inverse-time characteristic because the fault-current magnitude depends mostly on the fault location and is practically unaffected by changes in generation or in the high-voltage transmission system. Not only may relays with extremely inverse curves be used for this reason but also such relays provide the best selectivity with fuses and reclosers. However, if ground-fault-current magnitude is severely limited by neutral-grounding impedance, as is often true in industrial circuits, there is little or no advantage to be gained from the inverse characteristic of a ground relay.

Inverse-time relaying is supplemented by instantaneous relaying wherever possible. Speed in clearing faults minimizes damage and thereby makes automatic reclosing more likely to be successful.

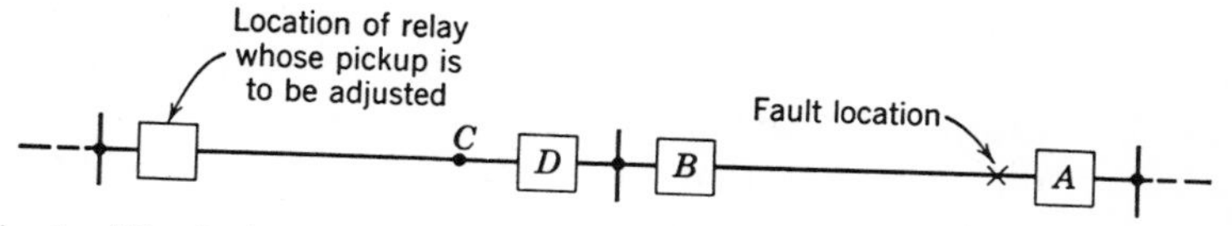

Fig. 1. The fault location for adjusting the pickup for back-up protection.

HOW TO SET INVERSE-TIME-OVERCURRENT RELAYS FOR COORDINATION

The first step is to choose the pickup of the relay so that it will (1) operate for all short circuits in its own line, and (2) provide back-up protection for short circuits in immediately adjoining system elements under certain circumstances. For example, if the adjoining element is a line section, the relay is set to pick up at a current somewhat less than it receives for a short circuit at the far end of this adjoining line section under minimum generating—or other—conditions that would cause the least current flow at the relay location. This is illustrated in Fig. 1.

For a phase relay, a phase-to-phase fault would be assumed since it causes less current to flow than does any other fault not involving ground. However, a phase relay must not be so sensitive that it will pick up under emergency conditions of maximum load over the line from which it receives its current. For a ground relay, a single-phase-to-ground fault would be assumed; load current is not a factor in the choice of a ground-relay's pickup except in a distribution system where there is ground current normally because of unbalanced loading. If there are two or more adjoining line sections, the fault should be assumed at the end of the section that causes the least current to flow at the location of the relay being adjusted.

Because of the effect of parallel circuits not shown, less current will flow at the relay location of Fig. 1 if breaker *A* is closed than if *A* is open. If satisfactory adjustment can be obtained with *A* closed, so much the better. However, the relay under consideration is being adjusted to operate if breaker *B* fails to open; it is not generally assumed that breaker *A* will also fail to open. There may be some occasions when one will wish to assume simultaneous equipment failures at different locations, but it is not the usual practice. Hence, it is permissible to assume that breaker *A* has opened, which is usually very helpful and may even be necessary.

Under certain circumstances, the relay will get less current for a phase-to-phase fault at *C* with breaker *D* closed and under minimum generating conditions than for the fault location shown in Fig. 1 with *A* open; the relay must be able to operate for this condition also.

In order to use the most inverse portion of the relay's time curves, the pickup in terms of primary current should be as high as possible and still be low enough so that the relay will operate reliably under the minimum fault-current condition. Under such conditions, the relay should operate at no less than about 1.5 times its pickup, but as near to that value as conveniently possible. The reason for this rule is that, closer to the pickup current, the torque is so low that a small increase in friction might prevent operation or it might increase the operating time too much. It may be that the CT ratio and the relay's range of adjustment do not permit adjusting for so low a multiple of pickup; in that event the only recourse, aside from changing the CT or the relay, is to use the highest possible pickup for which the relay can be adjusted.

To assure selectivity under all circumstances, the pickup of a given relay should be somewhat higher than that of other relays nearer to the fault and with which the given relay must be selective.

Because the impedance of generators increases from subtransient to synchronous as time progresses from the instant that a short circuit occurs, the question naturally arises as to which value of impedance to use in calculating the magnitude of short-circuit current for protective-relaying purposes where overcurrent relaying is involved. The answer to this question depends on the operating speed of the relay under consideration, on the amount by which generator impedance affects the magnitude of the short-circuit current, and on the particular relay setting involved. Usually, the impedance that limits the magnitude of the short-circuit current contains so much transformer and line impedance that the effect of changing generator impedance is negligible; one can always determine this effect in any given application. For relays near a large generating station that furnishes most of the short-circuit current, synchronous impedance would be best for determining the pickup of a relay for back-up purposes—particularly if the operating time of the relay was to be as long as a second or two. On the other hand, the pickup of a high-speed relay near such a generating station would be determined by the use of transient—or possibly even subtransient—impedance. Ordinarily, however, transient impedance will be found most suitable for all purposes—particularly for subtransmission or distribution circuits where overcurrent relays are generally used; there is enough transformer and line impedance between such circuits and the generating stations so that the effect of changing generator impedance is negligible. In fact, for distribution circuits, it is frequently sufficiently accurate to assume a source impedance that limits the current to the source-breaker interrupting capacity on the high-voltage side of a power transformer feeding such a circuit; in other words, only slightly more total impedance than that of the transformer itself and of the circuit to be protected is assumed.

Whether to take into account the effect of arc and ground resistance depends on what one is interested in. Arc resistance may or may not exist. Occasionally, a metallic fault with no arcing may occur. When one is concerned about the maximum possible value of fault current, he should assume no arc resistance unless he is willing to chance the possibility of faulty relay operation should a fault occur without resistance. Thus, as will be seen later, for choosing the pickup of instantaneous overcurrent relays or the time-delay adjustment for inverse-time relays, it is more conservative to assume no arc resistance.

When one is choosing the pickup of inverse-time relays, the effect of arc resistance should be considered This is done to a limited extent when one arbitrarily chooses a pickup current lower than the current at which pickup must surely occur, as recommended in the foregoing material; however, this pickup may not be low enough. In view of the fact that an arc may lengthen considerably in the wind, and thereby greatly increase its resistance, it is a question how far to go in this respect. At least, one should take into account the resistance of the arc, when it first occurs, whose length is the shortest distance between conductors or to ground. Beyond this, what one should do depends on the operating time of the relay under consideration and the wind velocity. The characteristics of an arc are considered later.

Ground resistance concerns us only for ground faults. It is in addition to arc resistance. This subject is considered later also, along with arc resistance.

For ground relays on lines between which there is mutual induction, this mutual induction should be taken into account in calculating the magnitude of current for single-phase-to-ground faults. In some studies, this will show that selectivity cannot be obtained when it would otherwise appear to be possible.

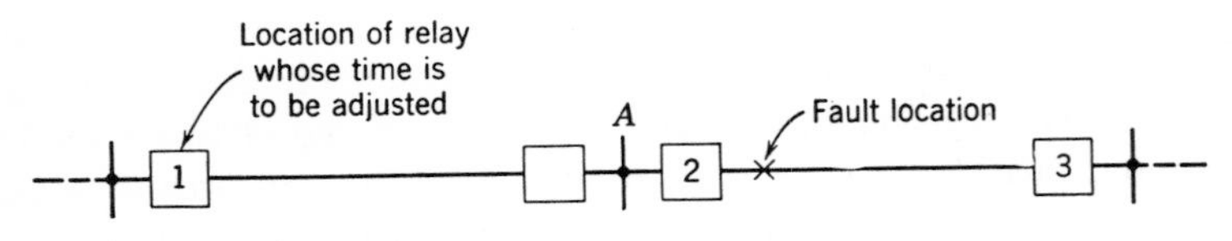

Fig. 2. The fault location for adjusting for selectivity.

The second step in the adjustment of inverse-time-overcurrent relays is to adjust the time delay for obtaining selectivity with the relays of the immediately adjoining system elements. This adjustment should be made for the condition for which the maximum current would flow at the relay location. This condition would exist for a short circuit just beyond the breaker in an adjoining system element. This is illustrated in Fig. 2. Under certain circumstances, more current will flow at the relay location if breaker 3 is open. For adjusting a phase relay, a three-phase fault would be assumed; and for a ground relay a single-phase-to-ground fault would be assumed. In either event, one would use the fault current for maximum generating conditions and for any likely switching condition that would make the current at 1 most nearly equal to the current at 2. For example, when there are other sources of short-circuit current connected to bus *A*, if it is considered a practical operating condition to assume them to be disconnected, they should be so assumed.

The adjustment for selectivity is made under maximum fault-current conditions because, if selectivity is obtained under such conditions, it is certain to be obtained for lower currents. This will be seen by examining the time-current curves of any inverse-time overcurrent relay, such as Fig. 3 of Chapter 3, and observing that the time spacing between any two curves increases as the multiple of pickup decreases. Hence, if there is sufficient time spread at any given multiple of pickup, the spread will be more than sufficient at a lower multiple. The foregoing assumes that relays having the same time-current characteristics are involved. Relays with different characteristics are to be avoided.

This brings us to the question of how much difference there must be between the operating times of two relays in order that selectivity will be assured. Let us examine the elements involved in the answer to this question by using the example of Fig. 2. For the fault as shown in Fig. 2, the relay located at breaker 2 must close its contacts, and breaker 2 must trip and interrupt the flow of short-circuit current before the relay at breaker 1 can close its contacts. Furthermore, since the relay at breaker 1 may "overtravel" a bit after the flow of short-circuit current ceases, provision should also be made for this overtravel. We can express the required operating time of the relay at 1 in terms of the operating time of the relay at 2 by the following formula:

$$T_1 = T_2 + B_2 + O_1 + F$$

where T_1 = operating time of relay at 1.
T_2 = operating time of relay at 2.
B_2 = short-circuit interrupting time of breaker at 2.
O_1 = overtravel time of relay at 1.
F = factor-of-safety time.

The overtravel time will be different for different overcurrent relays and for different multiples of pickup, but, for the inverse-time types generally used, a value of about 0.1 second may be used. The factor-of-safety time is at the discretion of the user; this time should provide for normally expected variations in all the other times. A value of 0.2 to 0.3 second for overtravel plus the factor of safety will generally be sufficient; lower values may be used where accurate data are available.

We are now in a position to examine Fig. 3 wherein time-versus-distance curves are shown for relays that have been adjusted as described in the foregoing. The time S, called the "selective-time interval," is the sum of the breaker, overtravel, and factor-of-safety times. A vertical line drawn through any assumed fault location will intersect the operating-time curves of various relays and will thereby show the time at which each relay would operate if the short-circuit current continued to flow for that length of time.

The order in which the relays of Fig. 3 are adjusted is to start with the relay at breaker 1 and work back to the relay at breaker 4. This will become evident when one considers that the selectivity adjustment of each relay depends on the adjustment of the relay with which it must select. For cases like that shown in Fig. 3, we can generalize and say that one starts adjustment at the relay most distant electrically from the source of generation, and then works back toward the generating source.

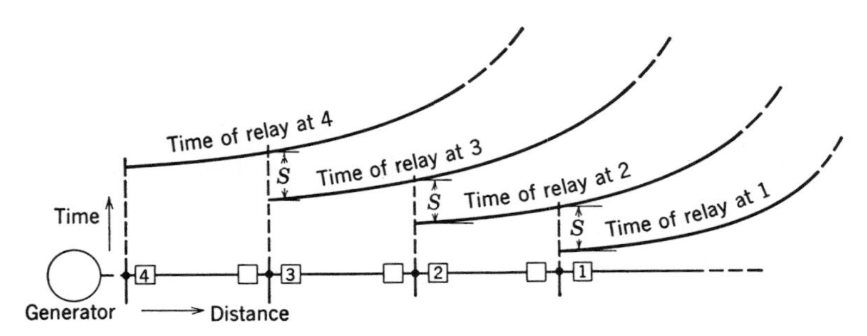

Fig. 3. Operating time of overcurrent relays with inverse-time characteristics.

ARC AND GROUND RESISTANCE

Although there is much difference of opinion on the interpretation of test data, the maximum value of rms volts per foot of arc length given by any of the data[1,2,3] for all arc currents greater than 1000 rms amperes is about 550. For currents below 1000 amperes, the formula $V = 8750/I^{0.4}$ gives the maximum reported value of rms volts per foot (V) for any rms value of arc current (I); from this formula, values considerably higher than 550 will be obtained at low currents. Actually, this formula gives a fairly good average of all the available data for any value of arc current, as will be seen by plotting superimposed the data of Fig. 1 of Reference 1, Fig. 5 of Reference 2, and the foregoing formula which is obtained from the formula given in Reference 3. However, because this average value is only about half of the maximum reported in Reference 1 for currents larger than 1000 amperes, it is more conservative not to use this average for such current values when one is interested in the maximum arc resistance.

To take into account the lengthening of the arc by wind, the approximate formula $L = 3vt + L_0$ may be used, where:

L = length of arc, in feet.
v = wind velocity, in miles per hour.
t = time, in seconds after the arc was first struck.
L_0 = initial arc length, i.e., the shortest distance between conductors or across insulator, in feet.

It will be evident that there are limits to which this formula may be applied because there are limits to the amount an arc may stretch without either restriking or being extinguished. Reference 2 gives several sets of data showing how the arc voltage increased during field tests.

Ground resistance is resistance in the earth. This resistance is in addition to that of an arc. When overhead ground wires are not used, or when they are insulated from the towers or poles, the ground resistance is the tower- or pole-footing resistance at the location where the ground fault has occurred plus the resistance of the earth back to the source. Electric utilities have measured data on such footing resistance. When overhead ground wires are connected to steel towers or to grounding connections on wood poles, the effect is somewhat as though all footing resistances were connected in parallel, which makes the resulting footing resistance negligible. Published zero-phase-sequence-impedance data do not include the effect of tower-footing resistance.

Occasionally, a conductor breaks and falls to the ground. The ground-contact resistance of such a fault may be much higher than tower-footing resistance where relatively low resistance is usually obtained with ground rods or counterpoises. The contact resistance depends on the geology of a given location, whether the ground is wet or dry, and, if dry, how high the voltage is; it takes a certain amount of voltage to break down the surface insulation.

Ground resistance can range over such wide limits that the only practical thing to do is to use measured values for any given locality. An example of extremely high ground resistance and the method of relaying is given in Reference 4.

In one system,[5] advantage was taken of ground resistance to decrease the rms magnitude of ground-fault current and to greatly shorten the time constant of its d-c component in order to reduce the interrupting stress on circuit breakers. However, such practice should be avoided in general.

EFFECT OF LOOP CIRCUITS ON OVERCURRENT-RELAY ADJUSTMENTS

Figure 3 best serves the purpose of illustrating how selectivity is provided with inverse-time-overcurrent relays. But, lest it mislead one by oversimplifying the problem, it is well to realize that, except for some parts of distribution systems, Fig. 3 does not truly represent most actual systems where loops are the rule and radial circuits are the exception. The principles involved and the general results obtained in the application and adjustment of overcurrent relays are correctly shown by reference to Fig. 3, but the difficulties in arriving at suitable adjustment in an actual system are minimized. This consideration is important because it is often the deciding factor that leads one to choose distance or pilot relaying in preference to overcurrent relaying.

When we studied the method of setting inverse-time-overcurrent relays, we saw that the relay most distant electrically from the generating source was adjusted first, and that one then worked back toward the generating source. The same procedure would be followed in the simple loop system illustrated in Fig. 4. The order in which the relays "looking" one way around the loop would be adjusted is

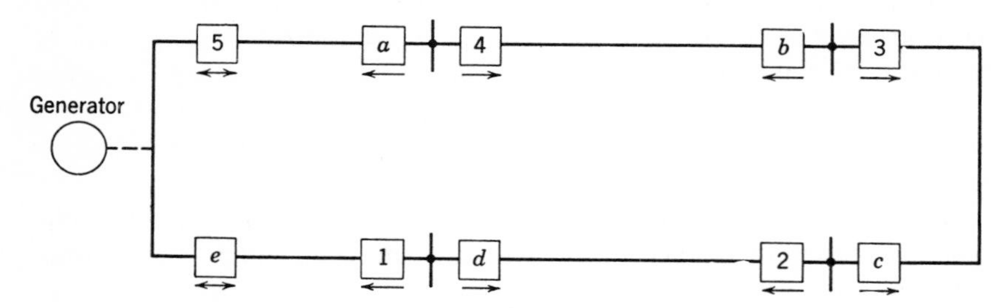

Fig. 4. The order for adjusting relays in a simple loop system.

1–2–3–4–5, and looking the other way, *a–b–c–d–e*. Directional overcurrent relays would usually be employed as indicated by the single-ended arrows that point in the direction of fault-current flow for which the relays should trip. Only at locations *e* and 5 can fault current flow only in the same direction as that for which tripping is desired, and the relays there may be non-directional as indicated by the double-headed arrows. The first relay to be adjusted in each of the two groups can be made as sensitive and as fast as possible because the current flow at the relay location will decrease to zero as faults are moved from the relay location to the generator bus, and hence there is no problem of selectivity for those relays. The phase relay at 1, for example, must receive at least 1.5 times its pickup current for a phase-to-phase fault at the far end of its line with the breaker at *e* open, and with minimum generation. Of course, no phase overcurrent relay should be so sensitive that it will pick up on maximum load.

Occasionally, the short-circuit current that can flow in the non-tripping direction is so small in comparison with the current that can flow in the tripping direction that certain relays need not be directional, the system itself having a directional characteristic. But, if a relay can pick up on the magnitude of current that flows in the non-tripping direction, it is wise to make the relay directional or else the problem of obtaining selectivity under all possible conditions is needlessly complicated; and a future change in the system or its operation may demand directional relays anyway.

The first complication in adjusting overcurrent relays in loop circuits arises when generators are located at the various stations around the loop. The problem then is where to start. And, finally, when circuits of one loop form a part of other loops, the problem is most difficult. The trial-and-error method is the only way to proceed with such circuits. In fact, some such systems cannot be relayed selectively by inverse-time-overcurrent relays without operating the system with certain breakers normally open, and closing them only in emergencies. Supplementary instantaneous overcurrent relaying will sometimes give relief in such cases, as will be described later. Of course, such systems are not created in their entirety; they develop slowly, and the relaying problems arise each time that a change is made.

EFFECT OF SYSTEM ON CHOICE OF INVERSENESS OF RELAY CHARACTERISTIC

The less change there is in the magnitude of short-circuit current with changes in connected generating capacity, etc., for a fault at a given location, the more benefit can be obtained from greater inverseness. This is particularly true in distribution circuits where short-circuit-current magnitude is practically independent of normal changes in generating capacity. In such circuits one can use the very inverse—or extremely inverse—overcurrent relays to advantage.

In systems, such as many industrial systems, where the magnitude of the ground-fault current is severely limited by neutral-grounding impedance, little or no advantage can be taken of the inverseness of a ground-relay's characteristic; the relays might just as well have definite-time characteristics. This would also be true even where no neutral impedance was used if the sum of the arc and ground resistance was high enough, no matter where the fault should happen to occur.

Later, under the heading "Restoration of Service to Distribution Feeders after Prolonged Outages," the need for the extremely inverse characteristic is described. This characteristic is also useful in areas of a distribution system adjacent to where fuses and reclosers begin to replace relays and circuit breakers, because the extremely inverse characteristic will coordinate with fuses and reclosers.

THE USE OF INSTANTANEOUS OVERCURRENT RELAYS

Instantaneous overcurrent relays are applicable if the fault-current magnitude under maximum generating conditions about triples as a fault is moved toward the relay location from the far end of the line. This will become evident by referring to Fig. 5 where the symmetrical fault-current magnitude is plotted as a function of fault location along a line for three-phase faults and for phase-to-phase faults, if we assume that the fault-current magnitude triples as the fault is moved

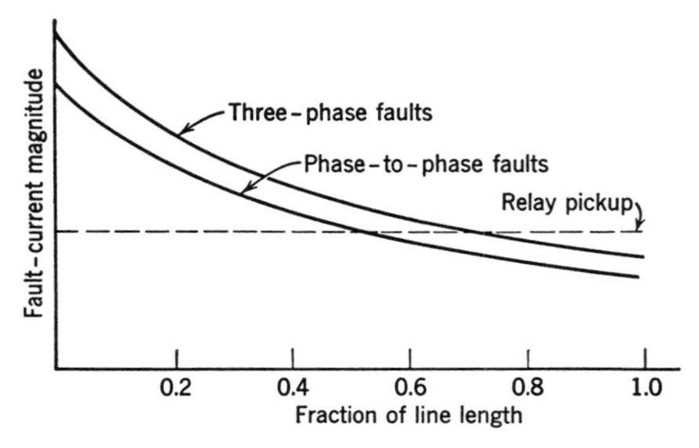

Fig. 5. Performance of instantaneous overcurrent relays.

from the far end of the line to the relay location. The pickup of the instantaneous relay is shown to be 25% higher than the magnitude of the current for a three-phase fault at the end of the line; the relay should not pick up at much less current or else it might overreach the end of the line when the fault-current wave is fully offset. In a distribution circuit, the relay could be adjusted to pick up at somewhat lower current because the tendency to overreach is less. For the condition of Fig. 5, it will be noted that the relay will operate for three-phase faults out to 70% of the line length and for phase-to-phase faults out to 54%. If the ground-fault current is not limited by neutral impedance, or if the ground resistance is not too high, a similar set of characteristics for ground faults would probably show somewhat more than 70% of the line protected; this is because the ground-fault current usually increases at a higher rate as the fault is moved toward the relay. The technique of Fig. 5 may be used for any other conditions to determine the effectiveness of instantaneous overcurrent relaying, including the effect on fault-current magnitude because of changes in generation, etc.

The shaded area of Fig. 6 shows how much instantaneous overcurrent relaying reduces the over-all relaying time for most faults. Even if such reduction is obtained only under maximum generating conditions, and if instantaneous relays were not even operable under minimum generating conditions, the use of supplementary instantaneous relays is considered to be worth while because they are relatively so inexpensive.

With instantaneous overcurrent relaying at both ends of a line, simultaneous tripping of both ends is obtained under maximum generating conditions for faults in the middle portion of the line. For faults near the ends of the line, sequential instantaneous tripping will often occur, i.e., the end nearest the fault will trip instantly and then the magnitude of the current flowing to the fault from the other end will usually increase sufficiently to pick up the instantaneous overcurrent relays there.

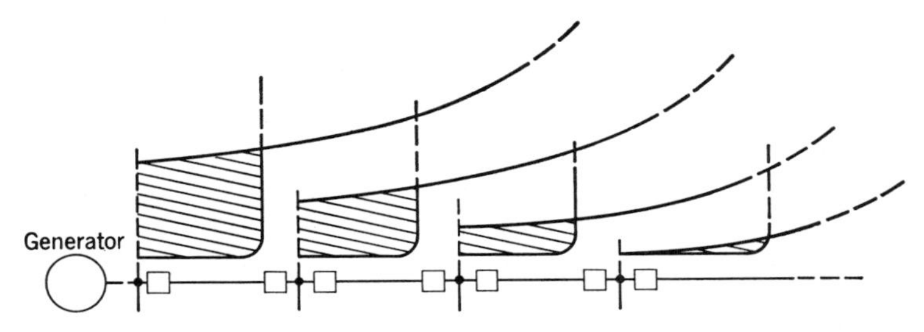

Fig. 6. Reduction in tripping time by the use of instantaneous overcurrent relays.

When the magnitude of fault current depends only on the fault location, instantaneous overcurrent relaying is like distance relaying[6] except that the overcurrent relays cannot generally be as sensitive as distance relays.

AN INCIDENTAL ADVANTAGE OF INSTANTANEOUS OVERCURRENT RELAYING

A useful advantage that can sometimes be taken of instantaneous overcurrent relaying is illustrated in Fig. 7. Without instantaneous overcurrent relaying at breaker 2, the inverse-time-overcurrent relays at 1 would have the dashed time curve so as to obtain the selective time interval *ab* with respect to the inverse-time relays at breaker 2. With instantaneous overcurrent relays at 2, the inverse-time relays at 1 need only be selective with the inverse-time relays at 2 for faults at and beyond the point where the instantaneous relays stop operating, as shown by *cd*. This permits speeding up the relays at 1 from the dashed to the solid curve, which is sometimes very useful, or even necessary when complicated loop circuits are involved. Of course, under minimum generating conditions when the instantaneous relays may not operate, one must be sure that selectivity between

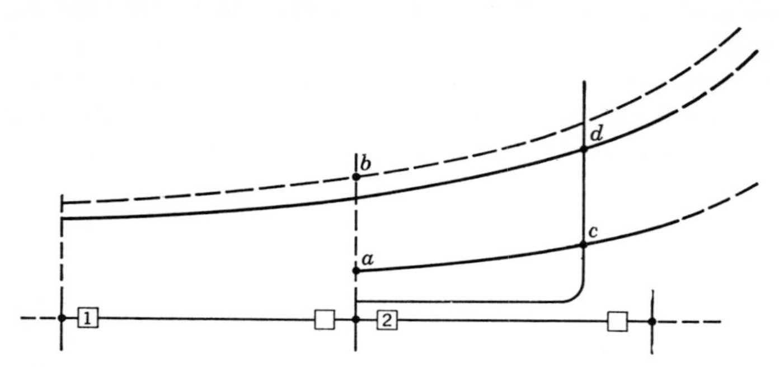

Fig. 7. Illustrating an additional advantage of instantaneous overcurrent relays.

the inverse-time relays is obtained. The fact that selectivity is not obtained between the inverse-time relays for faults just beyond the relay at 2 for maximum generating conditions is unimportant so long as selectivity is obtained down to the pickup current of the instantaneous relays.

OVERREACH OF INSTANTANEOUS OVERCURRENT RELAYS

"Overreach" is the tendency of a relay to pick up for faults farther away than one would expect if he neglected the effect of offset in the fault-current wave. Magnetic-attraction relays are more affected by offset waves than induction relays, and some induction relays are more affected than others. Certain induction relays can be designed to be unaffected by offset waves.

"Percent overreach" is a term that describes the degree to which the overreach tendency exists, and it has been defined as follows:

$$\text{Percent overreach} = 100\left(\frac{A - B}{A}\right)$$

where A = the relay's pickup current, in steady-state rms amperes.
B = the steady-state rms amperes which, when fully offset initially, will just pick up the relay.

In relays that have a tendency to overreach, the percent overreach increases as the ratio of reactance to resistance of the fault-current-limiting impedance increases, or, in other words, as the time constant of the d-c component of the fault current increases. The slower the decay of the d-c component, the sooner will the integrated force acting on the relay cause it to pick up; and the sooner the relay tends to pick up, the lower may be the rms component of the fault current and still cause pickup. If the fault-current wave were continually fully offset, the rms component for pickup would be the smallest. It may be evident from the foregoing that, other things being equal, the faster a relay is, the greater will its percent overreach be. The maximum percent overreach would be 50% for a relay that was fast

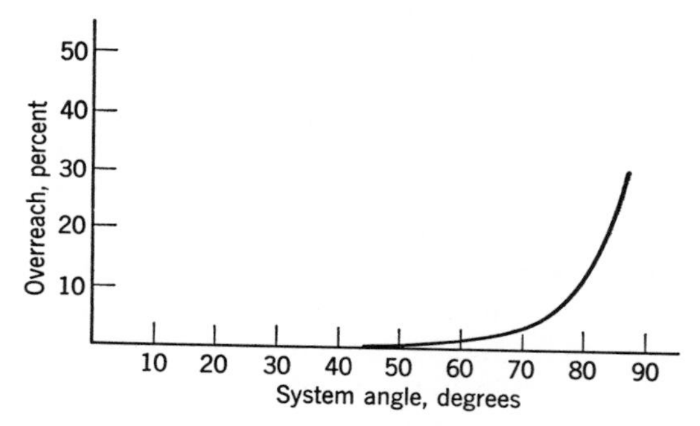

Fig. 8. Overreach characteristic of a certain instantaneous overcurrent relay.

enough to respond to the instantaneous magnitude of current. Since the rms value of a fully offset sine wave is $\sqrt{3}$ times that if the wave were symmetrical, the maximum value of percent overreach is 42% for relays that are not fast enough to respond to the instantaneous magnitude of current. Figure 8 shows how the percent overreach of a certain relay increases as the system angle ($\tan^{-1} X/R$) increases. Because X/R in a distribution circuit is only about 1 or 2, the tendency to overreach is negligible.

To allow accurately for overreach when choosing the pickup of an instantaneous overcurrent relay, one must have a percent-overreach curve for the relay, like that of Fig. 8. Then, solving the foregoing equation for A, we obtain:

$$A = \frac{100B}{(100 - \text{Percent overreach})}$$

Thus, with an overcurrent relay that has a 15% overreach for a fault whose steady-state component of current is 10 amperes, if the relay is not to operate for that fault, A must exceed:

$$\frac{100 \times 10}{100 - 15} \text{ amperes} = 11.8 \text{ amperes}$$

When percent-overreach data are not available, it will usually be satisfactory to make the pickup about 25% higher than the maximum value of symmetrical fault current for which the relay must not operate. This will provide for overreach and also for some error in the data on which the setting is based.

The tendency of an instantaneous overcurrent relay to overreach on offset waves can be minimized by a so-called "transient shunt."[7] This device is described in Chapter 14 in connection with its use with distance relays for the same purpose.

THE DIRECTIONAL FEATURE

Overcurrent relaying is made directional to simplify the problem of obtaining selectivity when about the same magnitude of fault current can flow in either direction at the relay location. It would be impossible to obtain selectivity under such circumstances if overcurrent relays could trip their breakers for either direction of current flow. The directional feature is not needed for a radial circuit with a generating source at only one end. Nor is it required where short-circuit current can flow in either direction if the magnitude of current that can flow in the tripping direction is several times that in the other direction; here, the system has a sufficiently directional characteristic. However, it is best to install directional relays, even if the directional feature is not presently needed, because system changes are likely to make directional relays necessary.

All directional-overcurrent relays should have the directional-control feature, as described in Chapter 3, whereby the overcurrent unit cannot begin to operate until the directional unit operates for current flow in the direction for which the overcurrent unit should operate. Here, again, this feature is not always required, but need for it may develop in the near future.

Occasionally, voltage-restrained directional units are desirable for use with phase overcurrent relays. This need arises when the magnitude of fault current in the direction for which an overcurrent relay must trip its breaker can be about the same as—or even somewhat less than—the maximum load current that might flow in the same direction. Voltage-restrained directional units also minimize the likelihood of undesired tripping when severe power swings occur. Such directional units should also provide directional control.

The terms "directional-overcurrent" and "directional-ground," as applied to ground relays with directional characteristics, are used by some people to denote two different types of relays. The term "directional-overcurrent" denotes a relay with separate directional and overcurrent units, and the term "directional-ground" denotes a directional unit with adjustable pickup and time-delay characteristics that combines the directional and overcurrent functions. The directional-overcurrent type is generally preferred because, although it imposes somewhat more burden on its CT's and also takes up somewhat more panel space, it is much easier to adjust. This is because only the line-current magnitude affects its operation both as to sensitivity and time delay; its directional unit is so sensitive and fast that its effect may be ignored. The sensitivity and speed of the directional-ground type is a function of the product of the line current, the polarizing current or voltage, and the phase angle between them. The two types are not simply interchangeable, and it is best to standardize on one or the other and not to mix them in a system, or else selectivity may be jeopardized.

USE OF TWO VERSUS THREE RELAYS FOR PHASE-FAULT PROTECTION

The consideration of two versus three relays for phase-fault protection arises because of a desire to save the expense of one CT and one relay, or at least of one relay, in applications where only the bare minimum expense can be tolerated for the protection of a certain line. The considerations involved in such practices are as follows.

Non-Directional Relaying. Non-directional-overcurrent phase-fault relaying can be provided by two relays energized from CT's in two of the three phases. However, protection will not necessarily be provided if the CT's in all circuits are not located in the same phases, as illustrated in Fig. 9. The system shown in Fig. 9 is assumed to be ungrounded. Simultaneous ground faults on different phases of two different circuits will constitute a phase-to-phase fault on the system, and yet neither overcurrent relay will operate.

Whether the system neutral is grounded or not, complete protection against phase and ground faults, even in the situation of Fig. 9, is provided if three CT's are used with two phase relays and one ground relay as illustrated in Fig. 10.

If a wye-delta or a delta-wye power-transformer bank lies between the relays and a phase-to-phase fault, the magnitude of the current in one of the phases at the relay location will be twice as great as in either of the other two phases. If only two relays are used, neither relay will get this larger current for a fault between one pair of the three possible pairs of phases that may be faulted on the other side of the bank. This fact should be taken into account in choosing the pickup and time settings.

If the fault-current magnitude for a phase-to-phase fault is of the same order as the load current, the effect of load current adding to

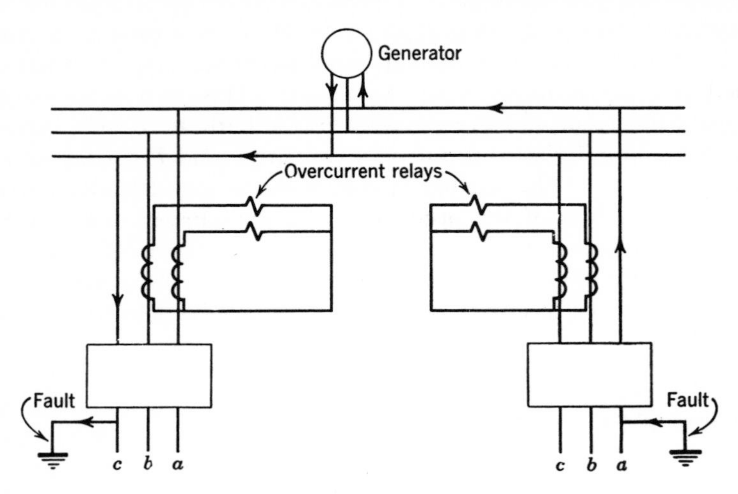

Fig. 9. A case of lack of protection with two overcurrent relays.

fault current in one phase and subtracting from it in another phase should be considered. This affects the pickup and time settings in a manner similar to that of an intervening power-transformer bank.

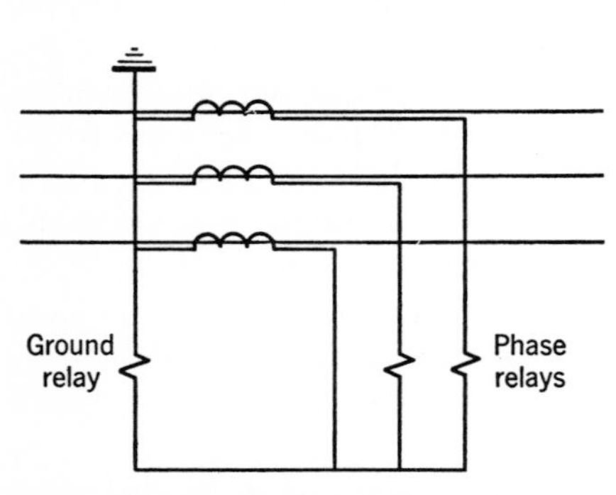

Fig. 10. Complete protection with two phase relays and one ground relay.

Three CT's and three phase relays are used wherever economically justifiable to avoid the foregoing difficulties because at least one relay will always operate for all interphase faults; and, except for the special conditions just described, two relays will operate, thereby giving double assurance of protection for much less than double the cost.

Directional Relaying. Directional-overcurrent phase-fault relaying is subject to the considerations described for non-directional-overcurrent relaying in so far as overcurrent units are concerned. In addition, there are the following considerations.

In a non-grounded system, two single-phase relays may generally be used if one is sure that the relays of all circuits are energized by currents from the same phases. Otherwise, grounds could occur on different phases of two different circuits, as in Fig. 9, thereby imposing a phase-to-phase fault on the system, and no protection would be provided. Directional-overcurrent relays for ground-fault protection are not usable on non-grounded systems, and, therefore, they could not alleviate this possible difficulty in the same way that non-directional ground-overcurrent relays do.

If directional phase relaying is to be used in two phases of a grounded-neutral system, ground relays must be provided for protection against ground faults. Then, the only question is if one or the other of the two phase relays will always operate for a phase-to-phase fault in the tripping direction.

If the magnitude of fault current for phase-to-phase faults is not several times the load-current magnitude, three single-phase directional-overcurrent relays should be used to assure tripping when desired. However, this problem involves more than just the number of relays required; the relays may operate to trip undesirably as well as to fail to trip when desired. Therefore, not only are three relays necessary but also voltage restraint on the directional units to keep them from operating undesirably. If three relays were not required for other reasons, they would be required as soon as voltage restraint is used. (This is why it is never the practice to use only two distance relays.)

With only two directional-overcurrent relays, the quadrature connection should be used. This is the best assurance that one of the two relays will always operate under the borderline conditions existing when faults occur close to the relay location.

SINGLE-PHASE VERSUS POLYPHASE DIRECTIONAL-OVERCURRENT RELAYS

Single-phase directional-overcurrent relays are generally preferred for protection against interphase faults. The main reason for this is that the very desirable feature of "directional control" is more simply and reliably obtained with single-phase directional-overcurrent relays than with a polyphase directional relay in combination with single-phase overcurrent relays. The directional-unit contact of a single-phase directional-overcurrent relay controls the operation of the overcurrent unit directly; an intermediate auxiliary relay is required when a polyphase directional unit is used.

Single-phase directional-overcurrent relays must be used when a-c tripping is involved, because separate contacts must be available for connection in each of the three CT secondary circuits. An auxiliary relay cannot be used with a polyphase directional relay to get the necessary contact separation, because there is no suitable voltage source to operate the auxiliary relay; the lack of such a voltage source is why a-c tripping is used.

A set of three single-phase directional-overcurrent relays can often be used to provide protection against single-phase-to-ground faults as well as against interphase faults. Polyphase directional relays may not be used for this purpose unless the minimum ground-fault current is more than 3 times the maximum load current.

Some users want to test the relays of each phase separately so that, if a fault occurs during testing, the relays of the other two phases can provide protection. This requires single-phase relays.

A minor advantage of single-phase relays is that they provide somewhat more flexibility in the layout of panels.

The advantage of a polyphase directional relay is that it is less subject to occasional misoperation than single-phase relays. For a certain fault condition, one of three single-phase relays may develop torque in the tripping direction when tripping would be undesirable; if the current in that one relay was high enough to operate the overcurrent unit, improper tripping would result. Since a polyphase directional relay operates on the net torque of its three elements, a reversed torque in one element can be overbalanced by the other two elements, and a correct net torque usually results. The subject of directional-relay misoperation is treated in the following section.

HOW TO PREVENT SINGLE-PHASE DIRECTIONAL-OVERCURRENT-RELAY MISOPERATION DURING GROUND FAULTS

Under certain circumstances, single-phase directional-overcurrent relays for phase-fault protection may cause undesired tripping for ground faults in the non-tripping direction. Such undesired tripping can be prevented if one only knows when special preventive measures are necessary.

The zero-phase-sequence components of ground-fault current produce the misoperating tendency. All these components are in phase, and, when wye-connected current transformers are used, these components always produce contact-closing torque in one of the three directional units no matter in which direction the current may be flowing. Generally, the other fault-current components are able to "swamp" the effect of the zero-phase-sequence components. But, when the fault current is largely composed of zero-phase-sequence component, misoperation is most likely.

Figure 11 shows the basic type of application where undesired tripping is most apt to occur. Let us assume that the directional units of the relays are intended to permit tripping only for faults to the left of the relay location, as indicated by the arrow. However, a ground fault to the right, as shown, will cause at least one directional unit to close its contact and permit tripping by its overcurrent unit. Whether the overcurrent unit will actually trip its breaker will depend on its pickup and time settings, and on whether it gets enough current to operate before the fault is removed from the system by some other relay that is supposed to operate for this fault. Failure to trip when tripping is desired is not a problem.

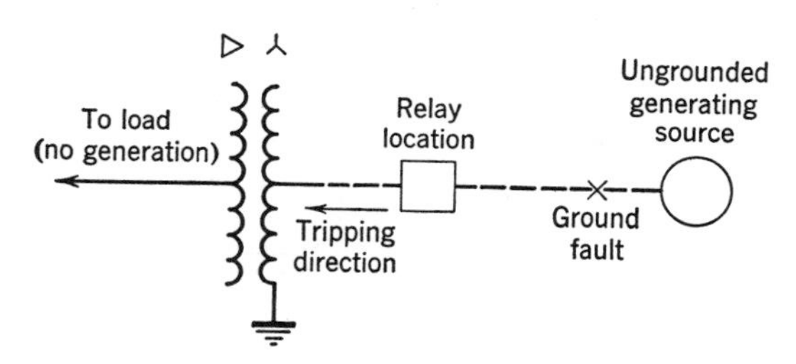

Fig. 11. A situation where single-phase directional-overcurrent relays may misoperate.

Misoperation can occur also for conditions *approaching* those of Fig. 11.[8] In other words, misoperation may still occur if there is a small source of generation at the load end of Fig. 11. In general, one should examine the possibility of misoperation whenever the zero-phase-sequence impedances from the ground-fault location back to the sources of generation on either side of the fault are not approximately in the same complex ratio as the corresponding positive-phase-sequence impedances. Thus, in addition to the situation of Fig. 11, if one side of a system is grounded through resistance and the other side through reactance, misoperation is possible.

The likelihood of misoperation is greatest when the phase relays are used also for ground-fault protection, and particularly when the relays have to be more sensitive because the ground-fault current is limited by neutral impedance.

To prevent misoperation for the situation shown in Fig. 11, the phase relays should be prevented from responding to the zero-phase-sequence component of current. This can be done with a zero-phase-sequence shunt using three auxiliary current transformers, as shown in Fig. 12. It is emphasized that the neutral of the phase relays should not be connected to the CT neutral or else part of the effectiveness of the shunt will be lost. Delta-connecting the secondaries of the main current transformers would also remove the zero-phase-sequence component, but it introduces other misoperating tendencies, and it does not provide the required source of energization for ground relays. The 90-degree or quadrature connection of the phase relays should be used.

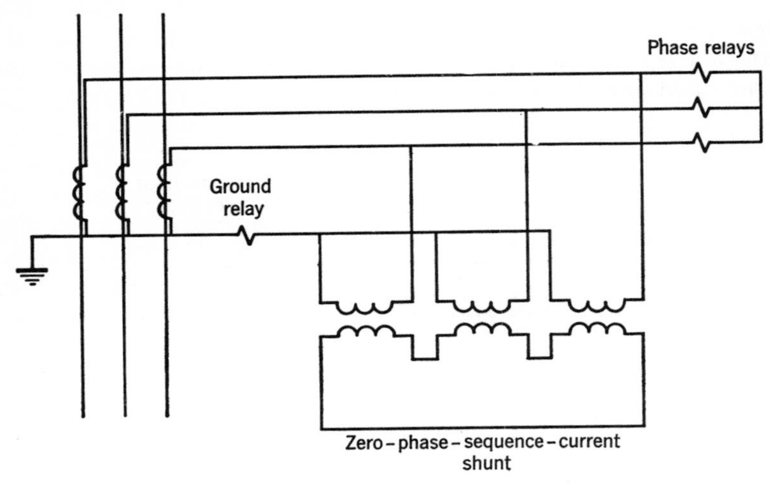

Fig. 12. Application of a zero-phase-sequence-current shunt.

The phase relays will still be able to respond to ground faults if the current magnitude is large enough, and if positive-phase-sequence components are present in the fault current. However, it is generally preferable to use separate directional-overcurrent ground relays for ground-fault protection because faster tripping can thereby be obtained. Where only zero-phase-sequence current can flow, as at the relay location of Fig. 11, separate ground relays must be used.

ADJUSTMENT OF GROUND VERSUS PHASE RELAYS

The satisfactory adjustment of ground overcurrent relays is generally easier to achieve than that of phase overcurrent relays in any system, including complicated loop systems. The principal reason for this is that the zero-phase-sequence impedance of lines (except single-phase cables) is approximately 2 to 5.5 times the positive-phase-sequence impedance, which provides two beneficial effects: (1) the magnitude of zero-phase-sequence current varies much more with fault location, and (2) the magnitude of the zero-phase-sequence current is not so much affected by changes in generating capacity.

These effects make it possible to take greater advantage of the inverse-time characteristic, and particularly of the very inverse characteristic; and also they aid the application of instantaneous overcurrent relays. Another simplification with ground relaying is that the pickup does not have to be higher than load current, because ground relays are not energized during normal load conditions, except in some distribution systems where ground current flows normally because of unbalanced phase-to-ground loading. (For other reasons to be discussed later there are limits as to how sensitive ground relays may be.) Finally, two-winding wye-delta or delta-wye power transformers are open circuits in the system so far as ground relays are concerned; in other words, except for the effect of CT errors, a ground-overcurrent relay cannot reach through such a transformer for any kind of fault on the other side. This minimizes the selectivity problem because it restricts the "reach" of the ground relays. The foregoing are the reasons why, when it is necessary to use distance relays for interphase faults, inverse-time and instantaneous overcurrent relaying are often satisfactory for ground faults.

An excellent coverage of the whole subject of ground-fault protection of transmission lines is in Reference 9.

EFFECT OF LIMITING THE MAGNITUDE OF GROUND-FAULT CURRENT

Current-limiting impedance in the grounded neutrals of generators or power transformers may be desirable from the standpoint of limiting the severity of a phase-to-ground short circuit. If carried too far, however, such practice tends to jeopardize the application of ground relaying where fast and selective operation is required.

Two problems must be considered. The first problem is that of obtaining sufficient sensitivity without danger of misoperation because of CT errors when large phase-fault currents flow. As explained later, an overcurrent relay in the neutral of wye-connected CT's will receive a current, even though a fault may not involve ground, if residual flux or d-c offset current in a CT of one phase causes it to have a different ratio error from a CT in another phase during a fault that involves large phase currents. The more sensitive the ground relay has to be in order to operate on the severely restricted phase-to-ground-fault currents, the more likely it is to operate on this CT error current.

The second problem when ground-fault current is limited by neutral impedance is to obtain prompt and selective tripping. The effect of limiting the ground-fault current by neutral impedance is to reduce the difference in magnitude of ground-fault current for faults at different locations. If there is little or no difference in the fault current for nearby or for distant faults, the inverse-time characteristic is of little use. The result is that selectivity must be obtained on the basis of time alone. Where several line sections are in series, the tripping time must be increased a fixed amount for each line section, the nearer a fault is to the source of fault current. As a consequence, faults adjacent to the source may not be cleared in less than several seconds, and during such a relatively long time the fault might involve other phases, thus causing a severe shock to the system.

It has been found from experience that a good rule of thumb is not to limit the ground-fault current of the generator or transformer to less than about its rated full-load current. Even this might be objectionable on some systems. The best procedure is to study the effect of limiting the current in each individual case.

Ground relays that are much more sensitive than phase relays may misoperate when simultaneous grounds exist at two different locations on different phases. Actually, this constitutes a phase-to-phase fault on the system, and the fault-current magnitude will be that for phase-to-phase faults. However, the ground relays of the circuits having these faults will operate as though a ground fault existed but at much higher speed than they would operate for ground faults because the current is so much higher. It is possible that these ground relays may operate faster than the phase relays that should operate, and the wrong breakers might be tripped because the ground relays might not be selective with each other at these higher currents.[10]

TRANSIENT CT ERRORS

The principal problem introduced by transient CT errors is the effect on fast and sensitive overcurrent ground relays. The troublesome effect, often called "false residual current," is that large transient currents flow through the ground-relay coils in the CT neutral when there is no actual ground-fault current in the CT primaries. This happens because the CT's have different errors because of unequal d-c offset in the primary currents or because of different amounts of residual magnetism. As a consequence, if ground-fault current is severely limited by neutral impedance and one needs to use very sensitive ground relays to detect ground faults reliably, the relays should have time delay or else they may operate undesirably on high-current interphase faults.

Such false residual current can be reduced appreciably by the addition of resistance to the CT neutral circuit if the relay burden is not already high enough to limit the current. The same mechanism applies as with current-differential relaying for bus protection, described in Chapter 12. The addition of such resistance tends to equalize the CT errors. However, the best solution is a CT whose magnetic circuit encircles all three phase conductors.[11]

Although not properly includable as consequences of transient CT errors, many other transient and steady-state conditions may cause ground-relay misoperation. Two of these—open phases and simultaneous ground faults at different locations—are described further in this chapter. The others are comprehensively treated in Reference 12. For these, as well as for transient CT errors, a solution that usually works is to make the ground relays less sensitive.

When none of the suggested solutions to false residual current can be used, it is possible to provide equipment that will permit a ground relay to close a circuit only if fault current is flowing in only one phase, such as, for example, by the relay described in Reference 13. However, this may be too heroic a solution for a distribution circuit.

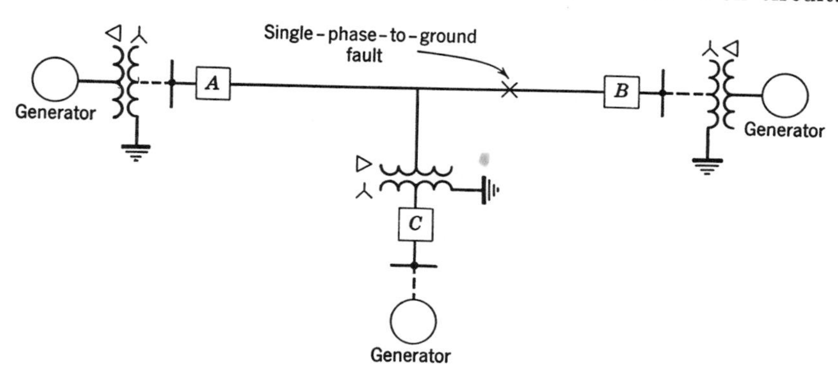

Fig. 13. Illustrating the possibility of ungrounded-neutral operation with a ground fault on the line.

DETECTION OF GROUND FAULTS IN UNGROUNDED SYSTEMS

The operation of other than distribution systems completely ungrounded is recognized as poor practice, and hence ground relaying in such systems is not usually a consideration. However, through circumstance, a portion of a system containing a source of generation may become disconnected from the rest of the system, and a condition of ungrounded operation may result. Consider the situation of Fig. 13 which is encountered quite frequently. Figure 13 shows a delta-wye power transformer tapped to a transmission line. A single-phase-to-ground fault on the line would cause the prompt tripping of breakers

A and *B*. If we assume that breaker *C* did not trip by the operation of its phase relays before breakers *A* and *B* had tripped, the line section would still be left connected to the tapped station with no means for opening breaker *C*. So long as the fault remained single-phase-to-ground, there would be no current of short-circuit magnitude flowing at *C* after breakers *A* and *B* had opened; only charging currents would flow. But, if the fault was of a persistent nature, as, for example, if a conductor had fallen to the ground, it is highly desirable that breaker *C* open automatically.

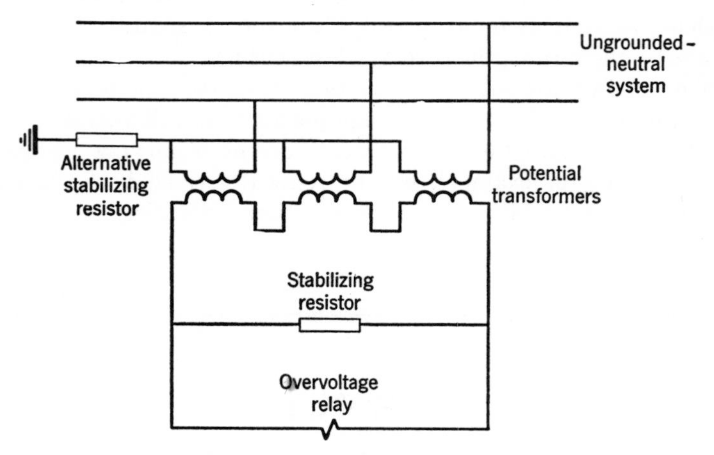

Fig. 14. A wye-broken-delta potential-transformer connection to detect a ground fault on an ungrounded-neutral system.

The presence of a ground fault on an ungrounded-neutral system can be detected through the use of a wye-broken-delta potential transformer with an overvoltage relay connected across the opening in the delta, as illustrated in Fig. 14, or the wye windings of Fig. 14 may be connected to the system indirectly through the secondary of a wye-wye potential transformer, both of whose neutrals are grounded.

This potential-transformer connection will be recognized as being that described in Chapter 8 for obtaining a polarizing voltage for ground directional relays. When such a connection is used on an ungrounded system, the burden placed across the break in the delta should be greater than a certain minimum value, or else an unstable neutral condition (ferroresonance) may develop that would indicate the presence of a ground fault when no fault existed.[14] However, it has been found that occasionally the required burden is higher than that recommended in Reference 14, and is so high that it will quickly thermally overload the potential transformers. A promising solution in such cases appears to be a resistor in series with the grounded neutral of the wye windings, as shown in Fig. 14.

An alternative to Fig. 14 is one high-voltage potential transformer connected between one phase and ground, and one induction-type voltage relay with double-throw contacts. For this method to work, it is necessary that there be sufficient and well-enough-balanced capacitance to ground to establish the neutral of the three-phase system approximately at ground potential. The relay would be provided with a stiff enough control spring so that neither contact would be closed under normal-voltage conditions. Should a ground fault occur on the phase to which the potential transformer was connected, the voltage on the relay would drop and the relay would close its undervoltage "*b*" contact. Should a ground fault occur on either of the other two phases, the relay voltage would rise and cause the overvoltage "*a*" contact to close. With such a relay arranged to trip a breaker on the closing of either contact, some provision must be made to permit reclosing the breaker if such reclosing is desired even though voltage has not been restored to the circuit to which the potential transformer is connected. Here, as in Fig. 14, sufficient stabilizing load should be connected to the potential-transformer secondary or inserted in series with the primary to prevent ferroresonance.[15] If a capacitance potential device were used, it would avoid this ferroresonance problem.

EFFECT OF GROUND-FAULT NEUTRALIZERS ON LINE RELAYING

Ground-fault neutralizers, or "Petersen coils," do not affect the choice of phase relays for line protection. When provision is made to short-circuit the neutralizer automatically after the expiration of a definite time if the ground fault is still present, ground relays are applied on the basis of solidly grounded-neutral operation

Ordinarily, the ground relays will have no tendency to operate until after the neutralizer has been shorted, but rarely there may be a tendency to misoperate at certain locations before the neutralizer is shorted. Figure 15 shows a one-line diagram of a portion of a system, and the corresponding zero-phase-sequence diagram, to illustrate the cause of such a misoperating tendency. Although the total zero-phase-sequence current (I_0) may be zero or very small, the magnitudes of the currents flowing in certain branches of the zero-phase-sequence network can be large. The current through the neutralizer will be nearly equal to the normal phase-to-neutral voltage divided by the neutralizer's impedance. The total capacitance current is approximately equal in magnitude to the neutralizer current, but reversed in phase.

Ground-relay misoperation is possible in the branches where large capacitance current flows, such as at location *P* of Fig. 15. All the capacitance current in the network to the right of location *P* flows through this location. A ground directional relay at *P* is arranged not

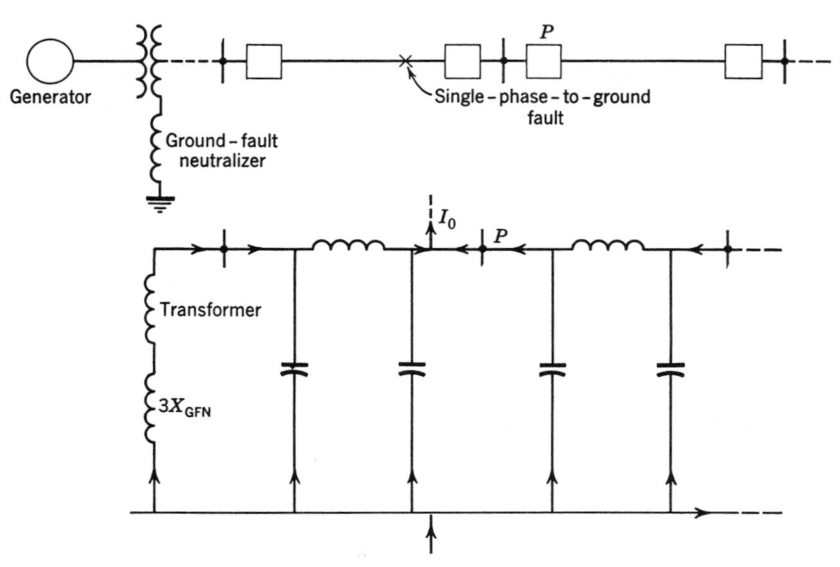

Fig. 15. System and zero-phase-sequence-network diagram illustrating tendency of ground relays to misoperate in the presence of a ground-fault neutralizer.

to operate when lagging current flows in the direction of the arrow. But capacitance, or leading, current flowing in the direction of the arrow will cause relay operation if the current is above the relay pickup. This tendency to misoperate can usually be corrected by increasing the relay's pickup slightly.

The more neutralizers there are at different points in the system, the less will be the charging current at any one location, and the tendency for ground relays to misoperate on charging current will be reduced.

The practice of attempting to use very sensitive relays to obtain selective ground-relay operation without shorting neutralizers is to be discouraged in general. Such practice may be successful with radial lines, but in loop circuits it is not reliable.[16] If one objects to shorting neutralizers on the basis that destructive short-circuit currents may flow, one can always insert sufficient resistance in the neutralizer short circuit to limit the magnitude of the current and still permit enough to flow for reliable selective relay operation.

An application problem is frequently encountered when ground-fault neutralizers have been applied to a system where the line conductors are not sufficiently transposed. The consequence is that, owing to unbalanced phase-to-neutral capacitances, a net phase-to-neutral charging current returns through the earth and the neutralizer coil, and flows continuously. In the zero-phase-sequence circuit, this current path is essentially a series-resonant circuit, and large zero-phase-sequence voltages will exist continuously for relatively low currents in the series-resonant circuit. If these voltages are large enough, they will tend to cause overheating of relay voltage-polarizing coils. (In the same system, but with solidly grounded neutrals, there is no series-resonant circuit, and the voltages are small enough to be negligible.)

Aside from not applying ground-fault neutralizers to such a system, the only solution is to be sure that the continuous ratings of ground-relay polarizing-coil circuits are high enough so that overheating will not occur, which may mean special relays or relays with decreased sensitivity. Incidentally, the ground-fault neutralizers must also be capable of withstanding the current that will flow through them continuously.

THE EFFECT OF OPEN PHASES NOT ACCOMPANIED BY A SHORT CIRCUIT

An open phase not accompanied by a short circuit may be caused by the blowing of a fuse, or by faulty contacts in a circuit breaker. Or an open-phase condition may exist for a short time because of non-simultaneous circuit-breaker-pole closing or opening, or because of single-phase switching.

This subject is too extensive to treat here in all its various aspects. In general, only ground relays are apt to operate undesirably, although sometimes the phase relays of balanced-current relaying may also operate undesirably. However, ground relays, being generally sensitive enough to operate on less than normal load current, are more apt to be the offenders. This subject is treated in more detail in Reference 12.

The operating tendencies of various types of relays can be determined from a study of the phase-sequence currents that flow during open-phase conditions. Consider the system diagram of Fig. 16. If one phase is open-circuited at the location shown in Fig. 16, the phase-sequence diagrams for the open phase are as shown in Fig. 17.

If two phases are open-circuited at the location shown in Fig. 16, the phase-sequence diagrams for the phase that is not open-circuited are shown in Fig. 18.

The arrows of Figs. 17 and 18 show the direction of current flow in the various phase-sequence networks. Note that all these currents are of load-current magnitude, their size depending on the magnitude of load current flowing before the open-phase condition occurred.

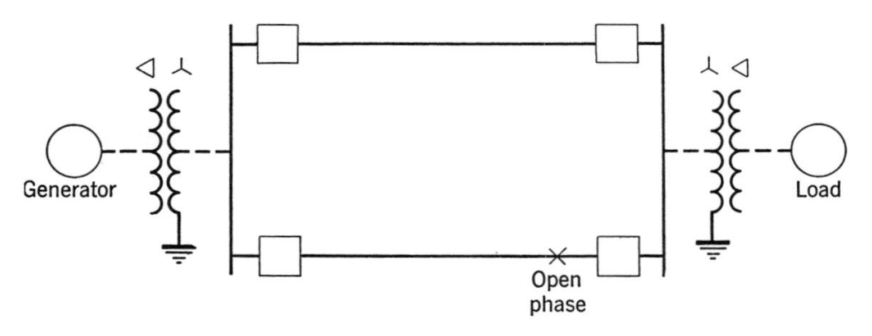

Fig. 16. Relay operation resulting from open phases.

The operating tendency of current-polarized directional-ground relays can be analyzed quite simply from Figs. 17 and 18 by remembering that the relays can trip when zero-phase-sequence current flows

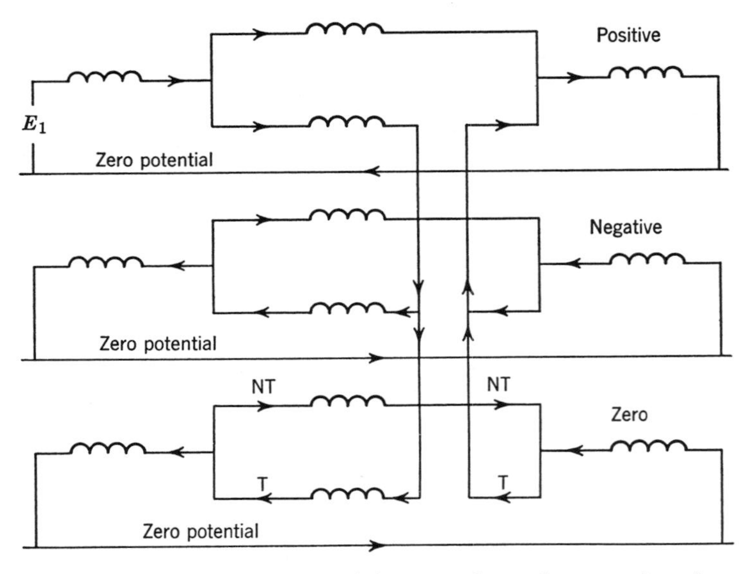

Fig. 17. Phase-sequence diagrams of the open phase when one phase is open.

out of the zero potential bus and into a line at a given location. A reversal of both of these directions also produces a tripping tendency. Based on this fact, Figs. 17 and 18 are labeled T (trip) and NT (not trip) at the ends of the lines to indicate the operating tendencies of the ground directional relays at those locations.

The operation of voltage-polarized ground directional relays obtaining voltage from a bus source will be the same as for current-polarized relays. However, if voltage is obtained from the line side of the

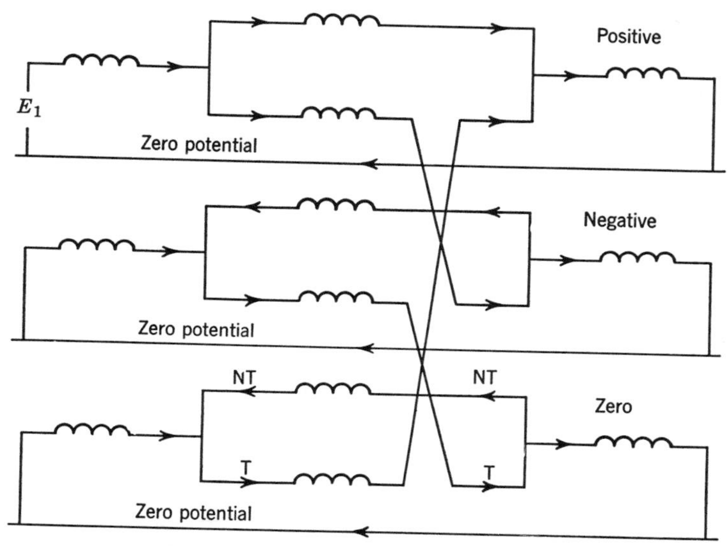

Fig. 18. Phase-sequence diagrams of the closed phase when two phases are open.

breakers, as with coupling-capacitor potential devices, an open circuit between a bus and the voltage source of a given relay will produce an operating tendency opposite to that shown on Figs. 17 and 18 for the relay at that location.

The effect that has been most troublesome, actually, is the transient effect on high-speed ground relays of non-simultaneous circuit-breaker-pole closing or opening. It has been necessary to employ induction types and sometimes to increase the pickup of the relays in order to avoid undesirable operation.

THE EFFECT OF OPEN PHASES ACCOMPANIED BY SHORT CIRCUITS

If a line conductor breaks, and one or both ends fall to the ground, we have simultaneous faults of two different types—an open-circuit and a phase-to-ground short circuit. Such faults can be analyzed by the method of symmetrical components.[17] However, it is rarely, if ever, that one has to make such an analysis because the presence of the short circuit takes command of the situation, and the proper protective relays operate to remove the short circuit from the system, the situation caused by the open circuit also being relieved.

POLARIZING THE DIRECTIONAL UNITS OF GROUND RELAYS

Directional units of ground relays may be polarized from certain zero-phase-sequence-current or voltage sources, or from both simultaneously.[18] Chapter 8 describes voltage-polarization sources utilizing potential transformers or capacitance potential devices.

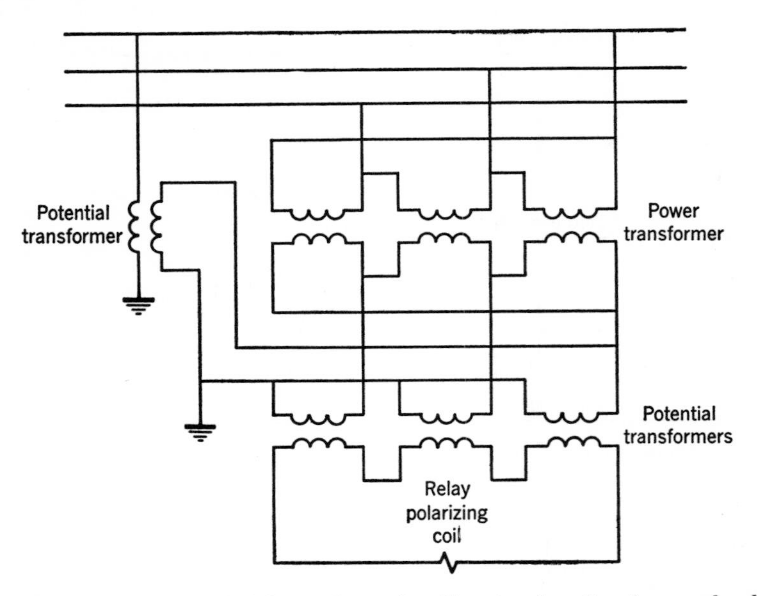

Fig. 19. Low-tension polarizing voltage for directional units of ground relays.

Figure 19 shows a method, not described in Chapter 8, for obtaining polarizing voltage from the low-voltage side of a delta-delta power-transformer bank, using only one high-voltage potential transformer to establish the neutral on the low-voltage side.[19] The same principle can be applied to a delta-wye transformer bank if the necessary auxiliary potential transformers are used to compensate for the 30° phase shift of voltages.

The voltage obtained by the method of Fig. 19 has an error caused by the voltage regulation in the transformer bank, owing to load current flowing from the high-voltage side to the low-voltage side, or caused by positive- and negative-phase-sequence currents flowing toward the fault if there is a source of generation connected to the low-voltage side. This error is not great enough to cause concern for large values of zero-phase-sequence voltage, but for low values it might cause a directional relay to misoperate. Magnetizing-current inrush to the bank also could cause misoperation if the relay was a high-speed type. For these reasons, this kind of a polarizing source is not considered suitable for high-speed directional relays. It may be used with inverse-time directional-overcurrent relays.

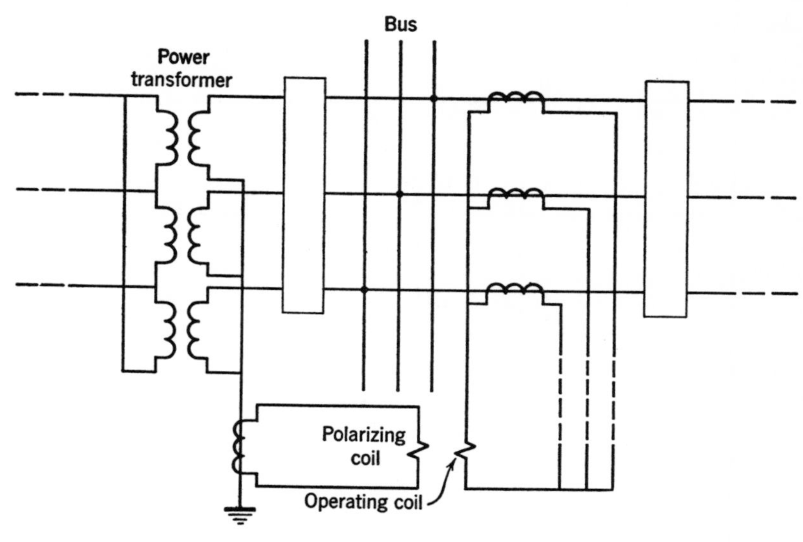

Fig. 20. Current polarization from power-transformer-neutral current.

Figure 20 shows how current polarization can be obtained from the grounded-neutral current of a three-phase power-transformer bank. As mentioned in Chapter 8 in connection with the use of voltage from the low-voltage side of a transformer bank, one should consider the reliability of a single transformer bank as a polarizing medium in view of the fact that it will be out of service occasionally for maintenance, or possibly for repair because of an internal fault. Polarizing current from paralleled CT's in the grounded neutrals of two or more transformer banks is considered sufficiently reliable if the banks have separate circuit breakers so that one bank will always be in service.

With a three-winding wye-delta-wye power-transformer bank, polarizing CT's should be put in the grounded neutrals of both wye windings and paralleled. The ratios of these two CT's should be inversely proportional to the voltage ratings of the wye windings.

As an alternative to the neutral CT's with either two- or three-winding transformers, a single CT in series with one of the delta windings may be used if these windings do not supply external load or are not connected to a generating source. If there are external connections to the delta, three CT's are required, one in each of the three windings. These CT's should be paralleled, as shown in Fig. 21, in such a way that their output is proportional to 3 times the zero-phase-sequence component of current circulating in the delta when a ground fault occurs.

As a second alternative to the neutral CT's, the neutral current of wye-connected CT's in series with the wye windings may be used,

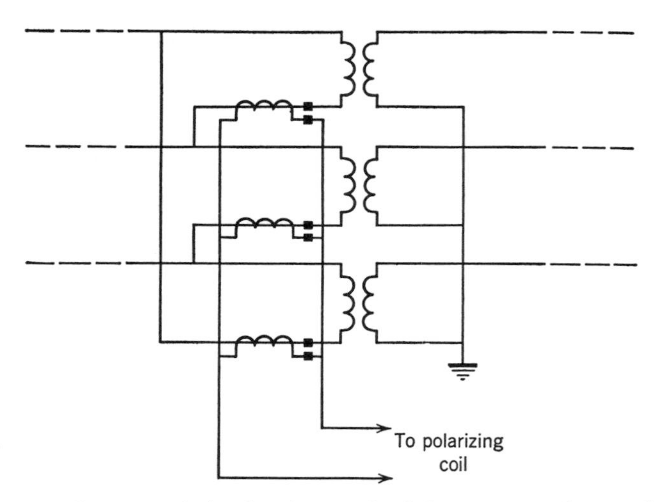

Fig. 21. Current polarization from a loaded power-transformer delta.

as in Fig. 22. In the three-winding power transformer, the ratios of these CT's should be inversely proportional to the voltage ratings of the wye windings, as when neutral CT's are used. This alternative is not exactly equivalent to neutral CT's because of the possibility of false residual current, and, therefore, it should not be used with high-speed relays.

In an autotransformer bank with a delta tertiary, either of the two alternatives to the neutral CT's may be employed. It is generally not permissible to use a CT in the neutral because the neutral current for a low-voltage fault may be reversed from the neutral current for a high-voltage fault. Infrequently, the distribution of fault currents is such that a neutral CT may be used; however, one should realize that the conditions might change as system changes are made.

The primary-current rating of a neutral or delta-winding CT used for polarizing directional units of ground relays should be such that the polarizing and operating coils of a directional unit get about the same current magnitudes for any fault for which it must operate.

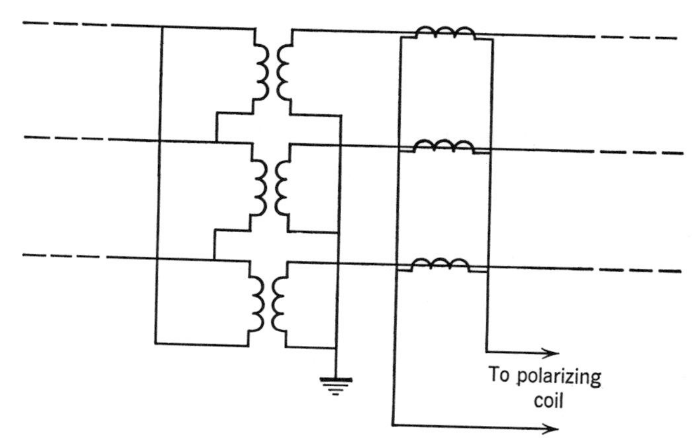

Fig. 22. Alternative to Fig. 20.

This is more important for the so-called "directional-ground" relays whose published characteristics hold only if one current does not differ too much from the other.

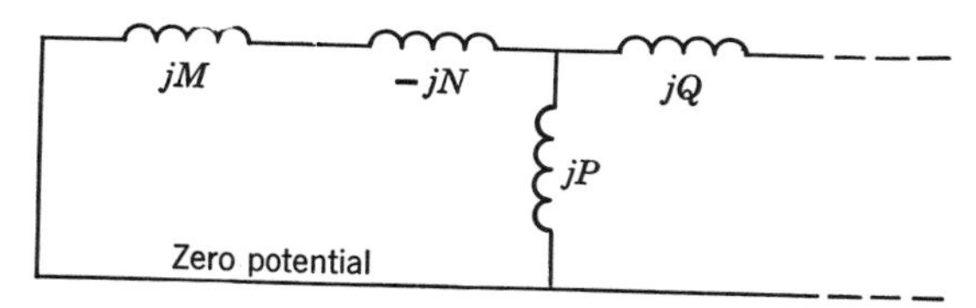

Fig. 23. Zero-phase-sequence diagram illustrating a case where polarization by the methods of Figs. 20 to 22 will cause misoperation.

In rare cases, such as that illustrated in Fig 23, none of the foregoing methods of current polarization may be used. Such circumstances may exist when one branch (N) of the equivalent circuit of a three-winding power transformer or autotransformer has negative reactance and when the zero-phase-sequence reactance of the system (M) connected to this negative branch is less than the reactance of the negative branch. In other words, on the side of this branch, the total zero-phase-sequence reactance including this branch of the transformer ($M + N$) is negative. Then, if this total negative reactance is smaller than the positive reactance of the branch representing the delta winding (P), all conditions are satisfied to make unsuitable any of the current-polarizing methods that have been described. When these circumstances exist, it becomes necessary either to use voltage polarization or, possibly, a special combination of currents.[20]

Directional relays are available that are arranged for polarization simultaneously by voltage and current.[18] Apart from simplifying the problem of stocking spare relays, "dual polarization," as it is called, has certain functional advantages. Sometimes, current or voltage alone is unsatisfactory because either source may sometime be disconnected from the system and thereby be rendered useless when it is still needed. With dual polarization, either source may be disconnected so long as the other is left in service. In other circumstances, either voltage or current polarization alone provides objectionably weak polarization but the two together assure strong polarization.

NEGATIVE-PHASE-SEQUENCE DIRECTIONAL UNITS FOR GROUND-FAULT RELAYING

When there is no zero-phase-sequence-current or voltage source for polarizing the directional unit of a ground relay, it is often possible to use a negative-phase-sequence directional unit if separate ground relaying is required. However, one must be sure that sufficient negative-phase-sequence current and voltage will be available to assure reliable operation of the directional unit for all conditions for which it must operate. In some systems that are grounded through impedance, the negative-phase-sequence quantities may be too small.

A negative-phase-sequence directional unit may be either a simple directional unit supplied with negative-phase-sequence current and voltage from filter circuits[21] or it may consist of two polyphase directional units with opposing torques, as described in Chapter 9.

Another advantage of negative-phase-sequence directional units is that they are not affected by mutual induction between paralleled circuits when ground faults occur. Chapter 15 shows that directional relays with zero-phase-sequence polarization may operate undesirably under such circumstances.

In spite of any advantages the negative-phase-sequence relay may have, it is used only as a last resort, because the zero-phase-sequence relay is simpler and easier to test, and because it produces more reliable torque under all conditions where it is applicable.

CURRENT-BALANCE AND POWER-BALANCE RELAYING

Prior to the introduction of high-speed distance and pilot relaying, current-balance and power-balance relaying were used extensively for the protection of parallel lines. Aside from instantaneous overcurrent relaying, they were the only available forms of instantaneous relaying for transmission lines. Their present-day usage for new installations is rare, but many old installations are still in service, and occasionally a new one is justified.

Current-balance relaying is illustrated in Fig. 24 for one phase of a pair of three-phase parallel lines. Two overcurrent-type current-

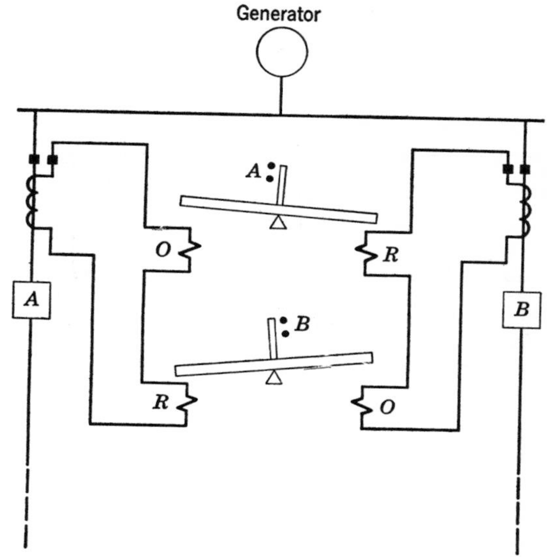

Fig. 24. Current-balance relaying of parallel lines.

balance units are shown schematically, having operating coils *O* and restraining coils *R*. One unit has contacts *A* that trip breaker *A*, and the other unit has contacts *B* for tripping breaker *B*. The trip circuits are not shown, but they are arranged so that tripping of either breaker can occur only when both breakers are closed. Furthermore, each phase-relay unit is restrained either by its control spring or by voltage (not shown) so that it will not operate with operating-coil current corresponding to maximum load and no restraining current (or, in other words, it will not operate when only one line is closed and carrying maximum load). This restraint is necessary with this type of relay to prevent immediate undesirable tripping when a line is being returned to service and one end is closed while the other end is open. Such restraint is unnecessary on a current-balance ground relay obtaining current from the CT neutrals, because normally there is no neutral current.

Power-balance-relaying is illustrated in Fig. 25. As shown, the CT's of each corresponding phase of the two parallel lines are cross-connected, and the current coil of a directional relay is differentially connected across the CT interconnections. While the currents flowing into the two lines are equal vectorially, no current will flow in the directional-relay coil, the currents merely circulating between the two CT's. If the current in one line becomes larger than that of the other, current will flow in one direction through the relay coil. If

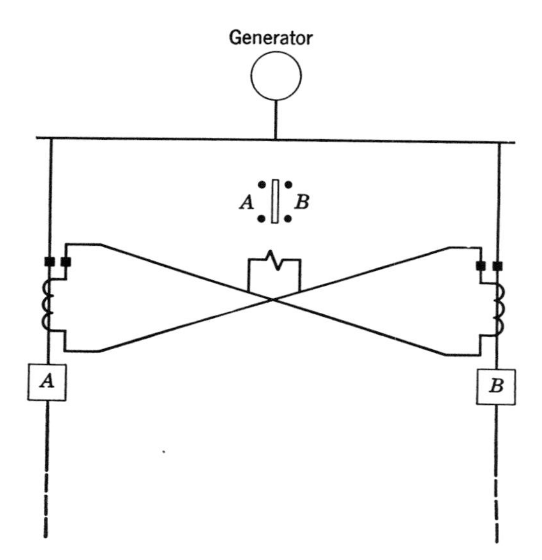

Fig. 25. Power-balance relaying of parallel lines.

the current of the other line is larger, the current will flow in the other direction. Thus, the relay will close one or the other of its double-throw contacts to trip the breaker of the line having the larger current. Instead of the single relay shown in Fig. 25, two relays may be used, contact *A* being on one, and contact *B* on the other. As described for current-balance relaying, the trip circuits are arranged so that both breakers must be closed before either can be tripped. Supplementary overcurrent relays connected in series with the directional-relay current coils provide a sensitivity adjustment.

Power-balance relaying can be used at either end of parallel lines, as contrasted with current-balance relaying, which can be used only where there is a connection to a source of short-circuit current. If there is no such source at one end of a pair of parallel lines, the magnitudes of the currents in the two lines are equal at that end for a fault on one line, and hence a relay that compares only the magnitudes of the currents would not operate. The power-balance equipment compares not only magnitude but also the phase relation of the currents in the two lines, and, since the direction of current flow in one line is reversed from that in the other line at the load end, this equipment will operate to trip the faulty line. The only reason for ever using current-balance relaying is that it is somewhat simpler and less expensive, especially if no voltage restraint is involved and since supplementary overcurrent relays are not required to establish the sensitivity.

Both current-balance and power-balance relaying are effective only while both lines are in service. For single-line operation, supplementary relaying is required. This supplementary relaying is also required to provide back-up protection for faults in adjoining lines or other system elements, since the current- or power-balance relaying will not operate for faults outside the two parallel lines. Unless relaying comparable to that provided by distance relays is used for single-line operation, faults during single-line operation will not be cleared nearly so quickly as when both lines are in service; and, if fast clearing is required for single-line operation, current-balance or power-balance relaying cannot be justified. The only exception to this is for protection against single-phase-to-ground faults for which distance relaying may not be economically feasible; then, current-balance or power-balance relaying will minimize the likelihood of a single-phase-to-ground fault on one line developing into a fault involving the other line, when both lines are close together.

AUTOMATIC RECLOSING

Experience has shown that 70% to 95% of all high-voltage transmission-, subtransmission-, and distribution-line faults are non-persisting if the faulty circuit is quickly disconnected from the system. This is because most line faults are caused by lightning, and, if the ensuing arcing at the fault is not allowed to continue long enough to badly damage conductors or insulators, the line can be returned to service immediately.[22] Where the fault persists after the first trip and closure, experience has shown it to be desirable to try as many as two or three more reclosures before keeping the line out of service until the trouble can be found and repaired.

Automatic reclosing is generally applied to all types of circuits. Subtransmission lines having overcurrent relaying usually have multi-reclosure equipment, with supplementary "synchronism-check" equipment at one end if it is likely that the line may sometime be the only tie between certain generating stations. Synchronism-check equipment is relay equipment that permits a circuit breaker to be closed only if the parts to be connected by the breaker are in synchronism. On radial lines, there is no need for synchronism check.

In distribution systems in which selectivity with branch-circuit fuses is involved, multireclosure is also used.[23] Instantaneous and inverse-time overcurrent relays are arranged so that, when a fault occurs, the instantaneous relays operate to trip the breaker before a branch-circuit fuse can blow, and the breaker is then immediately reclosed. However, after the first tripout, the instantaneous relays are automatically cut out of service so that if the fault should persist the inverse-time relays would have to operate to trip the breaker. This gives time for the branch-circuit fuse of the faulty circuit to blow, if we assume that the fault is beyond this fuse. In this way, the cost of replacing blown branch-circuit fuses is minimized, and at the same time the branch-circuit outage is also minimized. If the breaker is not tripped within a certain time after reclosure, the instantaneous relays are automatically returned to service.

When industrial loads are to be fed from lines with automatic reclosing, certain problems exist that must be solved before automatic reclosing can be safely permitted. They have to do with the possible loss of synchronism between the utility and the industrial plant that makes it necessary to cut the plant loose from the utility before the utility's breakers reclose. At the same time, the industrial plant may have to resort to some "load shedding" of unessential loads so as to be able to supply its essential load from its own generating source. These problems, as well as other problems of industrial-plant protection, are discussed at length and solutions are given in the material under Reference 24.

RESTORATION OF SERVICE TO DISTRIBUTION FEEDERS AFTER PROLONGED OUTAGES

Large economies can be effected in the design and operation of distribution circuits and related equipment by taking advantage of the diversity between loads. Under normal operating conditions, the actual load on a distribution feeder will generally be considerably less than the total rating of the power-utilization equipment served by the feeder. When such a feeder is quickly reclosed after the clearing of a fault, the inrush current will not greatly exceed the normal load current, and the current will quickly return to normal. But after such a feeder has been out of service long enough so that the normal "off" period of all loads, such as electric furnaces, refrigerators, water heaters, pumps, air conditioners, etc., has been exceeded, all these loads will be thrown on together with no diversity between them. The total inrush current to all the electric motors involved, added to the current of other loads, may be several times the normal peak-load current, decaying to approximately 1.5 times the normal peak-load current after several seconds.[25] This subject has been called "cold-load restoration" or "cold-load pickup."

Protective relaying for such feeders is faced with the problem of selecting between such inrush currents and the current to a fault at or beyond the next automatic sectionalizing point. An induction-type extremely inverse-time-overcurrent relay, having a time-current curve closely matching that of a fuse down to an operating time of approximately 0.1 second, has given some relief for such an application. However, the best solution seems to be automatic sectionalizing on prolonged loss of voltage, and time-staggered automatic reclosing on the return of voltage.

COORDINATING WITH FUSES

When overcurrent relays must be selective with fuses, the relay that will provide such selectivity and still be the fastest for faults for which it must operate is the extremely inverse type just described for cold-load restoration. Occasionally it may be necessary to add a delay of a few cycles, in the form of an auxiliary relay, to maintain selectivity at very high fault currents.

A-C AND CAPACITOR TRIPPING

Some installations, particularly in distribution systems, cannot justify the expense and the maintenance of a storage battery and its charging equipment solely for use with protective-relaying equipment of one or two circuits. Then "a-c tripping" or "capacitor tripping" may be used if they are applicable.

In one type of a-c tripping equipment, a reactor, called a "tripping reactor," is permanently connected in series with the coil of each overcurrent relay in the CT secondary circuit. The circuit breaker has a separate trip coil for each overcurrent relay. Each trip coil is connected across a tripping reactor through the contact of the corresponding overcurrent relay. Figure 26 shows the connections when three relays are involved. This method of tripping is applicable when the rms voltage drop through a reactor is sufficiently high during a short circuit to actuate the reactor's corresponding trip-coil mechanism should its overcurrent relay close its contact. Reactors of various ratings are available for producing the required voltage whenever the current is high enough to close the relay contacts. However, owing to limitations in the ability of CT's to produce sufficient rms voltage at low primary currents, this method of tripping is effective only for short-circuit currents of the order of rated load current or higher. Saturation of the tripping reactor limits the rms voltage to a safe value at high fault currents, and a shunt resistor, not shown in Fig. 26, holds down the crest value of voltage.

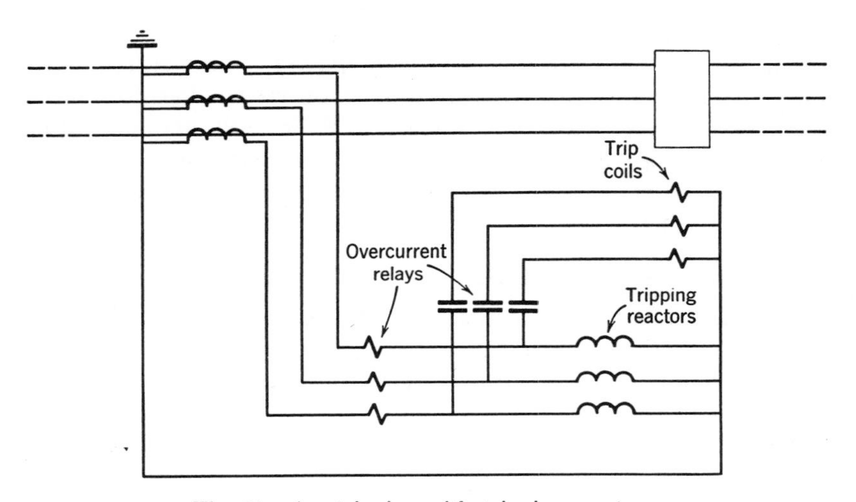

Fig. 26. A-c tripping with tripping reactors.

When the fault current for a single-phase-to-ground short circuit is less than the pickup of the phase overcurrent relays, phase-to-phase voltage from the secondary of a potential transformer may be used for tripping by a ground overcurrent relay in conjunction with a separate voltage trip coil. For a single-phase-to-ground fault, the phase-to-phase voltage will usually be large enough so that reliable tripping will be assured. A-c voltage can be used for tripping by other types of protective relays that operate for other than short circuits, or, in other words, when the voltage is not badly affected.

A-c tripping is inferior principally in its limited sensitivity for interphase faults, and also because of more severe relay contact duty which requires more maintenance. Also, the burden imposed on the CT's is high, and, unless the CT's are good enough to support this burden, the relaying sensitivity will be poor. Because of the high burden, accurate metering cannot be provided from the same CT's. A minor limitation is that single-phase relays must be used; a polyphase directional relay having but one contact could not be used because there would not be a suitable source for operating an auxiliary relay to provide the necessary additional contacts.

Capacitor tripping uses the stored energy of a charged capacitor to actuate a trip-coil mechanism. The capacitor is charged through a rectifier from a-c voltage obtained from the system. Its principal advantage over a-c tripping is greater sensitivity for interphase faults. Also, it permits the use of polyphase relays that do not need auxiliary relays, although an additional capacitor tripping device has been used to operate an auxiliary tripping relay. Its principal disadvantage is that the circuit-breaker trip mechanism gets only one impulse and not a steady force as with a-c tripping; this fact makes some people feel that the breaker requires a more "hair-triggered" trip mechanism, and consequently that it is apt to be less reliable.

Problem

Given a simple loop system as shown in Fig. 27, where the impedances are in percent based on the rating of one of the generators. Adjust overcurrent relays having time characteristics of Chapter 3, Fig. 3, for the protection of all the lines

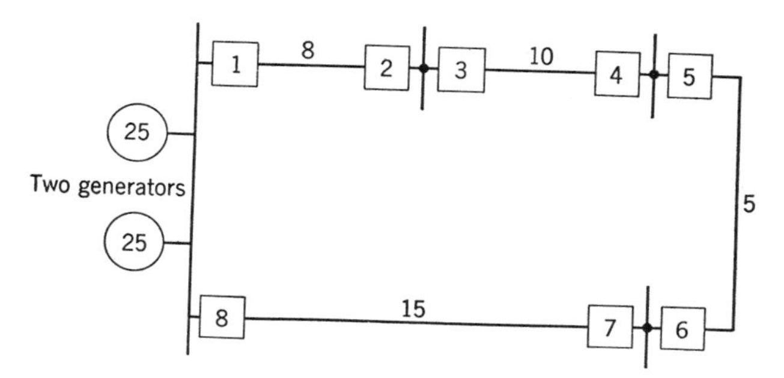

Fig. 27. Illustration for problem.

against phase faults, assuming and indicating directional elements, wherever necessary, by an arrow pointing in the tripping direction. Neglect the fact that the pickup should be well above load current. Make your adjustments on the basis of per unit primary current, assuming that any desired pickup can be obtained.

Bibliography

1. "Long 60-Cycle Arcs in Air," by A. P. Strom, *AIEE Trans., 65* (1946), pp. 113–117. Discussions, pp. 504–507.

2. "Power System Faults to Ground, Part II Fault Resistance," by C. L. Gilkeson, P. A. Jeanne, and E. F. Vaage, *AIEE Trans., 56* (1937), pp. 428–433, 474.

3. "Application of the Ohm and Mho Principles to Protective Relays," by A. R. van C. Warrington, *AIEE Trans., 65* (1946), pp. 378–386. Discussions, pp. 490–491.

"Reactance Relays Negligibly Affected by Arc Impedance," by A. R. van C. Warrington, *Elec. World, 98,* No. 12 (Sept. 19, 1931), pp. 502–505.

4. "Sensitive Ground Protection for Radial Distribution Feeders," by Lloyd F. Hunt and J. H. Vivian, *AIEE Trans., 59* (1940), pp. 84–87. Discussions, pp. 87–90.

"Some Recent Relay Developments," by Lloyd F. Hunt and A. A. Kroneberg, *AIEE Trans., 53* (1934), pp. 530–535.

5. "Effect of Fault Resistance on Ground-Fault Current," by Martin J. Lantz, *AIEE Trans., 72,* Part III (1953), pp. 1016–1019.

6. "Instantaneous Overcurrent Relays for Distance Relaying," by C. H. Frier, *AIEE Trans., 54* (1935), pp. 404–407. Discussions, pp. 883–884.

7. *Silent Sentinels,* Westinghouse Electric Corp., Meter Division, Newark, N. J.

8. "A Study of Directional Element Connections for Phase Relays," by W. K. Sonnemann, *AIEE Trans., 69,* Part II (1950), pp. 1438–1450. Discussions, pp. 1450–1451.

9. "Ground Fault Relay Protection of Transmission Lines," by J. L. Blackburn, *AIEE Trans., 71* (1952), pp. 685–691. Discussions, pp. 691–692.

10. "Ground Relays Set for Simultaneous 1-Ph Faults," by John M. Showman, *Elec. World,* Sept. 6, 1954, p. 102.

11. "Earth-Leakage Relays; with Use of Ring-Type Current Transformers," by E. T. B. Gross, *Electrician* (*London*), *123* (July, 1939), pp. 93–95.

"Sensitive Ground Protection," AIEE Committee, *AIEE Trans., 69,* Part I (1950), pp. 473–475. Discussions, pp. 475–476.

12. "Some Utility Ground-Relay Problems," by H. C. Barnes and A. J. McConnell, *AIEE Trans., 74,* Part III (1955), pp. 417–428. Discussions, pp. 428–433.

13. "A New Single-Phase-to-Ground Fault-Detecting Relay," by W. K. Sonnemann, *AIEE Trans., 61* (1942), pp. 677–680. Discussions, pp. 995–996.

14. "Criteria for Neutral Stability of Wye-Grounded-Primary Broken-Delta Secondary Transformer Circuits," by H. S. Shott and H. A. Peterson, *AIEE Trans., 60* (1941), pp. 997–1002. Discussions, pp. 1394–1395.

15. "Neutral Inversion of a Single Potential Transformer Connected Line-to-Ground on an Isolated Delta System," by Lyle L. Gleason, *AIEE Trans., 70,* Part I, pp. 103–110. Discussions, pp. 110–111.

16. "Sensitive Ground Protection for Transmission Lines and Distribution Feeders," E. T. B. Gross, *AIEE Trans., 60* (1941), pp. 968–971.

17. *Circuit Analysis of A-C Power Systems,* Vol. I, by Edith Clarke, John Wiley & Sons, New York, 1943.

18. "Dual-Polarized Directional-Ground Relays," by W. C. Morris, *Distribution,* April, 1952, pp. 8, 9.

19. United States Patent 2378800.

20. "Ground Relay Polarization," by J. L. Blackburn, *AIEE Trans., 71* (1952), pp. 1088–1093. Discussions, pp. 1093–1095.

21. "Directional Overcurrent Ground Relaying," by B. V. Hoard and S. C. Leyland, *Elec. J., 36,* No. 3, March, 1939, pp. 97–101.

22. "Faster Reclosing Breakers Needed," by J. T. Logan, *Elec. World, 103,* No. 16 (April 21, 1934), pp. 571–573.

23. "Progress Report on Coordination of Construction and Protection of Dis-

tribution Circuits Based on Operating Data for Year 1949," by AIEE Committee, *Technical Paper 52–13*.

"Reclosure Sequence and Coordination," by R. C. Kirk, *Elec. Light and Power*, Dec., 1954, pp. 99–100, 148.

24. "Relay Protection of 33-Kv Tie Feeders to a Large Industrial Customer," by W. H. O'Connor and H. D. Ruger, *Elec. Eng.*, Feb., 1952, p. 165.

"Protective Relaying on Industrial Power Systems," by Carl M. Lathrop and Charles E. Schleckser, *AIEE Trans., 70,* Part II, pp. 1341–1344. Discussions, pp. 1344–1345.

"Protection for an Industrial Plant Connected to a Utility System with Reclosing Breakers," by D. F. Langenwalter, *Distribution, 15,* No. 5 (Nov., 1953), pp. 17–20.

"Parallel Operation of Industrial Power Plants with Public Utilities," by L. G. Levoy, *Distribution, 15,* No. 5 (Nov., 1953), pp. 10–16.

"Frequency Relays Disconnect Synchronous Motors," by A. J. Meyers, *Power Generation,* July, 1949, p. 82.

"Application and Tests of Frequency Relays for Load Shedding," by L. L. Fountain and J. L. Blackburn, *AIEE Trans., 73,* Part III (1954), pp. 1660–1664. Discussions, pp. 1664–1668.

"Selecting A-C Overcurrent Protective Device Settings for Industrial Plants," by F. P. Brightman, *AIEE Trans., 71,* Part II (1952), pp. 203–211.

"More about Setting Industrial Relays," by F. P. Brightman, *AIEE Trans., 73* (1954), pp. 397–405. Discussions, pp. 405–406.

"Inertia Relay 'Anticipates' to Facilitate Reclosure," by J. T. Logan and J. H. Miles, *Elec. World, 113* (April, 1940), pp. 1061–1064.

Industrial Power Systems Handbook, McGraw-Hill Book Co., New York, 1955.

25. "A New Approach to Cold Load Restoration," by Oliver Ramsaur, *Elec. World,* Oct. 6, 1952, pp. 101–103.

"New Relay Assures Feeder Resumption after Outage," by M. H. Pratt, L. J. Audlin, and A. J. McConnell, *Elec. World,* Part I, Sept. 10, 1949, pp. 99–103; Part II, Sept. 24, 1949, pp. 95–98.

"Diversity a New Problem in Feeder Pickup," by C. E. Hartay and C. J. Couy, *Elec. Light and Power,* Oct., 1952, pp. 142–146.

14 LINE PROTECTION WITH DISTANCE RELAYS

Distance relaying should be considered when overcurrent relaying is too slow or is not selective. Distance relays are generally used for phase-fault primary and back-up protection on subtransmission lines, and on transmission lines where high-speed automatic reclosing is not necessary to maintain stability and where the short time delay for end-zone faults can be tolerated. Overcurrent relays have been used generally for ground-fault primary and back-up protection, but there is a growing trend toward distance relays for ground faults also.

Single-step distance relays are used for phase-fault back-up protection at the terminals of generators, as described in Chapter 10. Also, single-step distance relays might be used with advantage for back-up protection at power-transformer banks, but at the present such protection is generally provided by inverse-time overcurrent relays.

Distance relays are preferred to overcurrent relays because they are not nearly so much affected by changes in short-circuit-current magnitude as overcurrent relays are, and, hence, are much less affected by changes in generating capacity and in system configuration. This is because, as described in Chapter 9, distance relays achieve selectivity on the basis of impedance rather than current.

THE CHOICE BETWEEN IMPEDANCE, REACTANCE, OR MHO

Because ground resistance can be so variable, a ground distance relay must be practically unaffected by large variations in fault resistance. Consequently, reactance relays are generally preferred for ground relaying.

For phase-fault relaying, each type has certain advantages and disadvantages. For very short line sections, the reactance type is preferred for the reason that more of the line can be protected at high speed. This is because the reactance relay is practically unaffected by arc resistance which may be large compared with the line impedance, as described elsewhere in this chapter. On the other hand, reactance-type distance relays at certain locations in a system are the most likely to operate undesirably on severe synchronizing-power surges unless additional relay equipment is provided to prevent such operation.

The mho type is best suited for phase-fault relaying for longer lines, and particularly where severe synchronizing-power surges may occur. It is the least likely to require additional equipment to prevent tripping on synchronizing-power surges.[1] When mho relaying is adjusted to protect any given line section, its operating characteristic encloses the least space on the R-X diagram, which means that it will be least affected by abnormal system conditions other than line faults; in other words, it is the most selective of all distance relays. Because the mho relay is affected by arc resistance more than any other type, it is applied to longer lines. The fact that it combines both the directional and the distance-measuring functions in one unit with one contact makes it very reliable.

The impedance relay is better suited for phase-fault relaying for lines of moderate length than for either very short or very long lines. Arcs affect an impedance relay more than a reactance relay but less than a mho relay. Synchronizing-power surges affect an impedance relay less than a reactance relay but more than a mho relay. If an impedance-relay characteristic is offset, so as to make it a modified-impedance relay, it can be made to resemble either a reactance relay or a mho relay but it will always require a separate directional unit.

There is no sharp dividing line between areas of application where one or another type of distance relay is best suited. Actually, there is much overlapping of these areas. Also, changes that are made in systems, such as the addition of terminals to a line, can change the type of relay best suited to a particular location. Consequently, to realize the fullest capabilities of distance relaying, one should use the type best suited for each application. In some cases much better selectivity can be obtained between relays of the same type, but, if relays are used that are best suited to each line, different types on adjacent lines have no appreciable adverse effect on selectivity.

THE ADJUSTMENT OF DISTANCE RELAYS

Chapter 9 shows that phase distance relays are adjusted on the basis of the positive-phase-sequence impedance between the relay location and the fault location beyond which operation of a given relay unit should stop. Ground distance relays are adjusted in the same way, although some types may respond to the zero-phase-sequence impedance. This impedance, or the corresponding distance, is called the "reach" of the relay or unit. For purposes of rough approximation, it is customary to assume an average positive-phase-sequence-reactance value of about 0.8 ohm per mile for open transmission-line construction, and to neglect resistance. Accurate data are available in textbooks devoted to power-system analysis.[2]

To convert primary impedance to a secondary value for use in adjusting a phase or ground distance relay, the following formula is used:

$$Z_{sec} = Z_{pri} \times \frac{\text{CT ratio}}{\text{VT ratio}}$$

where the CT ratio is the ratio of the high-voltage phase current to the relay phase current, and the VT ratio is the ratio of the high-voltage phase-to-phase voltage to the relay phase-to-phase voltage—all under balanced three-phase conditions. Thus, for a 50-mile, 138-kv line with 600/5 wye-connected CT's, the secondary positive-phase-sequence reactance is about $50 \times 0.8 \times \frac{600}{5} \times \frac{115}{138{,}000} = 4.00$ ohms.

It is the practice to adjust the first, or high-speed, zone of distance relays to reach to 80% to 90% of the length of a two-ended line or to 80% to 90% of the distance to the nearest terminal of a multi-terminal line. There is no time-delay adjustment for this unit.

The principal purpose of the second-zone unit of a distance relay is to provide protection for the rest of the line beyond the reach of the first-zone unit. It should be adjusted so that it will be able to operate even for arcing faults at the end of the line. To do this, the unit must reach beyond the end of the line. Even if arcing faults did not have to be considered, one would have to take into account an underreaching tendency because of the effect of intermediate current sources, and of errors in: (1) the data on which adjustments are based, (2) the current and voltage transformers, and (3) the relays. It is customary to try to have the second-zone unit reach to at least 20% of an adjoining line section; the farther this can be extended into the adjoining line section, the more leeway is allowed in the reach of the third-zone unit of the next line-section back that must be selective with this second-zone unit.

The *maximum* value of the second-zone reach also has a limit. Under conditions of maximum overreach, the second-zone reach should be short enough to be selective with the second-zone units of distance relays on the shortest adjoining line sections, as illustrated in Fig. 1.

Transient overreach need not be considered with relays having a high ratio of reset to pickup because the transient that causes overreach will have expired before the second-zone tripping time. However, if the ratio of reset to pickup is low, the second-zone unit must be set either (1) with a reach short enough so that its overreach will not extend beyond the reach of the first-zone unit of the adjoining line

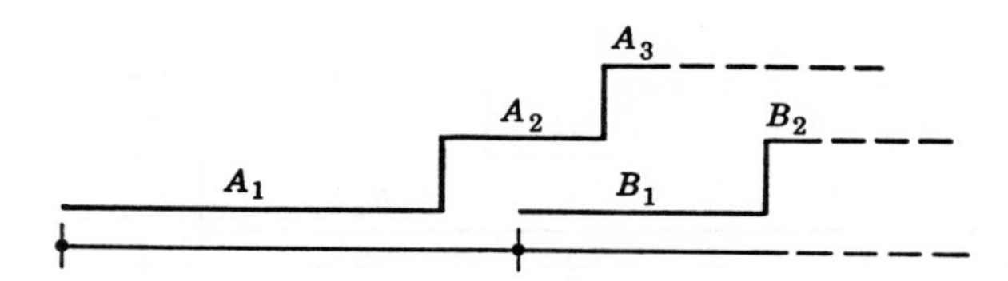

Fig. 1. Normal selectivity adjustment of second-zone unit.

section under the same conditions, or (2) with a time delay long enough to be selective with the second-zone time of the adjoining section, as shown in Fig. 2. In this connection, any underreaching tendencies of the relays on the adjoining line sections must be taken

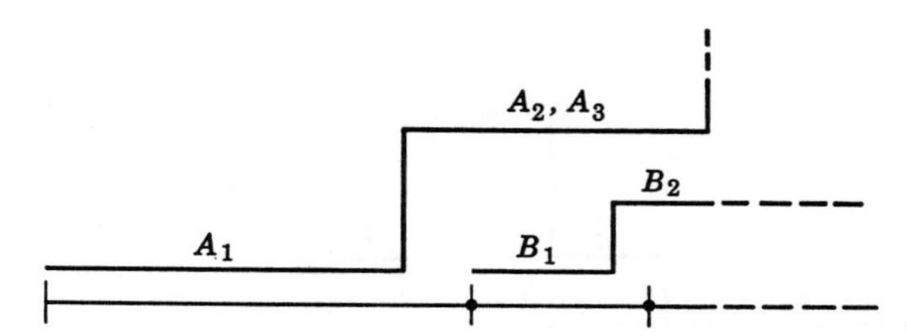

Fig. 2. Second-zone adjustment with additional time for selectivity with relay of a very short adjoining line section.

into account. When an adjoining line is so short that it is impossible to get the required selectivity on the basis of reach, it becomes necessary to increase the time delay, as illustrated in Fig. 2. Otherwise, the time delay of the second-zone unit should be long enough to provide selectivity with the slowest of (1) bus-differential relays of the bus at the other end of the line, (2) transformer-differential relays of transformers on the bus at the other end of the line, or (3) line relays of adjoining line sections. The interrupting time of the circuit breakers of these various elements will also affect the second-zone time. This second-zone time is normally about 0.2 second to 0.5 second.

The third-zone unit provides back-up protection for faults in adjoining line sections. So far as possible, its reach should extend beyond the end of the longest adjoining line section under the conditions that cause the maximum amount of underreach, namely, arcs

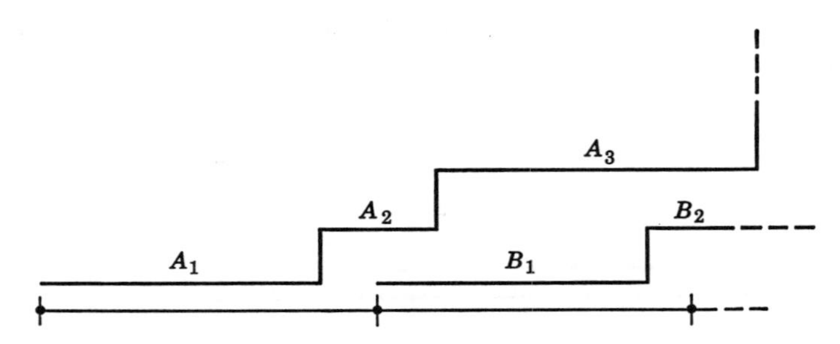

Fig. 3. Normal selectivity adjustment of third-zone unit.

and intermediate current sources. Figure 3 shows a normal back-up characteristic. The third-zone time delay is usually about 0.4 second to 1.0 second. To reach beyond the end of a long adjoining line and

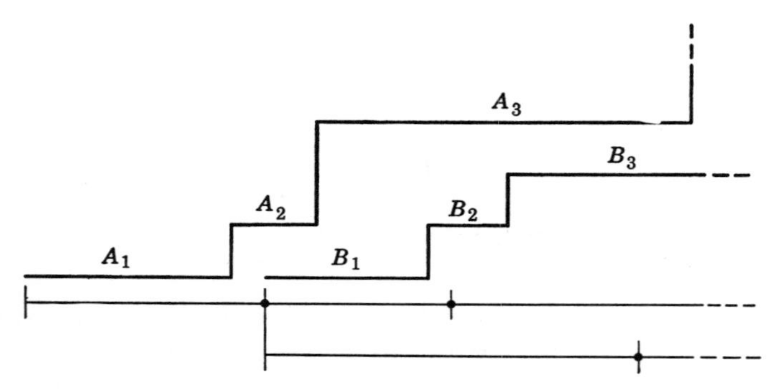

Fig. 4. Third-zone adjustment with additional time for selectivity with relay of a short adjoining line and to provide back-up protection for a long adjoining line.

still be selective with the relays of a short line, it may be necessary to get this selectivity with additional time delay, as in Fig. 4.

When conditions of Fig. 4 exist, the best solution is to use the type of back-up relaying described later under the heading "The Effect of Intermediate Current Sources on Distance-Relay Operation." Then, one has only the problem of adjusting the first- and second-zone units.

Under no circumustances should the reach of any unit be so long that the unit would operate for any load condition or would fail to reset for such a condition if it had previously operated for any reason. To determine how near a distance relay may be to operating under a maximum load condition, in lieu of more accurate information, it is the practice to superimpose the relay's reset characteristic on an *R-X* diagram with the point representing the impedance when the equivalent generators either side of the relay location are 90° out of phase. This is done by the method described in Chapter 9 for drawing the loss-of-synchronism characteristic. Stability can be maintained with somewhat more than a 90° displacement, but 90° is nearly the limit and is easy to depict, as described in Chapter 9.

THE EFFECT OF ARCS ON DISTANCE-RELAY OPERATION

Chapter 9 shows the effect of fault or arc resistance on the appearance of different kinds of short circuits when plotted on an *R-X* diagram in terms of the voltages and currents used by distance relays. Chapter 13 gives data from which arc resistance can be estimated for plotting such fault characteristics on the *R-X* diagram. It is only necessary, then, to superimpose the characteristic of any distance relay in order to see what its response will be.

The critical arc location is just short of the point on a line at which a distance relay's operation changes from high-speed to intermediate time or from intermediate time to back-up time. We are concerned with the possibility that an arc within the high-speed zone will make the relay operate in intermediate time, that an arc within the intermediate zone will make the relay operate in back-up time, or that an arc within the back-up zone will prevent relay operation completely. In other words, the effect of an arc may be to cause a distance relay to underreach.

For an arc just short of the end of the first- or high-speed zone, it is the initial characteristic of the arc that concerns us. A distance relay's first-zone unit is so fast that, if the impedance is such that the unit can operate immediately when the arc is struck, it will do so before the arc can stretch appreciably and thereby increase its resistance. Therefore, we can calculate the arc characteristic for a length equal to the distance between conductors for phase-to-phase faults, or across an insulator string for phase-to-ground faults. On the other hand, for arcs in the intermediate-time or back-up zones, the effect of wind stretching the arc should be considered, and then the operating time for which the relay is adjusted has an important bearing on the outcome.

Tending to offset the longer time an arc has to stretch in the wind when it is in the intermediate or back-up zones is the fact that, the farther an arcing fault is from a relay, the less will its effect be on the relay's operation. In other words, the more line impedance there is between the relay and the fault, the less change there will be in the total impedance when the arc resistance is added. On the other hand, the farther away an arc is, the higher its apparent resistance will be because the current contribution from the relay end of the line will be smaller, as considered later.

A small reduction in the high-speed-zone reach because of an arc is objectionable, but it can be tolerated if necessary. One can always use a reactance-type or modified-impedance-type distance relay to minimize such reduction.[3] The intermediate-zone reach must not be

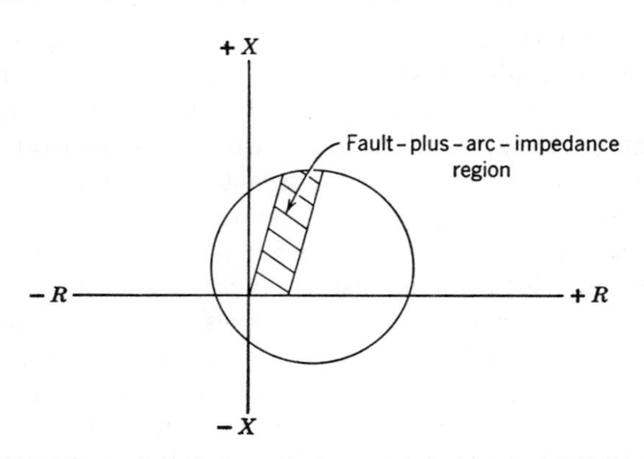

Fig. 5. Offsetting relay characteristic to minimize susceptibility to arcs.

reduced by an arc to the point at which relays of the next line back will not be selective; of course, they too will be affected by the arc, but not so much. Reactance-type or modified-impedance-type distance relays are useful here also for assuring the minimum reduction in second-zone reach. Figure 5 shows how an impedance or mho characteristic can be offset to minimize its susceptibility to an arc. One can also help the situation by making the second-zone reach as long as possible so that a certain amount of reach reduction by an arc is permissible. Conventional relays do not use the reactance unit for the back-up zone; instead, they use either an impedance unit, a modified-impedance unit, or a mho unit. If failure of the back-up unit to operate because of an arc extended by the wind is a problem, the modified-impedance unit can be used or the mho—or "starting"—unit characteristic can also be shifted to make its operation less affected by arc resistance. The low-reset characteristic of some types of distance relay is advantageous in preventing reset as the wind stretches out an arc.

Although an arc itself is practically all resistance, it may have a capacitive-reactance or an inductive-reactance component when viewed from the end of a line where the relays are. The impedance of an arc (Z_A) has the appearance:

$$Z_A = \frac{(I_1 + I_2)}{I_1} R_A = R_A + \frac{I_2}{I_1} R_A$$

where I_1 = the complex expression for the current flowing into the arc from the end of the line where the relays under consideration are.

I_2 = the complex expression for the current flowing into the arc from the other end of the line.

R_A = the arc resistance with current $(I_1 + I_2)$ flowing into it.

If I_1 and I_2 are out of phase, Z_A will be a complex number. Therefore, even a reactance-type distance relay may be adversely affected by an arc. This effect is small, however, and is generally neglected.

Of more practical significance is the fact that, as shown by the equation, the arc resistance will appear to be higher than it actually is, and it may be very much higher. After the other end of the line trips, the arc resistance will be higher because the arc current will be lower. However, its appearance to the relays will no longer be magnified, because I_2 will be zero. Whether its resistance will appear to the relays to be higher or lower than before will depend on the relative and actual magnitudes of the currents before and after the distant breaker opens.

THE EFFECT OF INTERMEDIATE CURRENT SOURCES ON DISTANCE-RELAY OPERATION

An "intermediate-current source" is a source of short-circuit current between a distance-relay location and a fault for which distance-relay operation is desired. Consider the example of Fig. 6. The true impedance to the fault is $Z_A + Z_B$, but, when the intermediate current I_2 flows, the impedance appears to the distance relays as $Z_A + Z_B + (I_2/I_1)\,Z_B$; in other words, the fault appears to be farther away because of the current I_2. This effect has been called the "mutual impedance" effect. It will be evident that, if I_1 and I_2 are out of phase, the impedance $(I_2/I_1)\,Z_B$ will have a different angle from Z_B.

If the distance relays are adjusted to operate for a fault at a given location when a given value of I_2 flows, they will operate for faults beyond that location for smaller values of I_2. Therefore, it is the practice to adjust distance relays to operate as desired on the basis of no intermediate current source. Then, they will not overreach and operate undesirably. Of course, when current flows from an intermediate source, the relays will "underreach," i.e., they will not operate for faults as far away as one might desire, but this is to be preferred to overreach.

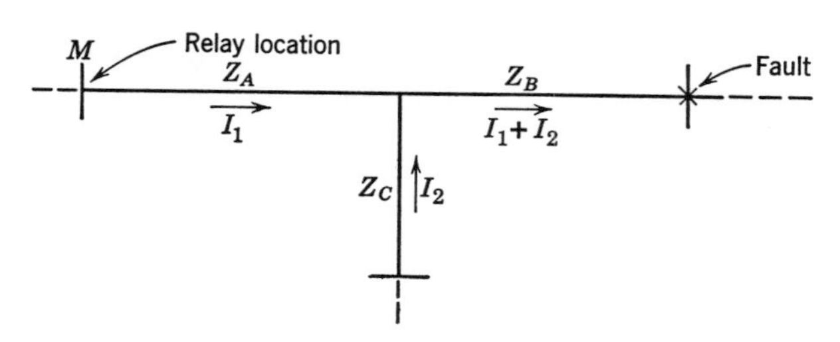

Fig. 6. Illustrating the effect of intermediate current sources on distance-relay operation.

Because of the effect of intermediate current sources, the full capabilities of distance relaying cannot be realized on multiterminal lines. It is the practice to adjust the high-speed zone of the relays at a given terminal to reach 80% to 90% of the distance to the nearest terminal, neglecting the effect of an intermediate current source. Thus, in Fig. 6, the maximum reach of the high-speed zone of the relays at M would be 80% to 90% of $Z_A + Z_B$ or of $Z_A + Z_C$, whichever was smaller. Neglecting the effect of an arc, if this maximum reach of the high-speed zone is less than Z_A, it will become evident that intermediate current cannot affect the high-speed-zone reach; if the maximum reach is greater than Z_A, intermediate current will cause the reach to approach Z_A as a minimum limit. If the second-zone reach is made to include double the impedance of the common branch, tripping will always be assured although it might be sequential.[4]

Back-up protection of the conventional type described in Chapter 1 is often impossible in the presence of intermediate current sources. In Fig. 7, consider the problem of adjusting relays at A to provide back-up for the fault location shown, in the event that breaker B fails to trip for any reason. The problem is similar whether inverse-time or distance relays are involved at A. The magnitude of the fault current flowing at A, or the impedance measured by a distance relay at A, may vary considerably, depending on the magnitude of fault current fed into the intermediate station from other sources. The range of such variations must be taken into account in determining the back-up adjustment. In some extreme cases, the apparent impedance between A and the fault may be such as to put the fault beyond the reach of the relays at *A*. Obviously, this problem may apply also to the relays in the other lines except the faulted one.

A solution that has been resorted to, when a fault can be beyond the reach of conventional back-up relays, is to have the back-up unit of the relaying equipment of each line at the intermediate station operate a timer which, after a definite time, will energize a multicontact auxiliary tripping relay to trip all the breakers connected to the bus of the intermediate station.[5] Admittedly, this solution violates one of the fundamental principles of back-up protection by assuming that the failure to trip is owing only to failure in the

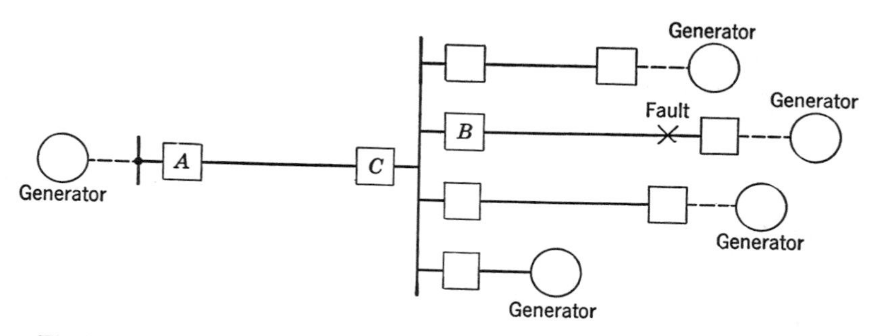

Fig. 7. A situation in which conventional back-up relaying is inadequate.

breaker or in the tripping circuit between the relay and the breaker; it assumes that the protective-relaying equipment or the source of tripping voltage will not fail. However, it is a practical solution, and it has been considered worth the risk. A more reliable arrangement is to use separate CT's and protective relays to energize the multicontact tripping relay, employing only the battery in common. It is also possible to have the back-up relay first try to trip the breaker of the faulty line before tripping all the other breakers.[5]

The back-up elements of mho-type distance relays can be made to operate for faults in the direction opposite to the conventional back-up direction. In fact, the reversed back-up direction, called "reversed third zone," is normally provided when mho-type distance relays are used with directional-comparison carrier-current-pilot relaying. This feature gives some relief in the problem of reaching far enough to provide back-up protection for adjoining line sections, since the back-up elements, being closer to these adjoining line sections, do not have to reach so far. Referring to Fig. 7, for example, if the back-up elements located at breaker *C* are arranged to operate for current flow toward the fault, their reach can be reduced by the distance from A to C as compared with back-up elements at A looking toward the fault.

Various other solutions have been resorted to, depending on how many different possibilities of failure one may wish to anticipate.[6]

OVERREACH BECAUSE OF OFFSET CURRENT WAVES

Distance relays have a tendency to overreach, similar to that described in Chapter 13 for overcurrent relays, when the fault current contains a d-c offset. Other things being equal, the tendency to overreach is greatest in magnetic-attraction types of distance relays, and particularly in the impedance type where the contact-closing torque is generated by current alone. The tendency is the least with induction-type relays.

"Percent overreach" for distance relays has been defined as follows:

$$\text{Percent overreach} = 100\left(\frac{Z_o - Z_s}{Z_s}\right)$$

where Z_o = the maximum impedance for which the relay will operate with an offset current wave, for a given adjustment.

Z_s = the maximum impedance for which the relay will operate for symmetrical currents, for the same adjustment as for Z_o.

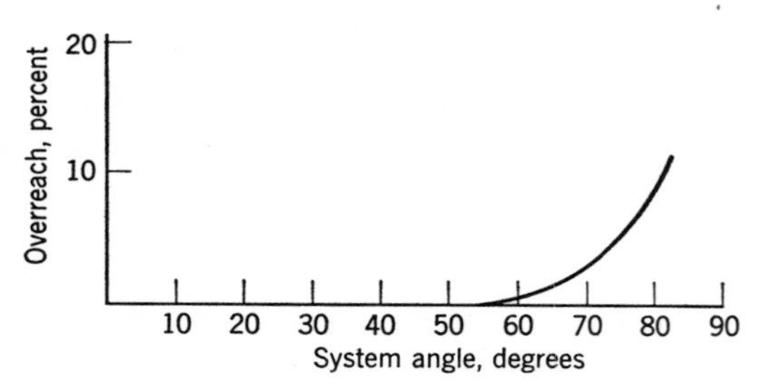

Fig. 8. Overreach characteristic of a certain distance relay.

As for overcurrent relays, the percent overreach increases as the system angle ($\tan^{-1} X/R$) increases. This angle increases with higher-voltage lines because the greater spacing between conductors makes the inductive reactance higher. Figure 8 shows a curve of percent overreach versus the system angle for one type of distance relay.

The overreach of a distance relay is of concern usually only for the first- or high-speed zone. The reaches of the intermediate and back-up zones are usually not nearly as critical as the reach of the first zone. Also, the intermediate and back-up-zone time delays are long enough for the offset to die out and to permit a distance unit to reset if it has overreached and if it is a type of unit whose reset is practically equal to its pickup.

The greater the first-zone overreach, the less of the line may the first zone be adjusted to protect. If the current wave were always fully offset, one could adjust the relay, relying on overreach, so as to protect the desired portion of the line at high speed. But the current wave will rarely be fully offset; it will usually have little offset. Therefore, one cannot depend on overreach, and less than the desired portion of the line will usually be protected at high speed. If the relays at both ends of the line are considered, and if each protects P percent of the line from its end at high speed, only the middle $(2P - 100)$ percent of the line is protected at high speed at both ends simultaneously.

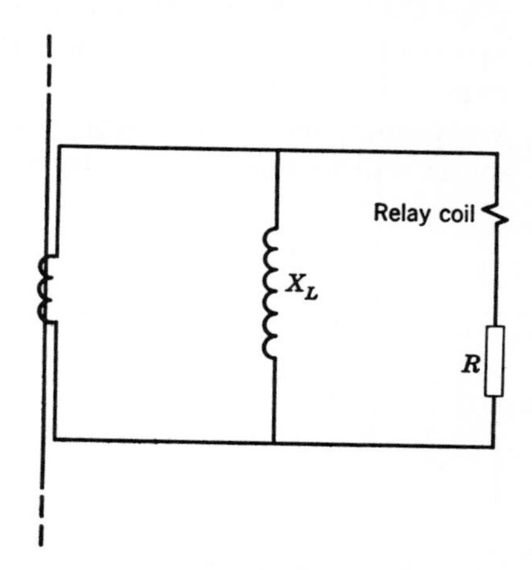

Fig. 9. A transient shunt to minimize overreach.

A device called a "transient shunt," shown in Fig. 9, has been used to minimize overreach with relays whose overreach might otherwise be objectionable.[7] The inductive reactor (X_L) is designed to have a very low resistance component so as to provide a low-impedance by-pass for the d-c component; the reactance is made high to block out most of the symmetrical a-c component. The resistor (R) in series with the relay coil may or may not be needed, depending on the impedance characteristic of the relay coil; the resistor is used, when necessary, to avoid a transient in the relay circuit that would occur if X/R of the relay circuit approached X/R of X_L.

OVERREACH OF GROUND DISTANCE RELAYS FOR PHASE FAULTS

Reference 8 shows that if phase-to-neutral voltage and compensated phase current are used, one of the three ground distance relays may overreach for phase-to-phase or two-phase-to-ground faults. This is not a transient effect like overreach because of offset waves; it is a consequence of the fact that such ground relays do not measure distance correctly for interphase faults. (Phase distance relays do not measure distance correctly for single-phase-to-ground faults, but, fortunately, they tend to underreach.) For this reason, it is necessary to use supplementary relaying equipment that will either permit tripping only when a single-phase-to-ground fault occurs or block tripping by the relay that tends to overreach.

USE OF LOW-TENSION VOLTAGE

Chapter 8 shows the connections of potential transformers for obtaining the proper voltages. It is emphasized that a low-tension source is not reliable unless there are two or more paralleled power-transformer banks with separate breakers; with only one bank, the source will be lost if this bank is taken out of service.

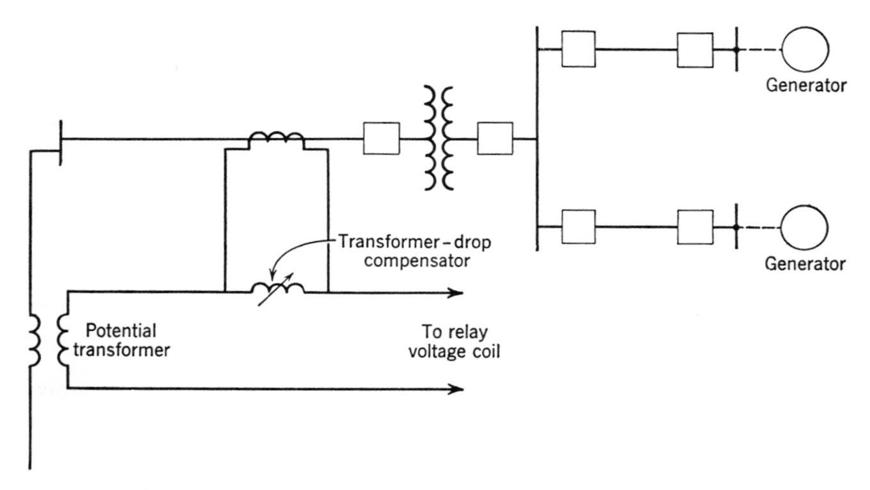

Fig. 10. Transformer-drop compensator for distance relays.

Whenever two or more high-voltage lines are connected to generating sources, as in Fig. 10, "transformer-drop compensation" should be used. For reasons to be given later, it is not sufficient merely to provide the relays with voltages that correspond in phase to the high-tension voltages that would be used if they were available; it is further necessary to correct for the voltage rise or drop in the transformer bank. In other words, by means of transformer-drop compensation, we take into account the fact that, in terms of per unit quantities:

$$V_H = V_L - I_T Z_T$$

where V_H = the high-tension voltage.

V_L = the low-tension voltage.
I_T = the current flowing from the low-tension side toward the high-tension side.
Z_T = the transformer impedance.

Figure 10 shows schematically how the low-tension current is used to produce a voltage that is added to the low-tension voltage in order to provide the relays with a voltage corresponding in phase and magnitude to that on the high-voltage side. Reference 9 gives detailed descriptions of actual connections.

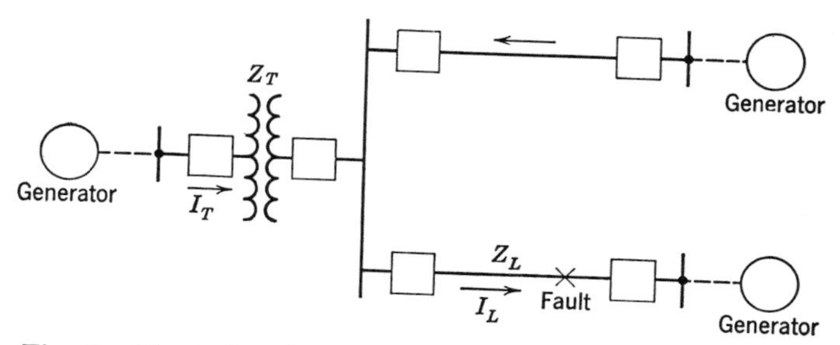

Fig. 11. Illustrating the need for transformer-drop compensation.

Transformer-drop compensation is required for Fig. 10 whether or not a generating source may be connected to the low-voltage side of the transformer bank. Synchronous motors may be considered such a source. Consider Fig. 11 in which a source is assumed to exist. Without transformer-drop compensation, and for a fault at the end of the line, the relays would see an impedance:

$$Z = \frac{I_T Z_T + I_L Z_L}{I_L} = \frac{I_T}{I_L} Z_T + Z_L$$

where Z_L is the per unit line impedance. It will be realized that I_T/I_L can be any value from zero to unity, depending on how the system is being operated; consequently, without transformer-drop compensation, the distance relays must be adjusted for $I_T/I_L = 0$ so as not to overreach. Then, for other values of I_T/I_L the relays will underreach, and underreaching is objectionable because less of the line is protected at high speed. With transformer-drop compensation, most of the first term is eliminated and the relays see practically the correct impedance Z_L, regardless of I_T/I_L, thus minimizing underreaching.

Even if there is never to be a source of generation on the low-voltage side, transformer-drop compensation is still necessary if there are two high-voltage lines either of which may be connected to a generating source. Consider the case of Fig. 12 in which a fault has occurred on the low-voltage side, and some load current continues to flow in line A on the high-voltage side. Without transformer-drop compensation, one or more of the relays of A would be likely to operate because there might be little or no voltage restraint but sufficient voltage for polarization to cause undesired tripping. Compensation would eliminate such a possibility. This possibility exists only when there are two or more high-voltage lines either of which may be connected to a generating source at its far end; otherwise, load current could not flow as in Fig. 12.

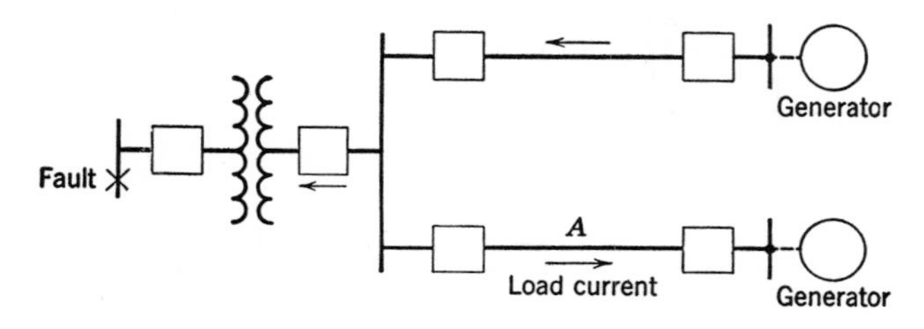

Fig. 12. Another instance in which transformer-drop compensation is needed.

Great care must be taken to avoid overcompensation. If the transformer drop is overcompensated, and if a short circuit that reduced the high-tension voltage to zero should occur on the protected line at the end where transformer-drop compensation is used, the relay voltage would be reversed in phase. This would prevent operation of the relays of the faulted line and would cause undesired operation of the relays of other lines supplying current to the fault. In order to avoid the possibility of overcompensation, the transformer drop should be undercompensated by an amount sufficiently large to take into account the effect of errors in the equipment and in the data on which the calculations are based.

Errors in compensation also affect the reach of the distance relays, and must be carefully considered so as to avoid overreach. In practice the high-speed reach of distance relays is adjusted to about 80% to 90% of the sum of the uncompensated part of the transformer impedance plus the line impedance for a fault at the far end of the line. In other words, if we use 90%,

$$R = 0.9\left[\frac{(1 - C)I_T Z_T + I_L Z_L}{I_L}\right]$$

where R = the high-speed reach of the distance relays.
C = the fraction of the total transformer drop that is intended to be compensated, neglecting any errors in the compensation.

I_T, Z_T, I_L, and Z_L are as previously defined. The *actual* uncompensated part of the effective transformer impedance, taking into account the effect of error, is:

$$Z_{TU} = [(1 - C) - EC]\frac{I_T Z_T}{I_L}$$

where E = the fractional error in C, being positive when the actual compensation is greater than C.

Therefore, the amount of the line (R_L) that is actually protected when the relay is adjusted for reach R is :

$$R_L = R - Z_{TU}$$

$$= [-0.1(1 - C) + EC]\frac{I_T Z_T}{I_L} + 0.9Z_L$$

Now, it would be undesirable for R_L to exceed $0.9Z_L$ because we need $0.1Z_L$ as a factor of safety against overreaching because of other reasons that are always applicable whether we have transformer-drop compensation or not. Therefore, the first term of the equation for R_L must be practically zero or be negative. If we let the first term be zero, we lose the significance of $I_T Z_T/I_L$; so let us assume that the first term is 10% of our factor of safety, or $0.01Z_L$. In other words:

$$0.01Z_L = [-0.1(1 - C) + EC]\frac{I_T Z_T}{I_L}$$

Solving for E we get:

$$E = \frac{Z_L I_L}{100C(Z_T I_T)} + 0.1\frac{(1 - C)}{C}$$

Let $Z_L I_L / Z_T I_T = N$. Then:

$$E = \frac{0.01N + 0.1(1 - C)}{C}$$

For any value of compensation, this equation gives the maximum positive error in the compensation that we can tolerate without having the relay's reach exceed 91% of the line length. An error of about ±3% is reasonable to expect, which permits about 80% to 90% compensation.

Negative compensation error will cause underreaching. This is objectionable also because less of the line will be protected at high speed, but it can be tolerated if necessary.

When a single line terminates in a power transformer with a low-voltage generating source, transformer-drop compensation offers no benefit unless it is so accurate that more than 90% of the transformer drop can be safely compensated, which is not apt to be the case. In practice compensation is not used, but the reach of the distance relays is adjusted for 80% to 90% of the sum of the transformer impedance plus the line impedance. Thus, if we adjust for 90%,

$$R = 0.9(Z_T + Z_L)$$

The amount of the line (R_L) that is protected with this reach is:

$$\begin{aligned} R_L &= 0.9(Z_T + Z_L) - Z_T \\ &= 0.9Z_L - 0.1Z_T \end{aligned}$$

If low-tension voltage is to be obtained from one low-tension side of a three-winding power-transformer bank having generating sources on both low-tension sides, it becomes necessary to use two sets of transformer-drop compensators. The details of this application are presented in Reference 9.

It will probably be evident from the foregoing that low-tension voltage for distance relays is an inferior alternative to high-tension voltage.[10] It will not permit the full capabilities of the relays to be realized, and, unless great care is taken in the adjustment of the equipment, it may even cause faulty operation.

USE OF LOW-TENSION CURRENT

Where a suitable current-transformer source of current for distance relays is not available in the high-voltage circuit to be protected, a source on the low-voltage side of an intervening power-transformer bank may be used. This practice is usually followed for external-fault back-up relays of unit generator-transformer arrangements. Low-tension current may infrequently be used where a line terminates in a power-transformer bank with no high-voltage breaker. In either of such circumstances, the possibility of losing the current source is not a consideration, as with a low-tension-voltage source, because the current source is not needed when the transformer bank is out of service.

When low-tension current is used where a line terminates in a transformer bank without a high-voltage breaker, it is theoretically possible that occasionally the distance relays might operate undesirably on magnetizing-current inrush. If such operation is possible, it can be avoided, if desired, by the addition of supplementary equipment that will open the trip circuit during the inrush period; such equipment uses the harmonic components of the inrush current in a manner similar to that of the harmonic-current-restraint relay described in Chapter 11 for power-transformer protection. However, there is really no need for concern. The probability of getting enough inrush current to operate a distance relay is quite low. In those infrequent cases in which a distance relay does operate to trip the transformer breaker, one may merely reclose the breaker and it probably will not trip again; this is permissible so long as the transformer-differential relay has not operated. As mentioned in Chapter 11, tripping on magnetizing-current inrush is objectionable only because one cannot be sure if it was actually an inrush or a fault that caused tripping; but, if the transformer-differential relay has not operated, one can be sure that it was not a transformer fault.

To use low-tension current, it is necessary to supply the relays with the same current components as when high-tension current is used. It will be seen from Chapter 9 that phase distance relays use the difference between the currents of the phases from which their voltage is obtained. (When high-tension current is used, this phase-difference —or so-called "delta"—current is obtained either by connecting the high-voltage CT's in delta or by providing two current coils on the magnetic circuits of each relay and passing the two phase currents through these coils in opposite directions. The two-coil type of relay has the advantage of permitting the CT's to be connected in wye; this is preferred because it avoids auxiliary CT's when the wye connection is needed for ground relaying.)

Figure 13 shows the current connections for one of the three distance relays used for inter-phase-fault protection; these connections are for a two-coil type of relay that uses the high-tension voltage V_{ab}. The connections A are the connections if high-tension CT's are available, and B and C are alternative low-tension connections. The power transformer is assumed to have the standard connections described in Chapter 8 for the voltage phase sequence a-b-c. The terminals of the two coils are labeled for each connection so that the three connections can be related. If we assume that each relay coil has N turns, the ampere-turns for each connection are as in the accompanying table. The significant thing about the three ampere-

Connection	Ampere-Turns
A	$(I_a - I_b)N$
B	$3(I_a - I_b)N$
C	$2(I_a - I_b)N$

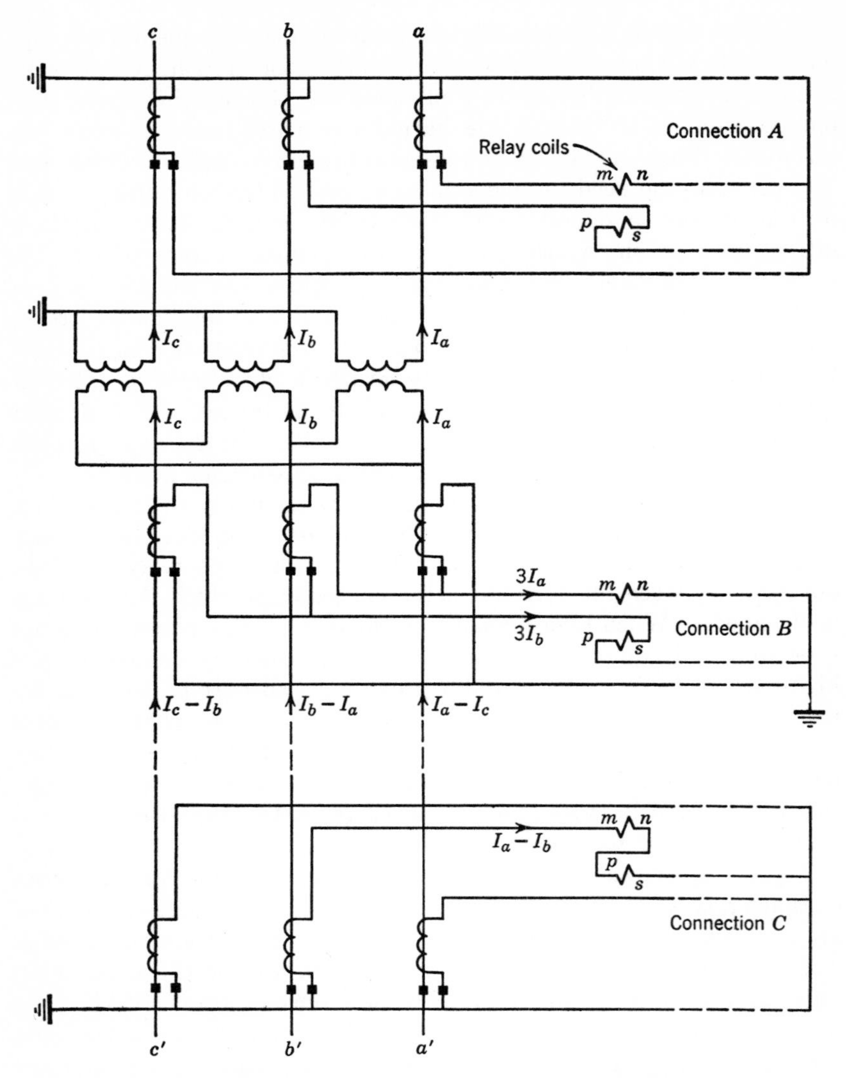

Fig. 13. Low-tension-current connections for a two-coil distance relay.

turns expressions is that all of them contain $(I_a - I_b)$ as required for proper distance measurement. That the ampere-turns of B and C are, respectively, 3 times and 2 times those of A is merely a consequence of the unity transformation ratios assumed for the power transformer and the CT's. However, connection C is different from the other two in that, if balanced three-phase currents of the same magnitude are supplied to the terminals of relays connected as in A (or B) and C, the ampere-turns of the relays of C will be $2/\sqrt{3}$ times the ampere-turns of the relays of A (or B). Since the adjustment procedure for such relays is usually given for the connections of A, the difference in ampere-turns can be taken care of by assuming that

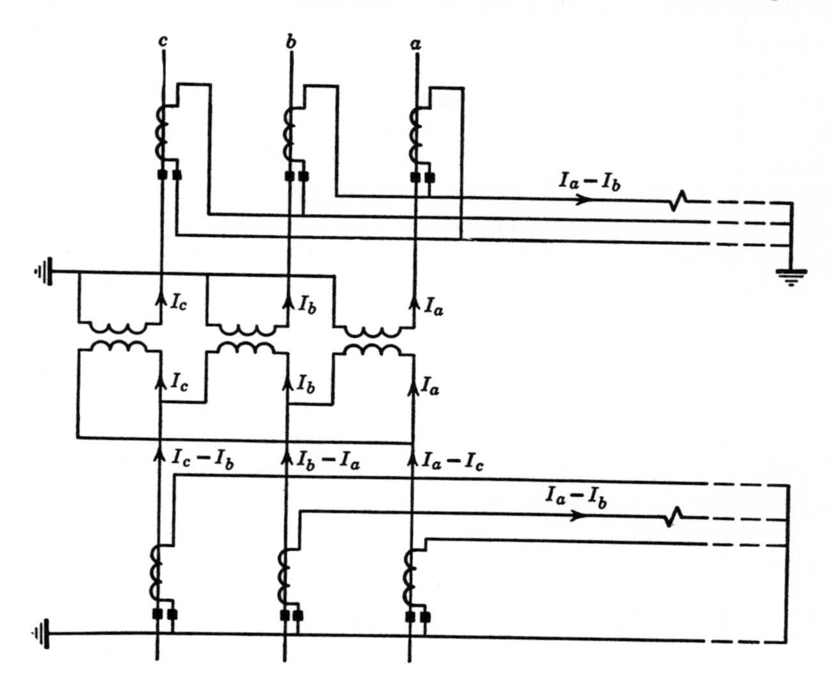

Fig. 14. Low-tension-current connections for a single-coil distance relay.

the magnitude of the current supplied to C is $2/\sqrt{3}$ times that supplied to A, or, in other words, that the CT ratio for C is $\sqrt{3}/2$—or 87%—of its actual value. The actual CT ratio for any connection is the ratio of the high-voltage-circuit phase-current magnitude to the relay phase-current magnitude under normal balanced three-phase conditions.

Figure 14 shows the high-tension and low-tension current connections for a single-coil type of distance relay. The CT ratio is the same as that defined in the preceding paragraph.

EFFECT OF POWER-TRANSFORMER MAGNETIZING-CURRENT INRUSH ON DISTANCE-RELAY OPERATION

The effect of power-transformer magnetizing-current inrush on distance-relay operation is also discussed under the heading "Use of

Low-Tension Current." It is the purpose here to consider the case of inrush to the HV windings of the transformers at the far end or ends of a line.

The writer does not know of any cases of tests, theoretical studies, or actual trouble in this respect. Therefore, it is concluded that, if the possibility of distance-relay misoperation exists, it must be extremely remote. The most severe inrush that can occur in existing conventional power transformers would be less likely to operate a distance relay than would a three-phase fault on the other side of the transformer bank. This is because the rms magnitude of the initial inrush current is generally less than that of the fault current. Furthermore, the inrush contains harmonic currents to which certain relays do not respond as well as to the fundamental components. The high-speed zone of a distance relay is not permitted to reach through transformers at the far ends of a line, and, therefore, if any relay unit is to operate it would have to be either the second- or third-zone unit. These units generally have enough time delay so that the inrush will have subsided considerably before a unit could operate, which further lessens the likelihood of their operation. And, finally, the distance relays in question will usually get only a part of the inrush current in those cases in which misoperation would be most objectionable, such as when energizing a transformer bank tapped to an important line. Therefore, there is little wonder that this subject is not cause for concern.

THE CONNECTIONS OF GROUND DISTANCE RELAYS

Reference 11 shows that for accurate distance measurement, a ground relay may be supplied with a phase-to-neutral voltage and the sum of the corresponding phase current and an amount proportional to the zero-phase-sequence current. If there is another line nearby that can induce voltage in the line under consideration when a ground fault occurs anywhere, there must also be added to the phase current an amount proportional to the zero-phase-sequence current of the other line. The addition of these zero-phase-sequence-current quantities is called "current compensation." Reference 11 also describes an alternative to current compensation called "voltage compensation," whereby the voltage is compensated by zero-phase-sequence-voltage quantities. Compensation is necessary because variations in the distribution of zero-phase-sequence current relative to the distribution of positive- and negative-phase-sequence current would otherwise cause objectionable errors in distance measurement.

Ground distance relays can also be energized by zero-phase-sequence-voltage drop and zero-phase-sequence current to measure distance by measuring the zero-phase-sequence impedance.[12]

OPERATION WHEN PT FUSES BLOW

Distance relays that are capable of operating on less than normal load current may operate to trip their breaker when a potential-transformer fuse blows. In one system, blown fuses caused more undesired tripping than any other thing until suitable fusing was provided.[13]

Since potential transformers are generally energized from a bus and supply voltage to the relays of several lines, it is advisable to provide separate voltage circuits for the relays of each line and to fuse them separately if fusing is to be used at all. With separate fusing, trouble in the circuit of one set of relays will not blow the fuses of another circuit. The principal objection to bus PT's is that "all the eggs are in one basket"; but, with separately fused circuits, this objection is largely eliminated.

Neon lamps should be used for pilot indication of the voltage supply to each set of distance relays.

When the relays do not have to operate on less than load current, instantaneous overcurrent units can be used to prevent tripping for a blown fuse during normal load conditions. The overcurrent-relay contacts are in series with the trip circuit. Undesired tripping is still possible should a fault occur before the blown fuse has been replaced; it is to minimize this possibility that the indicating lamps are recommended. If the ground-fault current is high enough, a single set of three instantaneous overcurrent relays can be used separately from the distance relays to prevent undesired tripping by either the phase or ground relays; otherwise, a single ground overcurrent relay would be used for the ground relays.

PURPOSEFUL TRIPPING ON LOSS OF SYNCHRONISM

When generators have gone out of synchronism, all ties between them should be opened to maintain service and to permit the generators to be resynchronized. The separation should be made only at such locations that the generating capacity and the loads on either side of the point of separation will be evenly matched so that there will be no interruption to the service.[14] Distance relays at those locations are sometimes suitable for tripping their breakers on loss of synchronism, and in some systems they are used for this purpose in addition to their usual protective functions. However, as mentioned in Chapter 9, additions or removals of generators or lines during normal operation will often change the response of certain distance relays to loss of synchronism. Therefore, each application should be examined to see if certain distance relays can always be relied on to trip.

A completely reliable method of tripping at preselected locations is available.[15] This relaying equipment contains two angle-impedance units as described in Chapter 4, an overcurrent unit, and several auxiliary relays. It is a single-phase equipment, which is all that is necessary because loss of synchronism is a balanced three-phase phenomenon. The two angle-impedance units use the same one of the three combinations of current and voltage used by phase distance relays for short-circuit protection. Figure 15 shows the operating

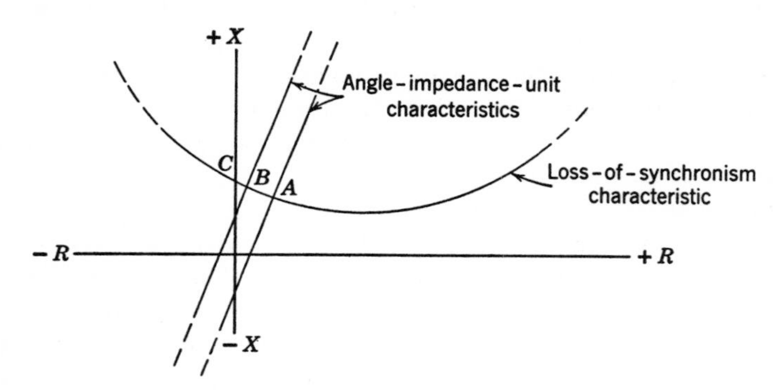

Fig. 15. Relay for tripping on loss of synchronism.

characteristics of the two angle-impedance units on an *R-X* diagram, together with a loss-of-synchronism characteristic for the relay's location on a tie line connecting the generators that have lost synchronism. The significant feature of the equipment is that the two angle-impedance units divide the diagram into the three regions *A*, *B*, and *C*. As the impedance changes during loss of synchronism, the point representing this impedance moves along the loss-of-synchronism characteristic from region *A* into *B* and then into *C*, or from *C* into *B* and then into *A*, depending on which generators are running the faster. As the point crosses the operating characteristic of an angle-impedance unit, the unit reverses its direction of torque and closes a contact to pick up an auxiliary relay. As the impedance point moves into one region after another, a chain of auxiliary relays picks up, one after the other. When the third region is entered, the last auxiliary relay of the chain picks up and trips its breaker. There are two such chains—one for each direction of movement—and the contacts of the two last auxiliary relays of the chains are connected in parallel so that either one can trip the breaker.

The purpose of the overcurrent unit is to prevent tripping during hunting between the generators at light load. This condition is represented by movement along a portion of the loss-of-synchronism characteristic diametrically opposite to the portion shown in Fig. 15. Such movement would also fulfill the requirements for tripping that have been described. Relatively very little current flows during hunting at light load compared with the high current flowing when generators pass through the 180° out-of-phase position which is in the B region on the portion of the loss-of-synchronism characteristic shown in Fig. 15. Therefore, the overcurrent unit's pickup can be adjusted so that the equipment will select between hunting and loss of synchronism.

No other condition can cause the impedance point to move successively through the three regions, and therefore the equipment is completely selective.

Changes in a system cannot cause the equipment to fail so long as there is enough current to operate the overcurrent unit when synchronism is lost. The loss-of-synchronism characteristic may shift up or down on the R-X diagram, or the characteristic may change from one of overexcitation to one of underexcitation, without adversely affecting the operation of the equipment.

When tripping is desired at a location where the current is too low to actuate a loss-of-synchronism relay, remote tripping is necessary over a suitable pilot channel from a location where a loss-of-synchronism relay can be actuated.[16]

When two or more loss-of-synchronism relays are used at different locations, one or more of these relays may need a supplementary single-step distance unit because the overcurrent units may not provide the desired additional selectivity.

The locations where tripping is desired on loss of synchronism may change from time to time as the relations between load and generation change. Under such circumstances, it is desirable to have installations of loss-of-synchronism-relay equipments at several locations so that the load dispatcher can select the equipments to be made operative during any period.

In lieu of complete freedom to choose the best locations to separate parts of a system when synchronism is lost, it may be necessary to resort to some automatic "load shedding." By such means, nonessential load can be dropped automatically either directly when the tie breakers are tripped or indirectly through the operation of relays such as the underfrequency type. This subject is treated in detail in Reference 24 of Chapter 13.

BLOCKING TRIPPING ON LOSS OF SYNCHRONISM

Tripping at certain locations is required when generators lose synchronism, but it should be limited to those locations only. If distance relays at any other locations have a tendency to trip, supplementary relaying equipment should be used to block tripping there. Also, wherever tripping can occur during severe power swings when a system is recovering from the effects of a shock, such as that caused

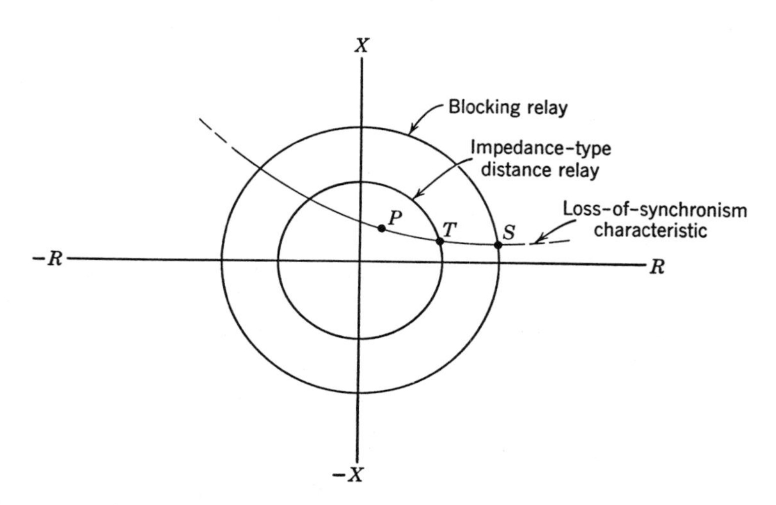

Fig. 16. Relay characteristics of equipment of Fig. 17.

by a short circuit, equipment to block tripping is most desirable; tripping a sound line that is carrying synchronizing power would very likely cause instability.

The method by which tripping on loss of synchronism can be blocked is very ingenious.[17] Consider the *R-X* diagram of Fig. 16. The point *P* represents the impedance for a three-phase fault well within the operating characteristic of an impedance-type distance relay. Assuming that the loss-of-synchronism characteristic passes through *P*, the problem, then, is to devise a selective relaying equipment that will permit tripping when the fault represented by *P* occurs, but not when loss of synchronism reaches a stage that is also represented by *P*. The way in which this selection is made is based on the fact that the change in impedance from the operating conditions just before the fault is instantaneous, whereas the change in impedance during loss of synchronism is relatively slow. The method used for recognizing this difference is to encircle the distance-relay characteristic with a blocking-relay characteristic such as that shown on Fig. 16. (For balanced three-phase conditions, such as those during loss of synchronism, all three phase distance relays see the same impedance; therefore, the one characteristic represents all three relays.) Then, a control circuit is provided so that, if the blocking relay operates sufficiently ahead of a distance relay, as when the impedance is changing from S to T, the distance-relay trip circuit will be opened. But, if the impedance changes instantly to any value such as P inside the distance-relay characteristic, the trip circuit will not be opened.

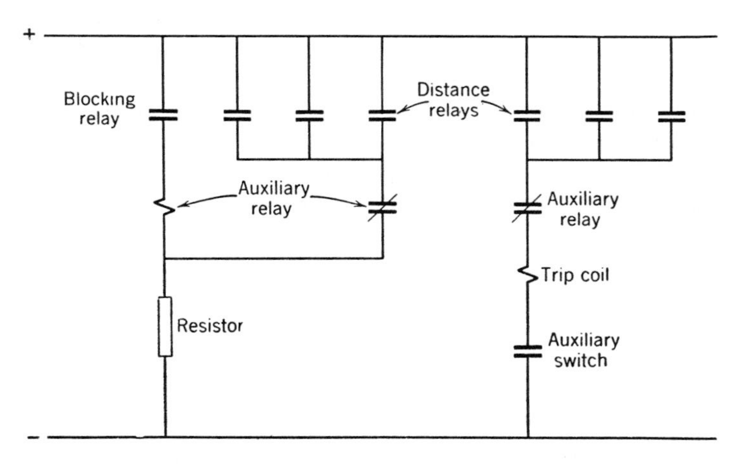

Fig. 17. Equipment for blocking tripping on loss of synchronism.

The circuit for blocking tripping is shown in Fig. 17. If the blocking relay closes its contact to energize the coil of the auxiliary relay before any distance relay closes its contact to short out the auxiliary-relay coil, the auxiliary relay will pick up and open its contact in the trip circuit. The subsequent closing of a distance-relay contact in the circuit around the auxiliary-relay coil will be ineffective because the auxiliary relay will have opened its contact in this circuit. The auxiliary relay is given a little time delay to pick up so as to be sure that it will not pick up when a fault occurs requiring tripping, and the distance and blocking relays close contacts practically together.

The blocking relay may be an impedance type or a mho type so long as its characteristic encircles the characteristic of the distance relays whose tripping is to be blocked. In practice the blocking relay is single phase. At one time, some considered three blocking relays necessary because a fault might still be on the system when blocking was required. Today, this is thought to be an unnecessary complication because of the general use of relatively high-speed relays. A single-phase blocking relay is permissible because loss of synchronism is a balanced three-phase phenomenon. The blocking relay uses the same voltage-and-current combination as one of the phase distance relays. Ground distance relays using phase-to-neutral voltage and compensated phase current could also tend to trip, and, therefore, their tripping should also be blocked.

AUTOMATIC RECLOSING

Chapter 13 introduces the subject of automatic reclosing and describes the practices with overcurrent relaying. Here, we are concerned with the practices with distance relaying. Lines protected by distance relays usually interconnect generating sources. Consequently, the problem arises of being sure that both ends are in synchronism before reclosing. "High-speed reclosing," defined here as reclosing the breaker contacts in about 20 cycles after the trip coil was energized to trip the breaker, cannot be used because of the inherent time delay of distance relaying for faults near the ends of a line. In order to be sure that an arc will not restrike when reclosing the line breakers, the line has to be disconnected at both ends for a time long enough for the ionized gas in the arc path to be dispersed. This takes from about 6 to 16 cycles, depending on the magnitude of the arc current and the system voltage, the average being about 8 to 10 cycles.[18] To provide this time with high-speed reclosing, both ends of a line must be tripped practically simultaneously. Since, with distance relays, one end may trip 6 to 12 cycles or more ahead of the other, depending on the intermediate-time adjustment, this additional time must be added to the reclosing cycle. In other words, about the fastest permissible reclosing time with distance relays is 26 to 32 cycles or longer. The only exceptions are lines with wye-delta power transformers at both ends; there simultaneous high-speed tripping is possible.

Because of the foregoing reclosing times and also the fact that inverse-time-overcurrent relays are generally used for ground-fault protection, such lines require synchronism check, as described in Chapter 13. Of course, one end of a line can be reclosed without synchronism check. Synchronism check is unnecessary if there are enough other interconnections between the generating sources that the line interconnects so that one can be sure that both sides will always be in synchronism. Automatic reclosing can be very harmful if it causes the connection of parts of a system that are out of synchronism;

this is known to have been the "last straw" that caused a major system shutdown.

It is usually the practice to provide one immediate (which is slower than high-speed) reclosure followed by 2 or 3 time-delay reclosures and then lockout if the fault persists.

When a line ends in a power-transformer bank with no high-voltage breaker, but with a grounding switch for remote tripping in the event of a transformer-bank fault, automatic reclosing of the remote breaker is affected. Some users adjust ground-distance relays' first-zone reach to 80% to 90% of the line (i.e., not to reach to the transformer) in order that the first-zone unit will not operate when the grounding switch is closed. Then, automatic reclosing is permitted only if a first-zone unit operates, thereby avoiding reclosing on the grounding switch and a transformer fault. If a three-phase grounding switch is used, the high-speed zone of the remote distance relays may be permitted to reach into the transformer bank; this will permit automatic reclosure on the grounding switch without harming the transformer, which may be permissible if the shock to the system is not too great.

EFFECT OF PRESENCE OF EXPULSION PROTECTIVE GAPS

It is generally necessary to delay tripping by the high-speed zone of distance relaying if a line is equipped with expulsion protective gaps. A minimum relay-operating time of 2 or 3 cycles is usually sufficient to prevent high-speed-relay tripping while an expulsion protective gap is functioning. This additional delay in the tripping time is provided by the addition of an auxiliary relay. The tripping circuit should be carried through the protective-relay contacts to avoid undesired tripping because of overtravel of the auxiliary relay.

EFFECT OF A SERIES CAPACITOR

A series capacitor can upset the basic premises on which the principles of distance and directional relaying are founded. These premises are (1) that the ratio of voltage to current at a relay location is a measure of the distance to a fault, and (2) that fault currents are approximately reversed in phase only for faults on opposite sides of a relay location. A series capacitor introduces a discontinuity in the ratio of voltage to current, and particularly in the reactance component of that ratio, as a fault is moved from the relay location toward and beyond a series capacitor. One can easily visualize the effect by plotting the impedance points on an *R*-*X* diagram. As a fault is moved from the relay side of the capacitor to the other side, the capacity reactance subtracts from the accumulated line reactance between the relay and the fault. As a consequence, the fault may appear to be much closer to the relay location or it may even have the appearance to some relays of being back of the relay location.

One way that series capacitors are used[19] minimizes their adverse effect on distance relays. A single capacitor bank is chosen to compensate no more than about half of the reactance of a given line section; if a higher degree of compensation is used, the capacitors are divided into two or more banks located at different places along a line. Also, a protective gap is provided across each capacitor bank, and this gap flashes over immediately when a fault occurs and effectively shorts out the capacitor bank while fault current flows. In other words, the capacitor banks are in service normally, are shorted out while a fault exists, and are returned to service immediately when the faulty line section is tripped.

Where capacitors are located otherwise, it will probably be necessary to add slight time delay to the distance-relay trip circuit. For fault currents that are too low to flash over the capacitor gap, it may be necessary to use phase-comparison relaying, probably with greater-than-normal sensitivity.

COST-REDUCTION SCHEMES FOR DISTANCE RELAYING

Many ways have been suggested for reducing the cost of high-speed distance relaying so that its use on lower-voltage circuits could be justified. What has been sought is a "class of protection somewhere in the middle ground between the cost, performance, and complexity of overcurrent and conventional 3-zone distance relays."[20] These schemes may be classified as follows:

(*a*) Abbreviated relays.[20]

(*b*) Three conventional relays for phase faults and three conventional relays for ground faults, except that certain units are used in common.[21]

(*c*) Three conventional relays for both phase and ground faults by means of "current and voltage switching."[22]

(*d*) One conventional relay for phase faults and/or one conventional relay for ground faults by means of current and voltage switching.[12,23]

(*e*) One conventional relay for both phase and ground faults by means of current and voltage switching.[22]

Current and voltage switching is a means for automatically connecting the relay to the proper current and voltage sources so that it will measure distance correctly for any fault that occurs.

With the possible exception of (*e*), all these schemes are in use in this country, although none of them very extensively. The greater the departure from the conventional arrangement of three phase and three ground relays, the poorer is the quality of relaying. They either have more time delay, are harder to apply, are less accurate, or require more frequent maintenance. Voltage switching may not be permitted with capacitance potential devices because of the errors that result from changing the burden. It is probably economically feasible to standardize on one intermediate arrangement, but there is little justification for much more.

To complete the list, other combinations of units should be included, most of which may be classified as "abbreviated" relays. They consist of various combinations of units like those used in conventional high-speed relays.[7] Such relays supplement existing relays for speeding up the protection. For example, three single-step directional-distance relays might supplement directional-overcurrent inverse-time relays; this would provide high-speed relaying for 80% to 90% of a line plus inverse-time-overcurrent relaying for the remainder of the line and for back-up protection.

ELECTRONIC DISTANCE RELAYS

Electronic distance relays have been extensively tested[24] that are functionally equivalent to conventional electromechanical distance relays, but that are faster and impose considerably lower burden on a-c sources of current and voltage. At the same time, they are more shock resistant. Their greater speed is most effective when they are used in conjunction with a carrier-current or microwave pilot. When they are used alone, it is still necessary to have some time delay for faults near the ends of a line, which tends to reduce the advantage of a higher-speed first-zone unit.

The lower-burden characteristic of electronic relays may contribute to less expensive current and voltage transformers, and therefore may increase the use of such relays even though the greater speed may not be required.

Apart from differences in physical characteristics, the basic principles of electromechanical distance relays and their application apply equally well to electronic distance relays.

Problems

1. Referring to Fig. 18, it has been determined that load transfer from *A* to *B* or from *B* to *A* prevents setting a distance relay at either *A* or *B* with a

greater ohmic reach than just sufficient to protect a 100-ohm, 2-terminal line with a 25% margin (i.e., third-zone reach = 125 ohms).

a. What is the apparent impedance of the fault at X_1 to the relay at A? Can the relay at A see the fault before the breaker at B has tripped?

b. Can the relay at B see the fault at X_1 before the breaker at A has tripped? Why?

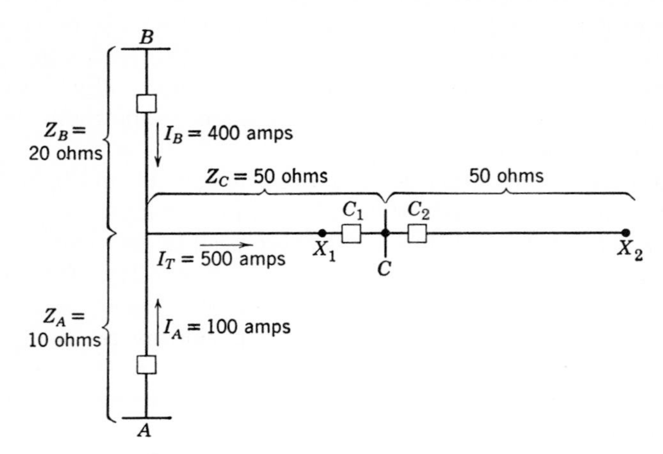

Fig. 18. Illustration for Problem 1.

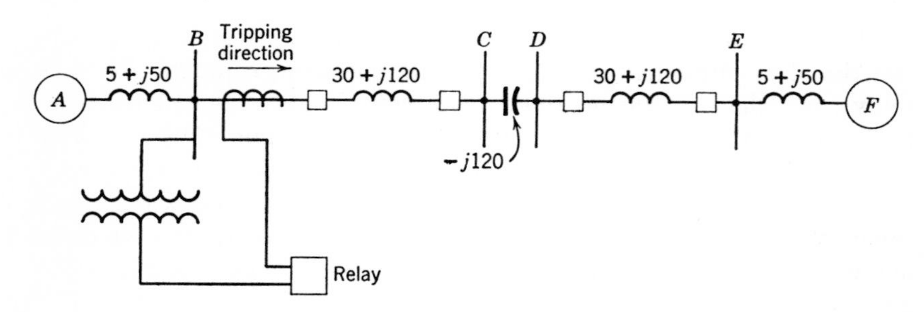

Fig. 19. Illustration for Problem 2.

c. If, for a fault at X_2, breaker C_2 fails to open, will the fault be cleared by relays at A and B? Assume the same relative fault-current magnitudes as for *a* and *b*.

d. By what means can the fault at X_2 be cleared?

2. Show the positive-phase-sequence impedances of the line sections and series capacitor of Fig. 19 on an R-X diagram as seen by phase distance relays at B, neglecting the effect of a protective gap across the capacitor. Comment on the use of distance and directional relays at B.

Bibliography

1. "New Distance Relays Know Faults from Swings," by W. A. Morgan, *Elec. World, 124* (Oct. 27, 1945), pp. 85–87.

2. *Circuit Analysis of A-C Power Systems,* Vols. I and II, by Edith Clarke, John Wiley & Sons, New York, 1943, 1950.

3. "Reactance Relays Negligibly Affected by Arc Impedance," by A. R. van C. Warrington, *Elec. World, 98,* No. 12 (Sept. 19, 1931), pp. 502–505.

"A High Speed Relay for Short Lines," by S. L. Goldsborough and W. A. Lewis, *Elec. Eng., 51* (1932), pp. 157–160.

4. "Relay Protection of Tapped Transmission Lines," by M. A. Bostwick and E. L. Harder, *AIEE Trans., 62* (1943), pp. 645–650. Discussions, pp. 969–972.

5. "Complete Your Back-Up Protection of Transmission Lines," by A. J. McConnell, *Elec. World, 137* (Feb. 11, 1952), pp. 90–91.

6. "Line Back-Up Protection without Additional Relays," by R. Bruce Shipley, *Elec. Light and Power, 31* (Sept., 1953), pp. 114–115.

"A Review of Backup Relaying Practices," by AIEE Committee, *AIEE Trans., 72,* Part III (1953), pp. 137–141. Discussions, p. 142.

7. *Silent Sentinels,* Westinghouse Electric Corporation, Meter Div., Newark, N. J.

8. "Graphical Method for Estimating the Performance of Distance Relays during Faults and Power Swings," by A. R. van C. Warrington, *AIEE Trans., 68,* 1949, pp. 608–620. Discussions, pp. 620–621.

9. "Distance Relaying with Low Tension Potentials," by J. L. Blackburn, *Westinghouse Engineer,* May, 1943, pp. 61–64.

"Voltage Drop Compensation," by J. Basta, *Elec. J., 30,* No. 12 (Dec., 1933), pp. 509–512.

"Compensators for Distance Relaying," by J. L. Blackburn, *Elec. World, 122* (July 22, 1944), pp. 146, 148, 150; (Aug. 19, 1944), pp. 112–113; (Sept. 16, 1944), p. 112.

10. "Operating Experience with Reactance Type Distance Relays," by E. E. George, *AIEE Trans., 50* (1931), pp. 288–293. Discussion, p. 293.

11. "Use of Impedance Relays for the Protection of Overhead Lines against Ground Faults," by Jean Fallou, *Bull. soc. franç. élec., 10,* Series 4 (January, 1930), pp. 82–94.

Appendix A of Reference 22.

"Fundamental Basis for Distance Relaying on Three-Phase Systems," by W. A. Lewis and L. S. Tippett, *AIEE Trans., 66* (1947), pp. 694–708. Discussions, pp. 708–709.

12. "A New Distance Ground Relay," by S. L. Goldsborough, *AIEE Trans., 67,* Part II (1948), pp. 1442–1446. Discussions, pp. 1446–1447.

13. "Operating Experience with Distance Ground Relays," by W. A. Wolfe, *AIEE Trans., 65* (1946), pp. 458–462. Discussions, pp. 1185–1187.

14. "Interim Report on Application and Operation of Out-of-Step Protection," by AIEE Committee, *AIEE Trans., 62* (1943), pp. 567–573.

15. "One Slip Cycle Out-of-Step Relay Equipment," by W. C. Morris, *AIEE Trans., 68,* Part II (1949), pp. 1246–1248.

16. "Operation of Synchronous Condensers on the Southern California Edison System," by C. R. Canady and J. H. Drake, *AIEE Trans., 71* (1952), pp. 1051–1058.

17. "Out-of-Step Blocking and Selective Tripping with Impedance Relays," by H. R. Vaughan and E. C. Sawyer, *AIEE Trans., 58* (1939), pp. 637–645. Discussions, pp. 645–646.

18. "Insulator Flashover Deionization Times As a Factor in Applying High-Speed Reclosing Breakers," by A. C. Boisseau, B. W. Wyman, and W. F. Skeats, *AIEE Trans., 68,* Part II (1949), pp. 1058–1066. Discussions, pp. 1066–1067.

19. "The Series Capacitor in Sweden," by Gunnar Jancke and K. F. Åkerström, *Elec. Eng.*, *71* (1952), pp. 222–227.

"Characteristics of a 400-Mile 230-Kv Series-Capacitor-Compensated Transmission Line," by B. V. Hoard, *AIEE Trans.*, *65* (1946), pp. 1102–1114. Discussions, pp. 1178–1180.

"Series Capacitors in High-Voltage Lines of the Bonneville Power Administration," by Alexander Dovjikov and E. C. Starr, *Elec. Eng.*, *71* (1952), pp. 228–232.

"Series Capacitors during Faults and Reclosing," by E. L. Harder, J. E. Barkle, and R. W. Ferguson, *AIEE Trans.*, *70* (1951). pp. 1627–1641. Discussions, pp. 1641–1642.

20. "A Simplified Unit for Distance Relaying," by A. W. Adams and F. R. Bergseth, *AIEE Trans.*, *72*, Part III (1953), pp. 996–998.

21. "Combined Phase and Ground Distance Relaying," by Warren C. New, *AIEE Trans.*, *69*, 1950, pp. 37–42. Discussions, pp. 42–44.

22. "Control of Distance Relay Potential Connections," by A. R. van C. Warrington, *AIEE Trans.*, *53* (1934), pp. 206–213. Discussions, pp. 465, 466, 617.

23. "Distance Relay Protection for Subtransmission Lines Made Economical," by L. J. Audlin and A. R. van C. Warrington, *AIEE Trans.*, *62* (1943), pp. 574–578. Discussions, pp. 976–979.

24. "Field Experience—Electronic Mho Distance Relay," by H. C. Barnes and R. H. Macpherson, *AIEE Trans.*, *72*, Part III (1953), pp. 857–864. Discussions, p. 865.

"Electronic Protective Relays," by R. H. Macpherson, A. R. van C. Warrington, and A. J. McConnell, *AIEE Trans.*, *67*, Part II (1948), pp. 1702–1707. Discussions, pp. 1707–1708.

"An Electronic Distance Relay Using a Phase-Discrimination Principle," by F. R. Bergseth, *AIEE Trans.*, *73*, Part III-B (1954), pp. 1276–1279. Discussions, p. 1279.

15 LINE PROTECTION WITH PILOT RELAYS

Pilot relaying is the best type for line protection. It is used whenever high-speed protection is required for all types of short circuits and for any fault location. For two-terminal lines, and for many multiterminal lines, all the terminal breakers are tripped practically simultaneously, thereby permitting high-speed automatic reclosing. The combination of high-speed tripping and high-speed reclosing permits the transmission system to be loaded more nearly to its stability limit, thereby providing the maximum return on the investment.

The continuing trend of increasing circuit-breaker-interrupting capabilities is increasing the allowable magnitude of short-circuit current. It is quite possible that damage rather than stability may dictate high-speed relaying in some cases.

Pilot relaying is used on some multiterminal lines where high-speed tripping and reclosing are not essential, but where the configuration of the circuit makes it impossible for distance relaying to provide even the moderate speed that may be required.

Some lines are too short for any type of distance relay. For such lines, it is not merely a matter of getting a distance relay with a lower minimum ohmic adjustment; the ohmic errors would be so high as compared with the ohms being measured that such relaying would be impractical.

Critical loads may require high-speed tripping beyond the capabilities of distance relays.[1]

For these reasons, it is the practice to use pilot relaying for most high-voltage transmission lines and for many subtransmission and distribution circuits. Therefore, it becomes necessary to choose between wire pilot, carrier-current pilot, and microwave pilot.[2] If either of the last two is indicated, one has to choose further between phase comparison and directional comparison, or a combination of the two. In addition to any of these, some form of remote tripping may be required. The application considerations of these various equipments will now be discussed.

Wire-Pilot Relaying

Chapter 5 tells why d-c wire-pilot relaying has largely given way to a-c types. In this chapter, we shall consider the application of only the a-c types.

Wire-pilot relaying is used on low-voltage circuits, and on high-voltage transmission lines when a carrier-current pilot is not economically justifiable. For the protection of certain power-cable circuits, wire pilot may be used because the cable-circuit attenuation is too high for carrier current. For short lines, a-c wire-pilot relaying is the most economical form of high-speed relaying.

Generally, wire pilots no longer than about 5 to 10 miles are used, but there are a few in service as long as about 27 miles.[3] As mentioned in Chapter 5, the technical limitations on the length of a pilot circuit are its resistance and shunt capacitance. Compensating reactors are sometimes used when the shunt capacitance is too high. A pilot circuit that is rented from the telephone company may be much longer than the transmission line for whose protection it is to be used, because such telephone circuits seldom run directly between the line terminals. Therefore, in borderline cases, one should find out the actual resistance and capacitance before deciding to use wire pilot. Other requirements imposed on the wire-pilot circuit are described in Chapter 5.

In general, wire-pilot relaying is not considered as reliable as carrier-current-pilot relaying, mostly because many of the wire-pilot circuits that are used are not very reliable. The pilot circuit represents so much exposure to the possibility of trouble that great care should be taken in its choice and protection.

OBTAINING ADEQUATE SENSITIVITY

Apart from making sure that associated equipment is suitable for the application, the principal step in the application procedure is to determine if the available adjustments of the relaying equipment are such that the necessary sensitivity and speed are assured. Manufacturers' bulletins describe how to do this when one knows the maximum and minimum fault-current magnitudes for phase and ground faults at either end of the line.

It is advisable not to adjust the equipment to have much greater sensitivity than is required or else, in so doing, the CT's may be burdened excessively. With excessive burden, the over-all sensitivity may be poorer, as illustrated by Problem 2 of Chapter 7.

If the phase-fault currents are high enough to permit it, it is advisable to adjust the phase-fault pickup to be at least 25% higher than maximum load current. Then, the equipment will not trip its breakers undesirably on load current should the pilot wires become open-circuited or short-circuited. Undesirable tripping could still occur for an external fault, unless supervising equipment is used to forestall such tripping.

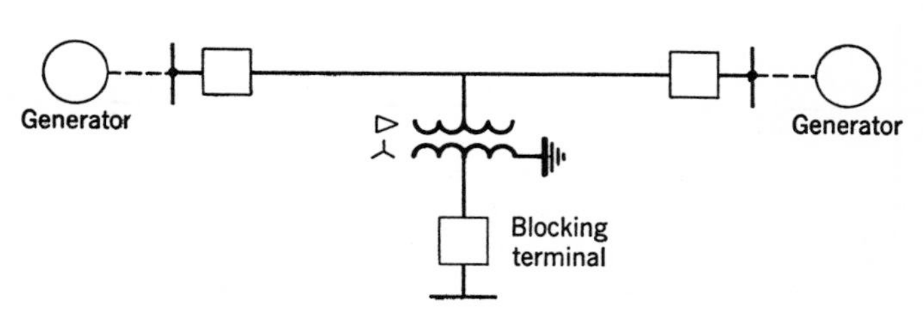

Fig. 1. Illustrating a blocking-terminal application.

THE PROTECTION OF MULTITERMINAL LINES

When there are sources of generation back of more than two terminals, or if there are grounded-neutral wye-delta power-transformer banks at more than two terminals, the application requires a careful study of the available short-circuit currents under various generating conditions to determine if the necessary sensitivity can be assured. The more source terminals there are, the less sensitive will the protection be.[4]

When there are sources of generation back of only two terminals as in Fig. 1, the problem is much simplified. A terminal that has no source of generation is treated as a so-called "blocking" terminal. Instantaneous overcurrent relays energized from CT's on the high-voltage side of each blocking terminal are connected either to open-circuit or to short-circuit the pilot wires, depending on the type of wire-pilot relaying used, to block tripping at the main terminals for a low-voltage fault at the blocking terminal. The operating time of the tripping relays at the main terminals must be coordinated with that of the blocking-terminal relays. The overcurrent-relay contacts operate in the secondary of an insulating transformer, as in Fig. 2. It is necessary to energize the blocking relays from high-voltage CT's so that tripping at the source terminals will also be blocked for magnetizing-current inrush. The blocking relays should not be sensitive enough to operate on the current fed back to high-voltage faults by motors on the low-voltage side of the blocking terminal, or else tripping at the main terminals will be delayed.

Blocking-terminal equipment will not trip the local breaker for high-voltage line faults; such tripping may be necessary if automatic reclosing is used at the source terminals of the line and if there are motors at the blocking terminal that might be damaged by such reclosing. If high-speed reclosing is used at the source terminals, the

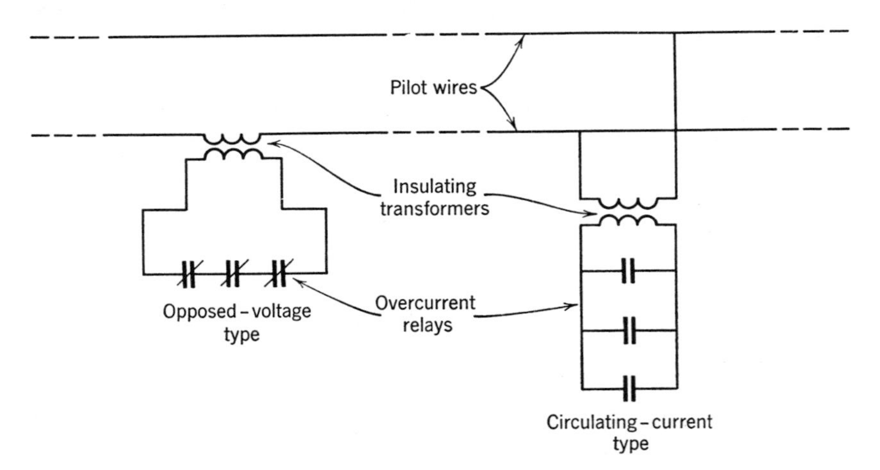

Fig. 2. Blocking-terminal technique.

breaker at the blocking terminal must be tripped, when necessary, by remote tripping from both source terminals. If the automatic reclosing is slow enough, the breaker could be tripped by local undervoltage or underfrequency relays.

If a blocking-terminal power-transformer bank is large enough to justify differential relaying, remote tripping of the source terminals for transformer faults would be used if there were no high-voltage breakers at the blocking terminal, as is usually true.[5] Otherwise, the bank would be protected only by fuses on the high-voltage side.

For the blocking-terminal technique to be permissible, the total load current of all blocking terminals on the line must be less than the current required to operate the wire-pilot relays at one source terminal of the line with the breaker at the other source terminal open.

If the power-transformer banks are small enough at the load terminals behind which there is no generation, the wire-pilot relays at the source terminals could be adjusted not to operate for low-voltage faults at the load terminals. This would also probably prevent operation on magnetizing-current inrush, particularly if the ground-fault pickup could be made high enough.[6] This would eliminate the need for any blocking-terminal equipment.

A multiterminal line can sometimes be well protected against ground faults with wire-pilot relaying, even though adequate phase-fault protection is impossible. This is because line taps are usually made through delta-wye power-transformer banks, and they are open circuits so far as zero-phase-sequence currents on the high-voltage side are concerned. Therefore, if the wire-pilot relays are arranged to receive only the CT neutral current, such a multiterminal line may be treated as a two-terminal line. Good protection against phase faults on such a line can often be provided by distance relays because the impedance of the transformer at each tap is so high that the distance relays can usually be adjusted to protect 80% to 90% of the line without reaching through any of the transformers.

CURRENT-TRANSFORMER REQUIREMENTS

Conventional a-c wire-pilot relays have variable percentage-differential characteristics that permit large CT ratio errors at high magnitudes of external-fault currents. Usually, it is only necessary to be sure that the CT's are able to supply the required current to operate the relays at high speed when internal faults occur. This is a matter involving the relay burdens for the sensitivity taps used, the pilot-wire resistance, and the characteristics of the CT's.

The equipment may also contain adjustment to compensate for the nominal CT ratio at one terminal differing from that at another terminal. In borderline cases, this adjustment might increase a tendency to operate undesirably for external faults because of transient differences between CT errors; therefore, in general, the same nominal CT ratios at all terminals are preferred.

If a line terminates in a power-transformer bank with no high-voltage breaker, the relaying equipment should be energized from high-voltage CT's—generally bushing CT's in the transformer bank. If low-voltage CT's were used, they would have to be connected so as to compensate for the phase shift caused by the power transformer and, possibly, to remove zero-phase-sequence components. The objection to low-voltage CT's is that the relaying equipment will operate to trip undesirably on magnetizing-current inrush either when the transformer bank is energized or when system disturbances occur. To avoid such objectionable operation would require additional relaying equipment. Also, some types of a-c wire-pilot relays would not respond to faults between a particular pair of phases.

BACK-UP PROTECTION

Wire-pilot relaying does not provide back-up protection. Separate overcurrent or distance relays are used for this purpose. When wire-pilot relaying is applied to an existing line, it is often the practice to use the existing relaying equipment for back-up protection.

Distance relays may be used for back-up protection even though the line is too short to use distance relays for primary protection. In such a case, the high-speed zone would be made inoperative.

When directional-overcurrent relays are used for back-up protection, the requirements on the voltage source are the least severe, and uncompensated low-tension voltage can be used. It will be noted that the conventional type of a-c wire-pilot-relaying equipment does not use any a-c voltage.

Carrier-Current-Pilot Relaying

Carrier-current-pilot relaying is the best and most commonly used kind of relaying for high-voltage lines. A report[7] showed that this kind of relaying is in service on lines whose voltage is as low as 33 kv. It is applicable in some form to any aerial line. Carrier-current-pilot relaying is preferred to wire-pilot relaying because it is somewhat more reliable and is more widely applicable. Consisting entirely of terminal equipment, it is completely under the control of the user, as contrasted with rented wire pilot. Also, the carrier-current pilot lends itself more conveniently to joint usage by other services such as emergency telephony and remote trip.

AUTOMATIC SUPERVISION OF THE CARRIER-CURRENT CHANNEL

When carrier-current-pilot relaying was first introduced, the reliability of vacuum tubes was not as good as it is now, and some users felt the need for automatic equipment to supervise the pilot channel. Today, users are content to rely on the manual tests that are made daily at various regular intervals, because the carrier-current channel has proved to be a very reliable element of the protective equipment.[7]

CARRIER-CURRENT ATTENUATION

Every proposed application should be studied to be sure that the losses, or attenuation, in the carrier-current channel will be within the allowable limits of the equipment.[8] Manufacturers' publications specify these limits and describe how to calculate the attenuation in each element of the channel.[9]

The protection of multiterminal lines requires very careful scrutiny of the attenuation. Depending on the length of line tapped from the main line, "reflections" from a tap may cause excessive attenuation unless the carrier-current frequency is very carefully chosen. If this length is 1/4, 3/4, 5/4, 7/4, 9/4, etc., wavelengths, excessive attenuation may be expected. Sometimes, only a test with carrier current of different frequencies will supply the required information.[10] In extreme cases, it is necessary to install line traps in the taps to eliminate reflections.

Power cable causes very high carrier-current attenuation, particularly where sheath-bonding transformers are used. Also, the so-called "mismatch," or discontinuity in the impedance characteristic of the channel where power cable connects to overhead line, causes high loss.[11] It is usually possible to use carrier current only on short lengths of cable, and then only with frequencies near the low end of the range. Because of the foregoing, wire pilot, or even microwave pilot, is sometimes used where carrier current might otherwise be preferred.

USE OF CARRIER CURRENT TO DETECT SLEET ACCUMULATION

The carrier-current channel provides a method for determining when sleet accumulation requires sleet melting to be started. This method has had varied reception among electric utilities.[12] It is generally agreed that the method indicates sleet accumulation, but it will also occasionally give false indication during fog, mist, or rain. Those who use this method of sleet detection feel that the extra cost incurred as a result of going through the sleet-melting process unnecessarily because of such false indications is small and is justified on a "fail-safe" basis.

The method of detecting sleet accumulation is based on the fact that the attenuation of a transmission line increases as sleet accumulates on the line. Figure 3 shows the effect of attenuation on the magnitude of the output from the carrier-current receiver. Normal operation is represented by the point *A*. The safety factor in the equipment, when it is properly applied, is sufficient so that under the most adverse atmospheric conditions, including sleet, the attenuation would not greatly exceed that represented by the point *B*. Therefore, the reliability of the equipment is assured under all conditions. Now, when it is desired to detect sleet accumulation, the operator at one end of the line causes carrier current to be transmitted, and the operator at one end or the other presses a button to introduce attenuation into the transmitter or receiver circuit so as to advance the normal operating position from A to B. Then, the receiver output decreases rapidly for any additional attenuation due to sleet. When conditions appear to be favorable for sleet, such a test at frequent intervals will detect increases in sleet accumulation. Such information must be

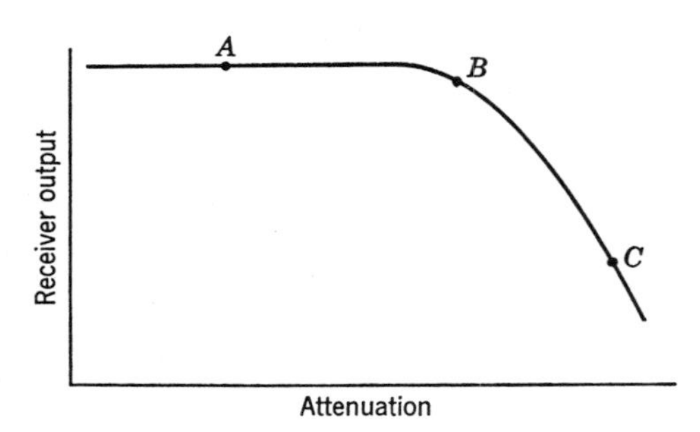

Fig. 3. Effect of attenuation on the strength of the receiver-output signal.

coordinated with visual observation and experience before the receiver-output readings have any useful meaning.

This sleet-detection feature also detects accumulation of dirt or salt on the line insulators, and deterioration of vacuum tubes. It is used for this purpose by many companies that do not use it for sleet detection.

TYPES OF RELAYING EQUIPMENT

The three types of carrier-current-pilot-relaying equipment in regular use are phase comparison, directional comparison, and combined phase and directional comparison. Each of them will be treated in the following material.

Phase Comparison

Phase-comparison relaying is much like a-c wire-pilot relaying. It is the simplest conventional type of carrier-current-pilot-relaying equipment. However, its best application is to two-terminal lines; multiterminal-line applications require very careful examination, and the sensitivity of the protection is quite inferior to that for two-terminal lines. Even for two-terminal lines, the phase-fault sensitivity of phase comparison is not as good as that of directional comparison.

The ideal application of phase comparison is to a two-terminal line that one is sure will not be tapped later, and where the fault-current magnitudes are high enough to assure high-speed tripping under all likely conditions of system operation.

The fact that phase-comparison relaying does not use a-c voltage (except for testing) may or may not be an advantage, depending on the type of back-up relaying that is used. If distance relays are used for back-up, the same quality of voltage source is required as for directional-comparison relaying. It is only when overcurrent relaying (possibly directional) is used for back-up protection that phase-comparison relaying enjoys any advantage from not using a-c voltages.

Phase-comparison relaying is unaffected by mutual induction from neighboring power circuits. This is an advantage over directional comparison. This subject is treated in more detail later under the heading "Combined Phase and Directional Comparison."

The fact that any back-up-relaying equipment that may be used is entirely separate from the phase-comparison equipment is an advantage of phase comparison. One equipment may be taken out of service for maintenance without disturbing the other in any way.

An excellent comparison of phase and directional-comparison relaying is given in Reference 13.

OBTAINING ADEQUATE SENSITIVITY

Apart from making sure that the carrier-current attenuation is not too high, and that associated equipment is suitable for the application, the principal step in the application procedure is to determine if the available adjustments of the relaying equipment are such that the necessary sensitivity and speed are assured. Manufacturers' bulletins describe how to do this when one knows the maximum and minimum fault-current magnitudes for phase and ground faults at either end of the line.

It is advisable not to adjust the equipment to have much greater sensitivity than is required or else, in so doing, the CT's may be burdened excessively. With excessive burden, the over-all sensitivity might be poorer, as illustrated by Problem 2 of Chapter 7. Or, if the ground-fault sensitivity is too high, the equipment might misoperate on "false residual currents"[6] caused by CT errors because of external-fault current having a large d-c offset, or because of differences in residual flux.

In practice the equipment is usually adjusted so that carrier current is not generated unless the phase-fault current exceeds the maximum load current. The purpose of this is to prolong the life of the vacuum tubes and to make the carrier-current channel available for other services when it is not required for relaying. Because the pickup of the tripping fault detectors must be higher than the pickup of the blocking fault detectors, the tripping pickup will be still higher above maximum load; in fact, the *reset* value of the tripping fault detectors must be a safe margin above the pickup of the blocking fault detectors. With such adjustment, failure of the carrier-current channel will not cause undesired tripping under load; however, undesired tripping could occur for an external fault if the channel failed.

THE PROTECTION OF MULTITERMINAL LINES

The more terminals there are with sources of generation back of them, the less sensitive will the protection be. This is illustrated with

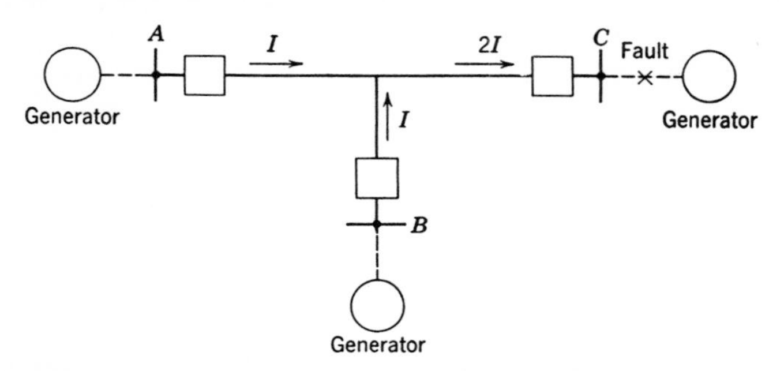

Fig. 4. Illustrating the reduction of sensitivity of phase-comparison relaying on multiterminal lines.

the help of Fig. 4 for a three-terminal line. Should equal magnitudes of fault current be fed into terminals *A* and *B* for an external fault beyond *C*, the pickup of the tripping fault detectors at *C* would have to be more than twice as high as the carrier-current-starting, or blocking, fault detectors at *A* and *B*. And, if the blocking fault detectors are adjusted to reset at more than the maximum full-load current, the tripping fault detectors at *C* will have to be adjusted to pick up at about 3 times maximum load.[14] Remember that phase-comparison relaying is not directional and that one terminal will operate to trip whenever its current is high enough unless it receives a blocking carrier-current signal from another terminal. The worst case is with equal currents entering at *A* and *B*. If one can be sure that these currents will not be equal, the tripping-fault-detector pickup at *C* can be lowered.

In general, the pickup adjustments of the fault detectors do not have to be the same at all terminals, but the pickup of the blocking fault detector having the highest pickup must be lower than the reset of the tripping fault detector having the lowest pickup.

Occasionally, in order to get the required tripping sensitivity, it may be considered justifiable to increase the blocking sensitivity to the point at which carrier current is transmitted continually when full-load current is flowing. In that event, vacuum-tube life will be shortened and the carrier-current channel cannot be used for any other services.

Another way of avoiding high pickup of the tripping fault detectors for situations like Fig. 4 is to use directional relays to control tripping at places like terminal C. However, this kind of solution will not work

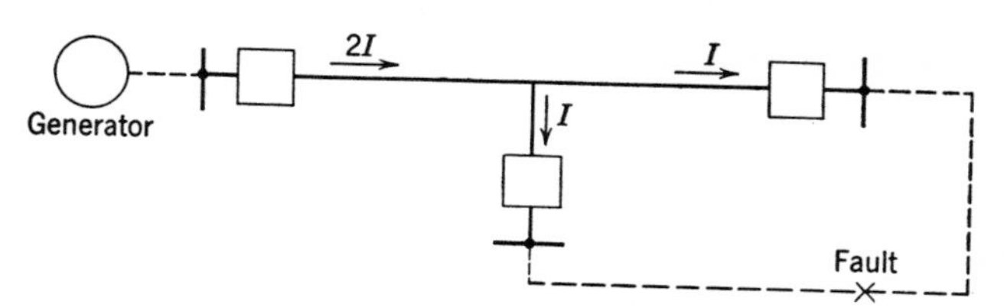

Fig. 5. A situation in which directional relaying will not prevent reduced sensitivity.

for the situation illustrated by Fig. 5 where the same problem exists, but at a terminal where the large current is flowing in the tripping direction.

At a load terminal, back of which there is no source of generation, and where there is no power-transformer neutral grounding on the high-voltage side, blocking-terminal equipment consisting of instantaneous overcurrent relays and a carrier-current transmitter can be used to block tripping at the main terminals for faults in the load circuits. Of course, this is necessary only if the main-terminal equipment is sensitive enough to operate for low-voltage faults at a load terminal. The phase-sequence network, comparer, etc., that are used in the equipment at the main terminals are not necessary, since no tripping function is provided at the blocking terminal. To block tripping at the main terminals, the overcurrent relays simply turn on carrier that is transmitted continuously and not every other half cycle. The blocking relays should be energized from CT's on the high-voltage side of the load-terminal power-transformer bank so that tripping will be blocked on magnetizing-current inrush. If the power transformer is wye-delta and grounded on the wye side, a ground overcurrent relay would be required to shut off carrier for ground faults on the high-voltage side.

If tripping at a blocking terminal is required to avoid damage to large motors when automatic reclosing is used at the main terminals, such tripping will probably have to be provided by local underfrequency relays. Remote tripping by carrier current over the protected line from the main terminals to such a load terminal could not be assured unless the tripping of the main terminals will extinguish an arcing phase-to-ground fault that might be on the phase to which the carrier-current equipment is coupled. Synchronous motors, acting as generators, might be able to generate sufficient voltage to maintain an arc through the capacitance to ground of the unfaulted conductors. Some users have installed equipment that relies on transmitting sufficient carrier current for remote tripping past an arcing ground fault on the coupling phase, but such operation cannot be assured in general.

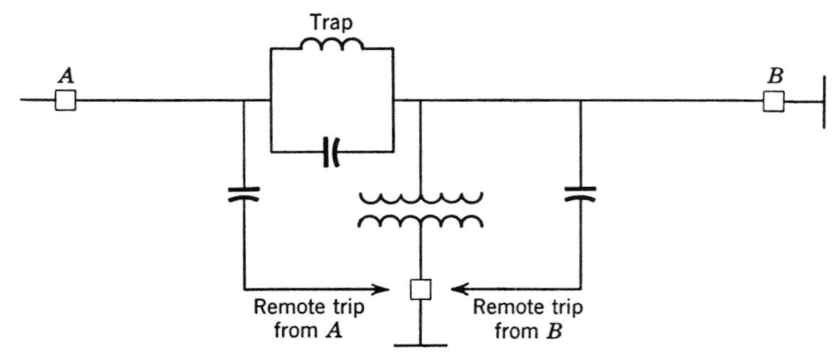

Fig. 6. Method of avoiding interference by a fault with remote tripping.

One solution is to use phase-to-phase coupling for the remote-trip signal, or to transmit this signal over another line section if the lines are parallel. Another solution that has been used is shown in Fig. 6; no matter where the fault is, one main terminal or the other can cause remote tripping.

Remote tripping from power-transformer differential relays at a load terminal to the breakers at the main terminals can be done over the carrier-current channel.[5] Whenever remote tripping for transformer faults is undertaken, a line trap should be inserted in the coupling phase between the coupling capacitor and the power transformer, so that power-transformer faults to ground on the coupling phase cannot short-circuit the carrier-current-transmitter output.

A multiterminal line can sometimes be well protected against ground faults even though adequate phase-fault protection is impossible. This is because line taps are usually made through delta-wye power-transformer banks, which are open circuits so far as zero-phase-sequence currents on the high-voltage side are concerned. Therefore, if only ground relaying equipment is used and arranged to receive only the CT neutral current, such a multiterminal line may be treated as a two-terminal line.

BACK-UP PROTECTION

Phase-comparison relaying does not provide back-up protection. This should be provided by phase distance relays and either overcurrent or distance ground relays. When phase-comparison relaying is applied to an existing line, it is often the practice to use the existing relaying equipment for back-up protection.

Conventional back-up relaying will be inadequate when intermediate current sources supply so much current to a fault that the fault is put beyond the reach of the back-up relays. Such a problem and its solution are described in Chapter 14 under the heading "The Effect of Intermediate Current Sources on Distance-Relay Operation." In such a situation it will be at the discretion of the user whether, in addition to the special back-up equipment, conventional back-up equipment is also applied to provide primary relaying while the phase-comparison equipment is being maintained or repaired.

Directional Comparison

Directional-comparison relaying is the most widely applicable type, and therefore it lends itself best to standardization programs. The only circumstance in which directional comparison is not applicable is when there is sufficient mutual induction with another line and when directional ground relays are used instead of ground distance relays; this is treated at greater length later under "Combined Phase and Directional Comparison."

In general, apart from considerations of carrier-current attenuation, the application of directional-comparison relaying is largely a matter of applying phase distance and directional-ground or ground distance relays. This is because, as mentioned in Chapter 6, conventional equipment uses certain units in common for carrier-current-pilot primary relaying and for back-up relaying. In fact, if a line is now protected by phase distance relays and ground overcurrent or distance relays, one may merely need to add some supplementary relays plus the carrier-current equipment to apply directional-comparison carrier-current-pilot relaying; the supplementary relays and carrier-current equipment provide the blocking function while the existing relays provide the tripping function. Of course, completely separate relaying could be used, but it would be more expensive.

RELATION BETWEEN SENSITIVITIES OF TRIPPING AND BLOCKING UNITS FOR TWO-TERMINAL LINES

The principal application procedure is to be sure that the correct relations are obtained between the operating ranges of the blocking and tripping units. This is seldom a problem for a two-terminal line. Figure 7 shows the relative operating ranges of the blocking and tripping units at both ends of a two-terminal line for, say, phase faults. The significant observation is that, for external faults beyond either end of the line, the blocking units must reach out farther than the

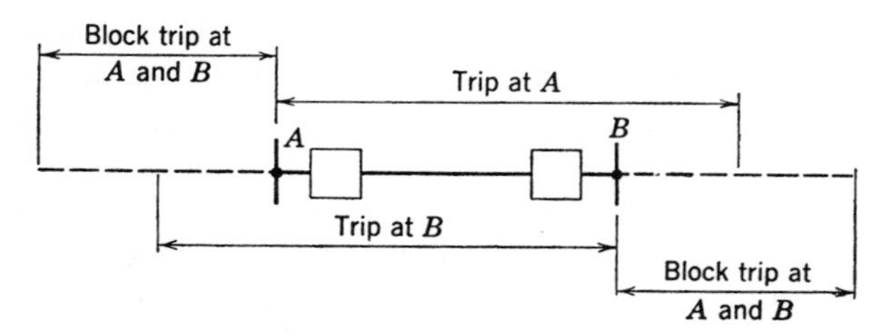

Fig. 7. Relative operating ranges of directional-comparison blocking and tripping units.

tripping units to be certain that, if there is any tendency to trip, it will surely be blocked. The tripping range for phase faults will be the operating range of the second- or third-zone distance-relay units, depending on the type of equipment.

The only time there is any problem adjusting the blocking units is when their range has to be so great that they might operate on load current, or that having operated for a fault they might not reset on load current. In such situations, it becomes necessary to use additional units called "blinders." These units are angle-impedance distance-relay units, one of which would be used with each blocking relay. The contacts of the two relays of each group would be connected in parallel so that both would have to open to start carrier. Figure 8 shows the operating characteristics of both relays and the point representing the load condition that makes blinders necessary. The resulting blocking region is shown cross-hatched. Blinders with impedance-type distance relays are shown because this type of relay is the most likely to require blinders for this purpose.

Incidentally, such blinders have been used also to prevent *tripping* on load current where distance relays have been applied to unusually long lines.

When low-tension voltage is used, transformer-drop compensation is not so necessary as when distance relays are used alone. The only time that transformer-drop compensation would be beneficial would be when the carrier-current equipment is out of service and complete reliance for protection is being placed on the distance relays. Such

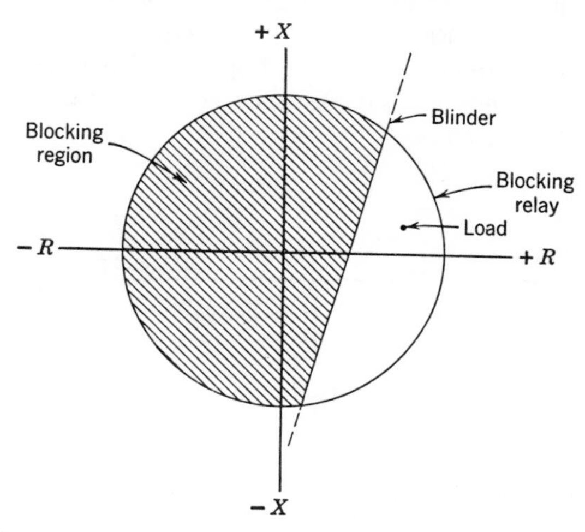

Fig. 8. A blinder to prevent blocking on load current.

circumstances occupy so little of the total time that the additional complications of transformer-drop compensation are not justified.

The problem of obtaining the correct relations between the tripping and blocking units for multiterminal-line applications, along with other related problems, is treated next.

THE PROTECTION OF MULTITERMINAL LINES

Directional-comparison relaying is applicable to any multiterminal line. However, under some circumstances proper operation will not be obtained without a very careful choice of the type of equipment and of the blocking- and tripping-relay adjustments.[4,15] And sometimes simultaneous high-speed tripping at all terminals will not be obtained. Therefore, one should be familiar with these circumstances so as to be able to avoid them, if possible, in the early stages of system planning. These circumstances will now be described.

Current Flow Out of One Terminal for an Internal Fault. In Fig. 9, directional-comparison relaying cannot trip for an internal fault if the current flowing out of the line at A is higher than the blocking-relay pickup there. This situation may exist for phase faults or ground faults or both. If it is not permissible to raise the blocking-relay pickup so as to avoid this situation, tripping must wait until the back-up relays at B trip their breaker, after which high-speed tripping can occur at the other two terminals. For phase faults, the distance relays at B will operate at high speed, and sequential high-speed tripping of all other terminals can follow if there is enough fault current, and if other features of the equipment do not introduce time delay. For ground faults, the tripping of breaker B will be delayed slightly unless ground distance or instantaneous overcurrent ground relays are used.

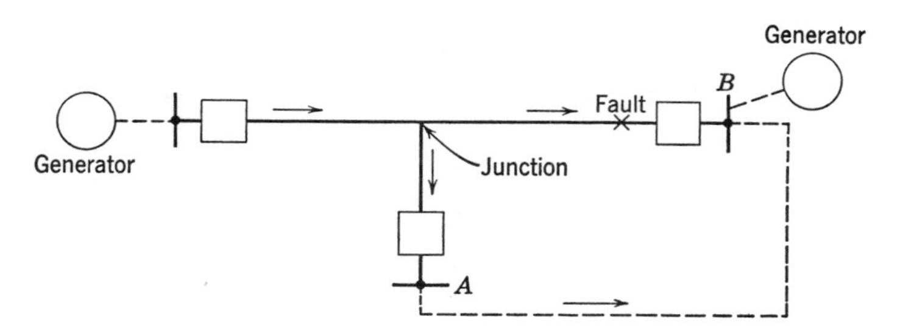

Fig. 9. Situation in which directional comparison blocks tripping for an internal fault.

Remote tripping from breaker B to the other terminals by means of carrier current over the protected line is not a reliable way to avoid sequential tripping, unless this type of difficulty occurs only for faults not involving ground. Or, if it occurs only for ground faults, phase-to-phase coupling could be used. Although some users are relying on getting sufficient remote-tripping signal past a phase-to-ground fault on the coupling phase, satisfactory results cannot be assured in general. Occasionally, another line section can be used for carrying the remote-tripping signal. Obviously, remote tripping would be practical if microwave were used instead of carrier current.

Insufficient Current for Tripping. Apart from there being too small a source of short-circuit current back of a terminal, other circumstances can make the current so low—or the apparent impedance to the fault so high—as to prevent or at least to delay tripping.

For the circumstance of Fig. 9, if the fault is closer to the junction, the current at A will be: (1) in the blocking direction but too low to operate a blocking relay, (2) zero, or (3) in the tripping direction but too low to operate a tripping relay. For any of these, tripping at the other terminals would not be blocked, but tripping at A would have to wait until the breakers at B had tripped, assuming that there would then be a redistribution of enough fault current at A to cause tripping there.

Another circumstance in which the fault current may be too low is shown in Fig. 10. Here, the intermediate current, or "mutual impedance effect," as it is sometimes called, may prevent tripping at both B and C. Furthermore, the tripping of the breakers at A may not relieve the inability to trip at breakers B and C, so that even sequential tripping may not be possible.

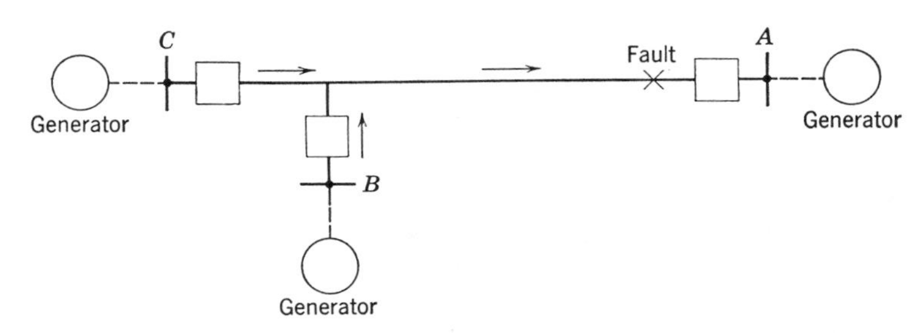

Fig. 10. Insufficient current for tripping.

As stated before, remote tripping is not a complete solution to the problem. Instantaneous undervoltage carrier-starting or tripping relays are a solution if their adjustments can be coordinated with those of the other relays.[2]

Shortcomings of Non-Directional Blocking Relays. Some directional-comparison equipments use non-directional overcurrent or impedance relays to start carrier. Occasionally, such equipments block tripping for an internal fault because the current or apparent impedance falls between the pickup of the blocking and tripping relays at the *same* terminal. Basically, the pickup adjustment of the blocking relays at a given terminal has to coordinate with the pickup adjustments of tripping relays at the other terminals. However, when non-directional blocking relays are used, coordination must also be obtained between the pickup adjustments of blocking and tripping relays at the *same* terminal. Such a circumstance can exist when there is insufficient current or too high an apparent impedance to cause tripping at a given terminal until after another terminal has tripped, as in one of the preceding cases. However, if a blocking relay at the given terminal should operate, it would block tripping at the other terminals; this situation would persist until a back-up relay operated to cause tripping at another terminal that would permit the given tripping relay to operate. The best solution to this problem is directional blocking relays.

Miscellaneous Problems of Coordinating Blocking and Tripping Sensitivities. It is necessary that the coordination between blocking

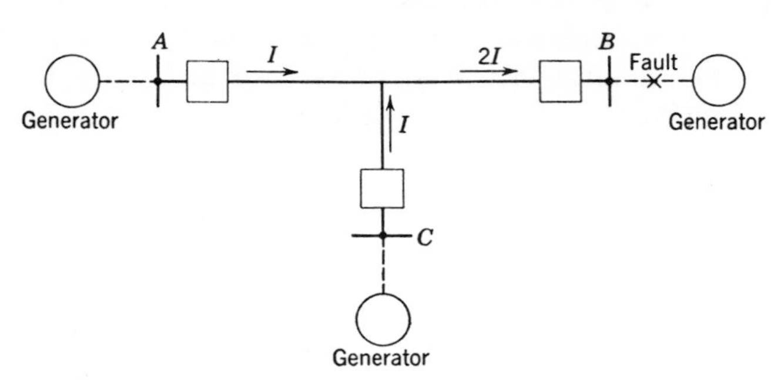

Fig. 11. Directional-comparison blocking relays get twice as much current as tripping relays.

and tripping sensitivities be carefully analyzed not only for multi-terminal operation but also when a line is operated with one or more terminals open. Owing to elimination of intermediate current sources,

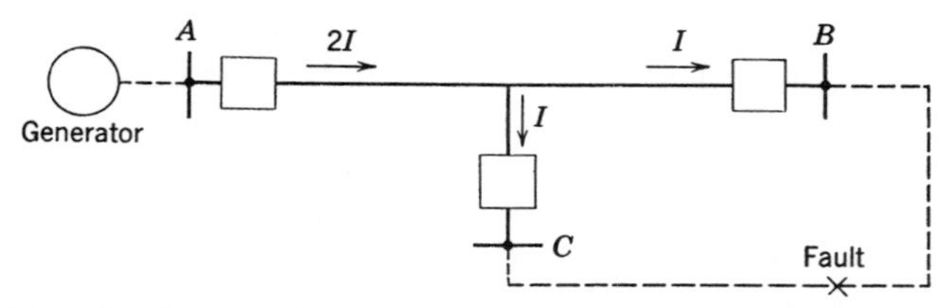

Fig. 12. Directional-comparison tripping relays get twice as much current as blocking relays.

such operation may increase the reach of the tripping relays at one terminal; one must be sure that these relays do not outreach the blocking relays at another terminal.

Figures 11 and 12 show two extreme operating conditions for a three-terminal line, so far as the relative blocking and tripping sensitivities are concerned. For Fig. 11, the blocking relays at terminal *B* get twice as much current as the tripping relays at *A* or *C*; for Fig. 12, the blocking relays at *B* and *C* get only half as much current as the tripping relays at *A*. Even greater extremes could exist for a line with more than three terminals.

Need for a Blinder on the Loss-of-Synchronism Blocking Relay. If the phase tripping relays at terminal *A* of Fig. 13 cannot operate to trip until after terminal B has tripped for the fault location P, a supplementary angle-impedance relay should be used to provide a "blinder" for the loss-of-synchronism blocking relay. The need for

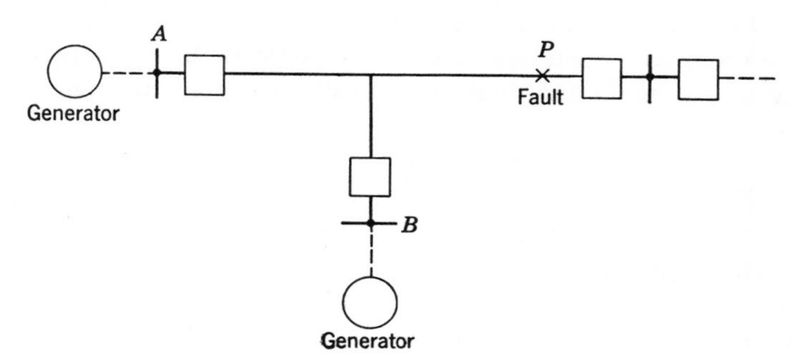

Fig. 13. Situation in which the loss-of-synchronism blocking relay will block tripping for an internal fault.

this blinder is shown in Fig. 14. The point P_1 represents the way the fault first appears to the tripping and blocking relays at A, and the point P_2 represents the appearance of the fault after B has tripped.

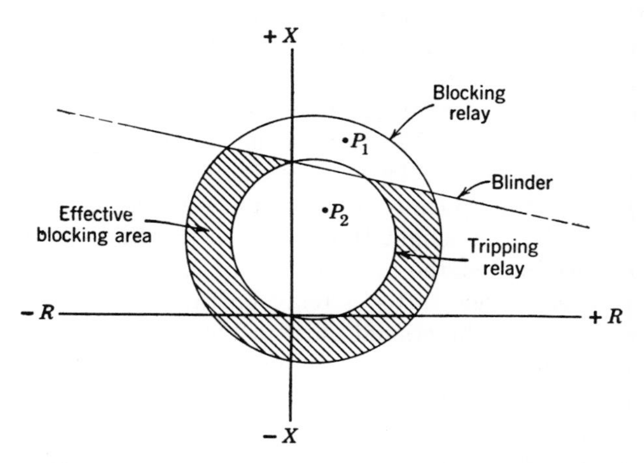

Fig. 14. R-X diagram of the conditions of Fig. 13.

It will be noted that this sequence will establish local blocking at A by the loss-of-synchronism relay, and, consequently, that tripping there can only occur in third-zone time.

Figure 14 shows how the blinder characteristic modifies the operating characteristic of the blocking relay so that when the fault first occurs it will not fulfill the requirement for the second step in the sequence of operations necessary to set up loss-of-synchronism blocking. The effective blocking area is shown cross-hatched.

Blocking-Terminal Equipment at Load Terminals. At terminals behind which there is no source of generation, only blocking equipment may be required. Such equipment is required if the high-speed relays at any of the main terminals are sensitive enough to operate for a low-voltage fault at such a load terminal. The blocking-terminal equipment consists of instantaneous overcurrent relays, energized from CT's on the high-voltage side of the power-transformer bank, and carrier-current transmitting and receiving equipment. The overcurrent relays start the transmission of carrier current to block tripping at the main terminals for low-voltage faults at the load terminal or for magnetizing-current inrush to the load-terminal power-transformer bank.

As described for phase-comparison relaying, remote tripping from a blocking-terminal power-transformer differential relay to the breakers at the main terminals can be accomplished over the carrier-current channel.[5]

EFFECT OF TRANSIENTS

Directional-comparison relaying using high-speed ground relays energized from zero-phase-sequence quantities is exposed to more possibilities of misoperation than is phase-comparison relaying. Reference 6 describes a host of things that tend to fool such ground relays. However, conventional directional-comparison-relaying equipments have certain features, developed as a result of experience, that minimize any tendency toward misoperation. Such features are: (1) limited sensitivity, (2) slight time delay in auxiliary relays, and (3) "transient blocking" or the prolongation of a carrier-current blocking signal for several cycles after a relay operates to try to shut it off. Also, induction-type directional units in both the carrier-starting and the tripping functions make the units unresponsive to transients in only one of the operating quantities.

Ground distance relays, that respond to positive-phase-sequence impedance, for controlling carrier-current transmission and for tripping eliminate the problem of misoperation on transients.

Combined Phase and Directional Comparison

Directional-comparison relaying using directional-ground relays may operate undesirably if there is sufficient mutual induction with a neighboring power circuit.[6] The directional-ground relays misoperate because their polarization is adversely affected, as will be described later. This would seem to indicate the desirability of phase-comparison relaying which would be unaffected by mutual induction. If phase comparison was completely applicable, it would be a good solution. However, occasionally it does not have sufficient sensitivity for phase faults, although it would be entirely satisfactory for ground faults. Under these circumstances combined phase- and directional-comparison relaying is chosen. The directional-comparison principle is used for phase faults and the phase-comparison principle is used for ground faults. Because the carrier-current transmitter and receiver are employed in common, the equipment is only a little more expensive than directional comparison alone. Incidentally, the phase-comparison ground-fault equipment is less affected by most of the transient conditions that affect directional-ground relays.

If ground distance relays were used in the directional-comparison equipment instead of directional-ground relays, it would be unnecessary to resort to combined phase- and directional-comparison equipment. However, it would be somewhat more expensive, but it would provide better back-up protection.

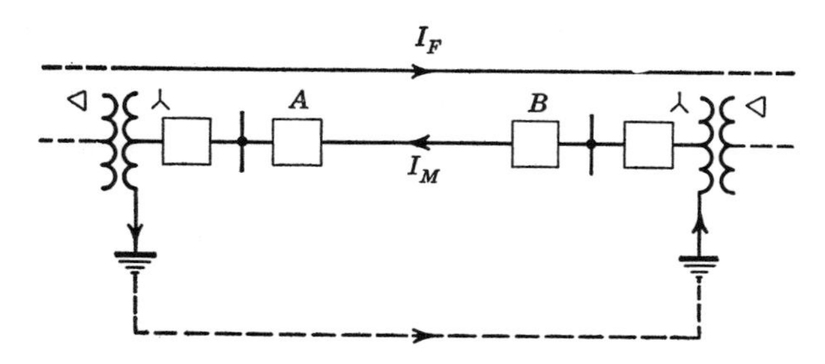

Fig. 15. Illustrating the cause of undesired directional-ground-relay operation resulting from mutual induction.

THE EFFECT OF MUTUAL INDUCTION ON DIRECTIONAL-GROUND RELAYS

Figure 15 illustrates the fundamental principle involved in the undesired operation of directional-ground relays. As shown in Fig. 15, fault current I_F flowing in a nearby line causes current I_M to flow by mutual induction in the line under consideration. The induced current circulates through grounded-neutral power-transformer banks at the ends of the line and the earth, as shown. The difficulty is that directional-ground relays at both ends of the line tend to operate under such circumstances. At location B, the polarizing current flows from the ground into the neutral of the grounded power transformer and from the bus into the line; this is the same as for a ground fault on the line for which directional-ground relays are intended to operate. The fact that, at end A, both the currents are reversed with respect to the directions at B also produces a tripping tendency. The same operating tendencies would exist if the relays were voltage-polarized. In

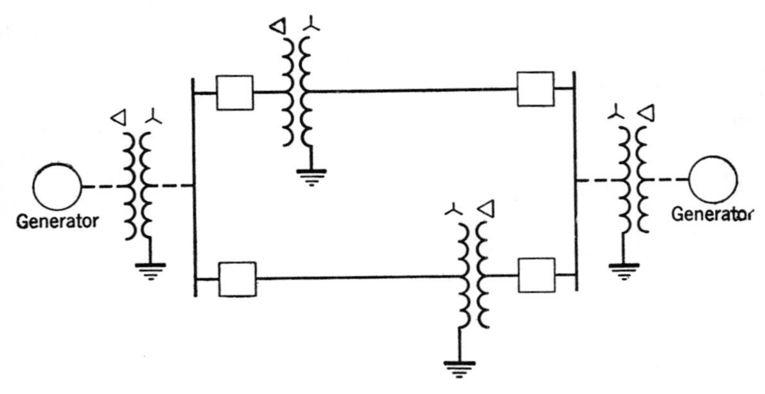

Fig. 16. Parallel circuits subject to mutual induction that are electrically independent so far as zero-phase-sequence currents are concerned.

other words, the phase of the polarizing quantity is not independent of the direction of current flow in the line, as it is when a short circuit occurs in the line or beyond either end.

One can immediately see the significance of the foregoing circumstances when directional-comparison pilot relaying is involved. Since the directional-ground relays at both ends of the circuit have a tripping tendency, if the induced current is high enough to pick up the relays, the circuit will be tripped undesirably.

The conditions of Fig. 15 are extreme in view of the fact that the line section in which induced current flows cannot directly contribute short-circuit current to the other circuit. However, it is not an impossible situation. It can exist whenever electrically independent circuits are closely paralleled, or in a case such as that illustrated in Fig. 16. Although these two circuits are paralleled at their ends, they are independent so far as zero-phase-sequence currents are concerned.

Situations that are more apt to be encountered are illustrated in Fig. 17. It is only necessary in either case of Fig. 17 that the breaker of the faulty line be open at the end where the lines are normally paralleled, in order to have the condition that can produce an undesirable tripping tendency of the directional-ground relays of the sound line. This breaker may be open either because it was tripped by its protective relays immediately when the fault occurred and before the breaker at the other end could trip, or because the line had been open at both ends and the breaker at the other end was reclosed first on a

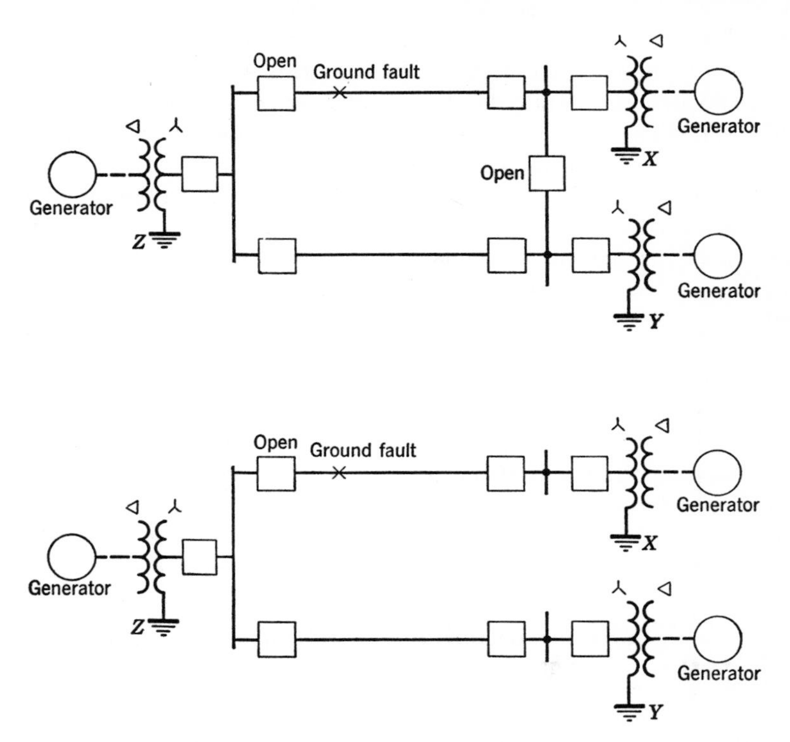

Fig. 17. Situations in which lines are directly paralleled at one end only.

persisting fault. Not only would the directional elements at both ends of the sound line permit tripping, but the current magnitudes may be large enough so that selectivity would not be obtained.

Sometimes it is even unnecessary that the breaker in the faulty line of Fig. 17 be open. If the effect of mutual induction is great enough, it can overcome the tendency of the unfaulted line to supply current to the fault, and actually reverse the direction of its current.

Apart from phase-comparison relaying or directional ground distance relays, other solutions to the problem can sometimes be found. The zero-phase-sequence current or voltage at the ends of the faulted circuit may be enough greater than at the corresponding ends of the unfaulted circuit so that a relay can be interposed that balances the corresponding quantities and permits only the circuit to trip that has the larger quantity. Another possible solution is to parallel the CT's in the neutrals of the power transformers, as, for example, at X and Y of Fig, 17, and to use the resulting current to polarize the directional-ground relays of both lines at that end. At the other end of the circuits of Fig. 17, a CT in the neutral of the one power-transformer bank shown at Z would suffice for the directional-ground relays of both lines; or voltage polarization might be used at this end.

Should the line terminals be too far apart at one end, as when two lines run close to one another for part of their length and then diverge, it would be impossible to employ the alternatives to phase-comparison relaying that have been described. One remaining alternative would be to determine if the magnitude of the zero-phase-sequence current or voltage at the ends of the lines could not be used alone to permit operation, subject to directional control, only if the magnitude was high enough. Another possibility is to take advantage of the fact that the phase-to-neutral voltages of the circuit in which the induced current flows are usually not as low during the induced-current condition as they are while a ground fault exists on the line itself.

It was mentioned in Chapter 13 that a negative-phase-sequence directional-ground relay would not be affected by mutual induction. However, such a relay has other disadvantages, as mentioned also, that make it desirable to seek some other alternative.

All-Electronic Directional-Comparison Equipment

All-electronic directional-comparison equipment, including electronic phase distance and directional-ground relays, has been in service since 1953.[16] The average operating time of this equipment is ⅝ cycle with a maximum of 1.0 cycle, as compared with 1.0 to 3.0 cycles for conventional electromechanical relay equipment.

Such operating speeds eventually become necessary, not only to maintain stability when faults occur but also to minimize the damage from ever-increasing concentrations of short-circuit current.

The application procedures and problems are the same as those described for the electromechanical equipment.

Microwave

A microwave pilot is used for relaying only when the relaying equipment can share the channel with enough other services; it is not economically justifiable for relaying alone if carrier current or wire pilot is applicable.[17]

Microwave is entirely suitable although it is not as reliable as carrier current for protective-relaying purposes; this is partly because of the complex circuitry and the large number of tubes involved, and also because of the large number of services on the same microwave channel. When repeater stations are necessary, the complexity practically doubles with further loss of reliability. Of course, one should realize that the requirements of protective relaying as to reliability are in certain respects more severe than the requirements of other services that use the microwave channel. Any lapse in the signal when a fault occurs is unacceptable.

Microwave has certain theoretical advantages over carrier current because it is dissociated from the power line,[18] but its only real advantage is in connection with remote tripping, which will be considered later. Occasionally, microwave is useful where the attenuation would be too high for carrier current, such as on a power-cable circuit, but even there microwave would probably not be selected unless there were many other uses in addition to protective relaying.

The same relaying equipments that are used with a carrier-current pilot are also used with a microwave pilot. Therefore, the application considerations are the same so far as the relaying equipment is concerned.

THE MICROWAVE CHANNEL

The microwave channel is a line-of-sight-radio system operating on a frequency band in the United States assigned by the Federal Communications Commission in the range from 950 to 30,000 megacycles.[19] Such a system requires that a straight line from one antenna to another be above intervening objects, preferably by about 50 feet. This usually limits the distance between antennae to about 20 to 50 miles, depending on the topography of the land. Where a longer channel is required, one or more "repeater stations" may be necessary. One repeater station doubles the base channel equipment, except that only one additional tower is necessary; hence, the cost of a microwave channel is dependent on its length.

It is the practice to use standby equipment automatically switched into service in the event that the regular equipment fails.

For protective relaying that cannot tolerate even a moment's outage when a fault occurs, operation from a power-system a-c source is not acceptable. It is necessary to provide an a-c generator operating from the station battery, or d-c-operated equipment. This becomes more of a problem at a repeater station where a suitable battery source would not otherwise be available.

For protective-relaying purposes, the practice is to modulate the microwave frequency directly by any of the usual methods, such as, for example, by a so-called "tone." Such a tone is a single-frequency voltage in the audio range or above. Tones above the audio range are preferred because the time constants of their filter circuits are shorter, and therefore it is unnecessary to delay tripping to allow time for the receiver output to build up sufficiently to block tripping.

REMOTE TRIPPING

The principal advantage of microwave for protective relaying is that the presence of a fault on the protected line will not interfere with the transmission of a remote-tripping signal. For the protection of three-terminal lines, there are circumstances when the relays at a given terminal cannot operate to trip their breakers until after the breakers trip at another terminal. With microwave, the first relays to operate can cause the transmission of a tripping signal to another terminal and thereby eliminate part of the time delay in the sequential tripping of this other terminal.[20]

This ability to perform remote tripping without hindrance by a fault makes possible the use of a different principle for line protection.[18] To apply this principle, it is first necessary that the high-speed-tripping zones of the relays at all terminals overlap for all types of fault in such a way that, for any fault, the relays of at least one terminal will always operate at high speed. Then, if each terminal is arranged to transmit a trip signal to each other terminal, practically simultaneous high-speed tripping will occur at all terminals; the remote tripping will be delayed about 2 to 3 cycles. Of course, each terminal is still free to trip at high speed independently of the remote-tripping equipment whenever a fault occurs within that terminal's high-speed-tripping zone. This principle eliminates the need for blocking relays, as required by directional comparison, but it often requires distance relays for phase- and ground-fault protection. Where remote tripping is required for multiterminal applications, this type of relaying would have its greatest application; otherwise, the added time delay for certain faults would discourage its general usage where simultaneous high-speed tripping is possible with directional comparison.

Incidentally, the foregoing principle can be applied to a wire-pilot system by the use of tones.

High-Speed Reclosing

High-speed automatic reclosing of transmission-line breakers after they have tripped to clear a fault is generally possible only with pilot relaying, because only pilot relaying is able to cause all line terminals to trip at high speed and practically simultaneously. With such high-speed tripping and reclosing, generators do not have time to swing very far out of phase, and therefore no synchronism check is necessary before reclosing. The experience with such high-speed reclosing (or "ultra-high-speed reclosing," as it is sometimes called) has been excellent.[21]

Generally, all three phases are tripped and reclosed for any kind of fault. Infrequently, however, such three-phase switching cannot be used, but it is possible to use single-phase switching to advantage.[22] Such a possibility exists when there is only one line connecting a hydroelectric generating station to its system. If about 25% or more of the load on the generating station is dropped when a line is tripped, the generators will speed up too rapidly to permit high-speed reclosing. But for single-phase-to-ground faults, if only the faulty phase is tripped and reclosed, stability can often be maintained; for any other kind of fault, all three phases are tripped but are not reclosed. Single-phase switching can be performed with conventional relaying equipment by the addition of "phase-selector" relays.[23]

High-speed reclosing is permitted only when high-speed tripping is caused by the operation of the pilot equipment or the first-zone units of distance relays. When tripping is caused by any other units, automatic reclosing is blocked until released locally by an operator or remotely by supervisory control.

Bibliography

1. "All-Electronic Carrier Relaying Reduces Fault-Clearing Time," by H. C. Barnes and L. F. Kennedy, *AIEE Trans., 73,* Part III-A (1954), pp. 170–173. Discussions, p. 173.

2. "Line and Transformer Bank Relaying," by J. L. Blackburn and G. D. Rockefeller, *AIEE Trans., 74,* Part III (1955), pp. 334–339. Discussions, pp. 339–343.

"Unique Protection Required for Midwest Interconnection," by A. J. Nicholson, *Elec. Light and Power,* Nov., 1950, pp. 90–96.

3. "Pilot-Wire Circuits for Protective Relaying—Experience and Practice, 1942–1950," by AIEE Committee, *AIEE Trans., 72,* Part III (1953), pp. 331–336. Discussions, p. 336.

4. "Relay Protection of Tapped Transmission Lines," by M. A. Bostwick and E. L. Harder, *AIEE Trans., 62* (1943), pp. 645–650. Discussions, pp. 969–972.

5. "Remote Tripping Schemes," by AIEE Committee, *AIEE Trans., 72,* Part III (1953), pp. 142–150. Discussions, pp. 150–151.

"Protection of Stations without High-Voltage Switching," by AIEE Committee, *AIEE Trans., 68,* Part I (1949), pp. 226–231. Discussions, pp. 231–232.

6. "Some Utility Ground-Relay Problems," by H. C. Barnes and A. J. McConnell, *AIEE Trans., 74,* Part III (1955), pp. 417–428. Discussions, pp. 428–433.

7. "Experience and Reliability of Carrier-Relaying Channels," by AIEE Committee, *AIEE Trans., 72,* Part III (1953), pp. 1223–1226. Discussions, p. 1227.

8. "Transmission Considerations in the Coordination of a Power Line Carrier Network," by G. E. Burridge and A. S. C. Jong, *AIEE Trans., 70,* Part II (1951), pp. 1335–40. Discussions, p. 1340.

"Applying Carrier Current to Power Lines," by H. J. Sutton, *Elec. World, 136* (Oct. 8, 1951), pp. 121–123.

"Propagation Characteristics of Power Line Carrier Links," *Brown Boveri Rev., 35* (Sept./Oct., 1948), pp. 266–275.

"Operation of Power Line Carrier Channels," by H. W. Lensner, *AIEE Trans., 66* (1947), pp. 888–893. Discussions, pp. 893–894.

9. "Application of Carrier to Power Lines," by F. M. Rives, *AIEE Trans., 62* (1943), pp. 835–844. Discussions, pp. 945–947.

10. "Measurement of Carrier Circuit Impedances," by W. H. Blankmeyer, *Elec. World, 126* (August 3, 1946), pp. 49–51.

"Report on Method of Measurements at Carrier Current Frequencies," by AIEE Committee, *AIEE Trans., 67,* Part II (1948), pp. 1429–1432. Discussions, p. 1432.

"A Method of Measurement of Carrier Characteristics on Power Cables," by B. J. Sparlin and J. D. Moynihan, *AIEE Trans., 74,* Part III, pp. 31–33.

11. "Loss Measurements Made on Underground-Cable Overhead-Conductor 132-Kv Transmission Line at Carrier Current Frequencies," by H. A. Cornelius and B. Wade Storer, *AIEE Trans., 68,* Part I (1949), pp. 597–601.

"Power Line Carrier Used on 110-Kv Cable," by R. H. Miller, *Elec. World,* November 5, 1949, p. 67.

12. "Sleet-Thawing Practices of the New England Electric System," by C. P. Corey, H. R. Selfridge, and H. R. Tomlinson, *AIEE Trans., 71,* Part III (1952), pp. 649–657. Discussions, p. 657.

"Sleet Melting on the American Gas and Electric System," by S. C. Bartlett, C. A. Imburgia, and G. H. McDaniel, *AIEE Trans., 71,* Part III (1952), pp. 704–708. Discussions, pp. 708–709.

"Forty-Two Years' Experience Combating Sleet Accumulations," by A. N. Shealy, K. L. Althouse, and R. N. Youtz, *AIEE Trans., 71,* Part III (1952), pp. 621–628.

"Ice-Melting and Prevention Practices on Transmission Lines," by V. L. Davies and L. C. St. Pierre, *AIEE Trans., 71,* Part III (1952), pp. 593–597.

"Sleet-Melting Practices—Niagara Mohawk System," by H. B. Smith and W. D. Wilder, *AIEE Trans., 71,* Part III (1952), pp. 631–634.

"Carrier Attenuation Discloses Glaze Formation," by G. G. Langdon and V. M. Marquis, *Elec. World, 112* (August 12, 1939), pp. 38–40, 100–101.

13. "Considerations in Selecting a Carrier Relaying System," by R. C. Cheek and J. L. Blackburn, *AIEE Trans., 71,* Part III (1952), pp. 10–15. Discussions, pp. 15–18.

14. "Phase-Comparison Carrier Relaying for 3-Terminal Lines," by H. W. Lensner, *AIEE Trans., 72,* Part III (1953), pp. 697–701. Discussions, pp. 701–702.

15. "Power-Line Carrier for Relaying and Joint Usage—Part II," by G. W. Hampe and B. Wade Storer, *AIEE Trans., 71,* Part III (1952), pp. 661–668. Discussions, pp. 668–670.

"Relaying of Three-Terminal Lines," by C. W. Cogburn, *Elec. Light and Power,* March, 1954, pp. 71–73.

16. "All-Electronic 1-Cycle Carrier Relaying Equipment—Relay Operating Principles," by M. E. Hodges and R. H. Macpherson, *AIEE Trans., 73,* Part III-A (1954), pp. 174–186.

"An All-Electronic 1-Cycle Carrier-Relaying System—Over-All Operating Principles," by H. T. Seeley and N. A. Koss, *AIEE Trans., 73,* Part III-A (1954), pp. 161–168. Discussions, pp. 168–169.

"Performance Evaluation of All-Electronic 1-Cycle Carrier-Relaying Equipment," by W. S. Price, R. E. Cordray, and R. H. Macpherson, *AIEE Trans., 73,* Part III-A, pp. 187–192. Discussions, pp. 192–195.

17. "Economics of Relaying by Microwave," by R. C. Cheek, *Elec. Light and Power,* May, 1951, pp. 82–84.

18. "Protective Relaying over Microwave Channels," by H. W. Lensner, *AIEE Trans., 71,* Part III (1952), pp. 240–244. Discussions, pp. 244–245.

19. "Microwave Channels for Power System Applications," by AIEE Committee, *AIEE Trans., 68,* Part I (1949), pp. 40–42. Discussions, pp. 42–43.

20. "Power System Stability Criteria for Design," by W. A. Morgan, *AIEE Trans., 71,* Part III (1952), pp. 499–503. Discussions, pp. 503–504.

21. "Five Years Experience with Ultra-High-Speed Reclosing of High-Voltage Transmission Lines," by Philip Sporn and C. A. Muller, *AIEE Trans., 60* (1941), pp. 241–246. Discussions, p. 690.

22. *Power System Stability—Vol. II,* by S. B. Crary, John Wiley & Sons, New York, 1947.

23. "Relays and Breakers for High-Speed Single-Pole Tripping and Reclosing," by S. L. Goldsborough and A. W. Hill, *AIEE Trans., 61* (1942), pp. 77–81. Discussions, p. 429.

Review Problems

1. A system short-circuit study is to be made. What quantities should be obtained for studying the application of protective relays? What other data are required to apply (*a*) overcurrent relays, (*b*) pilot relays, and (*c*) distance relays?

2. Name and describe briefly the various methods for obtaining selectivity.

3. Discuss the factors governing the choice of transmission-line-relaying equipment. Which type would lend itself best to standardization, and why?

4. Given a breaker with 1200/5 bushing CT's having the secondary-excitation characteristic of Fig. 3 of Chapter 7. What would the ASA accuracy classification be for the 40-turn tap (ratio 200/5)? Assume the tap winding to be fully distributed.

5. On what basis is it permissible to superimpose system and relay characteristics on the same *R-X* diagram for the purpose of studying relay response?

6. Under what circumstance is distance relaying affected by short-circuit-current magnitude?

7. Show the complete CT and relay connections for applying percentage-

differential protection to the power transformer of Fig. 18. Draw the three-phase voltage vector diagrams for both sides of the power transformer.

8. Discuss the virtues of high-speed relaying.

9. Write "true" or "false" after each of the following:

(*a*) Overcurrent relays with more-inverse curves are better to use when the generating capacity changes more from time to time.

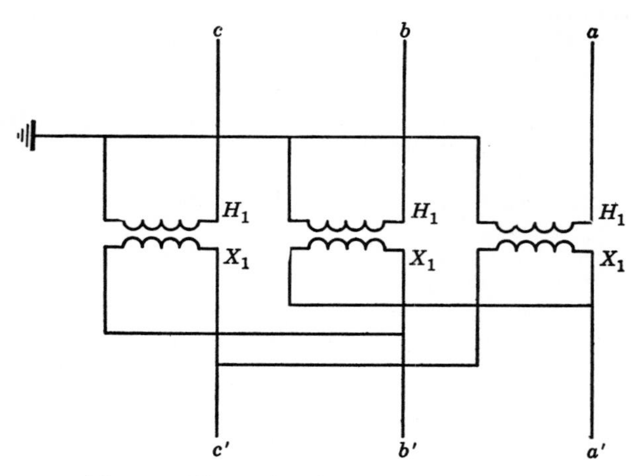

Fig. 18. Illustration for Review Problem 7.

(*b*) Instantaneous overcurrent relays are more applicable on the longer lines.

(*c*) Mho-type distance relays are more likely to operate undesirably on power swings than other types of distance relays.

(*d*) Any distance relay—no matter where it is located—will trip on loss of synchronism.

(*e*) Sensitive, high-speed bus protection can always be obtained with differentially connected overcurrent relays.

(*f*) In attended stations, it is not the practice to let transformer overload relays trip the transformer breakers.

(*g*) Back-up relaying is not a substitute for good maintenance.

INDEX

Abnormal conditions other than short circuits, 8
A-c tripping, 335
Angle-impedance relay, 79
  for tripping on loss of synchronism, 362
Angle of maximum torque, adjustment, 57
  of power relays, 52, 55
  of short-circuit relays, 55
Arcs, effect on distance relays, 345
  resistance, 302
Arc-furnace-transformer protection, 270
ASA accuracy classification, of current transformers, 121
  of potential transformers, 133
Attenuation, carrier-current, 378
Automatic reclosing, *see* Reclosing, automatic
Auxiliaries, station, protection of, 229
  see also Motor protection

Back-up relaying, defined, 6
  effect of intermediate current sources, 347
  for bus protection, 275, 291
  generator external fault, 227
  reversed third zone, 175, 349
  transformer external fault, 264, 268
  with pilot relaying, 378, 385
Blind spot in pilot relaying, 94
Blinder with directional-comparison relaying, 386, 390
Blocking pilot, 88, 91
Blocking terminal, with a-c wire-pilot relaying, 375
  with directional-comparison pilot relaying, 392
  with phase-comparison pilot relaying, 383

Broken-delta connection, burden calculation, 140
  for detecting grounds in ungrounded systems, 320
  for polarizing directional-ground relays, 151
  of capacitance potential devices, 135, 140
Buchholz relay, 261
Burden, current-transformer, 114
  potential transformer, 133
Bus protection, automatic reclosing, 292
  by back-up relays, 275, 291
  circuit-breaker by-passing, 292
  combined power transformer and bus, 287
  current differential with overcurrent relays, 278
  current differential with overvoltage relays, 286
  current differential with percentage-differential relays, 284
  directional comparison, 277
  fault bus, 275
  grounding secondaries of differentially connected CT's, 291
  partial-differential, 282
  ring-bus, 290
  testing, 292
  voltage differential with linear couplers, 284
Bushing potential device, *see* Capacitance potential device

Capacitance potential device, broken-delta burden, 140
  comparison with instrument PT's, 144
  coupling-capacitor insulation coordination, 142
  effect of overloading, 138

Capacitance potential device, equivalent circuit, 135
non-linear burdens, 139
standard accuracy, 137
standard rated burdens, 136
Capacitor, series, effect on distance relaying, 367
Capacitor tripping, 335
Carrier-current attenuation, 378
Carrier-current-pilot relaying, *see* Pilot relaying, carrier-current
Circuit breaker, by-passing, 292
standard capacities, 3
Circulating-current pilot relaying, 92, 94
Cold-load pickup, 334
Cold-load restoration, 334
Compensated voltage, *see* Transformer-drop compensation
Constant-product characteristic, 38
Contact definitions, 17
Contact races in pilot relaying, 91
Continuous pilot, 109
Control spring, 17
Conventions, vector, 53
Corrosion, effect of polarity on, 19
Coupling-capacitor description, 134
Coupling-capacitor insulation coordination, 142
Coupling-capacitor potential device, *see* Capacitance potential device
Current-balance relay, directional type, 62
for line protection, 330
overcurrent type, 58
Current biasing, for mho relay, 82
for offset impedance relay, 77
to avoid distance-relay misoperation on arcs, 346
Current compensation for ground distance relays, 360
Current switching for distance relays, 368
Current transformers, accuracy calculations, 113
ASA accuracy classification, 121
burden, 114
for generator differential relaying, 198
ratio-correction-factor curves, 116
secondary-excitation curves, 117
grounding the secondaries of differentially connected CT's, 291
overvoltage in secondary, 124
polarity and connections, 126
proximity effect, 126
Current transformers, requirements for pilot relaying, 377
secondary leakage reactance, 117, 119, 120
transient errors, 124, 278, 318
types, 113
zero-phase-sequence-current shunt, 130, 249

D-c offset, effect on induction relays, 32, 39
overreach of distance relays, 82, 350
overreach of overcurrent relays, 308
time constant, 279
D-c relays, single-quantity, 22
directional, 24, 49
Differential relays, 63
see also Percentage-differential relays
Directional-comparison relaying, for bus protection, 277
principle of operation, 106
see also Line protection with pilot relays
Directional control, of electromagnetic-attraction relays, 23
of single-quantity induction relays, 32, 57, 310, 313
Directional-overcurrent relay, 57
Directional relays, a-c types, 33, 52
connections, 52
power, 52
short circuits, 55
characteristics on R-X diagram, 74
d-c types, 49
use of shunts, 52
effect of mutual induction on ground relays, 393
electromagnetic-attraction type, 24
ground-relay polarization, 151, 326
misoperating tendencies, 314
negative-phase-sequence type, 330
operating characteristics, 37
response of polyphase relays to positive and negative phase sequence, 183
response of single-phase relays to short circuits, 187
Distance relays, current and voltage switching, 368
effect of power swings and loss of synchronism, 181
effect of wye-delta transformer between relay and fault, 172
electronic type, 369
ground-relay connections, 360
impedance seen during faults, 167

Distance relays, use of low-tension voltage, 145, 148
see *also* Line protection with distance relays
Distribution-circuit protection, *see* Line protection with overcurrent relays
Drop-out defined, 17

Electric arc-furnace-transformer protection, 270
Electromagnetic-attraction relay, directional, 24
general characteristics, 16
single-quantity, 22
Electronic relay, directional-comparison pilot, 396
distance, 369
Evaluation of protective relaying, 12
Expulsion protective gaps, effect of, on distance relays, 367
External-fault back-up relaying, *see* Back-up relaying

Failures, electrical, *see* Faults
False residual current, 318
Fault bus, 275
Faults, mitigation of effects of, 2
prevention of, 2
probability of, effect on practice, 11
see also Short circuits
Fire, protection against, 230
Fire-pump-motor protection, 230
Footing resistance, tower-, 303
Frequency, compensation of relays for changes in, 49
effect on induction relays, 32, 39
Frequency-converter protection, *see* Generator protection
Fundamental principles of protective relaying, 4
Fuse, coordinating with a, 335
Fuse blowing, potential-transformer, effect on distance relays, 361
effect on generator relays, 228

Generator protection, bearing overheating, 228
external-fault back-up, 227
field ground, 218
loss of excitation, 223
loss of synchronism, 218
miscellaneous, 228
motoring, 225
open circuits, 215
overexcitation, 225
overspeed, 226
Generator protection, overvoltage, 217
potential-transformer fuse blowing, 228
prime mover, 230
station auxiliary, 229
stator overheating, 216
stator short circuit, 195
calculation of CT errors, 198
ground faults, sensitive, 208
ground faults in unit generators, 209
overcurrent relays for, 215
turn-to-turn faults, 204
unbalanced phase currents, 221
vibration, 225
Ground-distance-relay connections, 360
Ground-fault neutralizer, effect on line relaying, 321
to mitigate the effect of a fault, 2
Grounding protective relay for transformer protection, 263
Ground preference, 91, 108
Ground resistance, 303
Grounding-transformer protection, 268

Harmonic-current restraint, for distance relays, 357
for transformer differential relays, 257
Holding coil, 18

Impedance diagram, *see* R-X diagram
Impedance relay, characteristic on R-X diagram, 72
for line protection, 340
general, 70
see also Distance relays
Induction-cup and induction-loop structures, 30, 31
Induction-type relay, directional, 31
general characteristics, 26
single-quantity, 31
structures, 29
torque production, 26
Insulating transformer for pilot-wire circuits, 98
Intermittent pilot, 109

Line protection with distance relays,
adjustment of distance relays, 341
arcs, effect of, 345
blocking tripping on loss of synchronism, 364
choice between impedance, reactance, and mho, 340
connections of ground distance relays, 360

Line protection with distance relays, current and voltage switching, 368
expulsion protective gaps, effect of, 367
fuse blowing, effect of, 361
electronic relays, 369
intermediate current sources, effect of, 347
low-tension current, use of, 356
low-tension voltage, use of, 352
magnetizing inrush, effect of, 359
overreach, 351, 360
purposeful tripping on loss of synchronism, 361
reclosing, automatic, 366
series capacitor, effect of, 367
see also Distance relays
Line protection with overcurrent relays, a-c and capacitor tripping, 335
adjustment of ground vs. phase relays, 316
adjustment of inverse-time-overcurrent relays, 297
arc and ground resistance, 302
directional feature, 310
fuses, coordination with, 335
ground faults in ungrounded systems, detection of, 319
ground-fault neutralizers, effect of, 321
instantaneous overcurrent relays, use of, 306
inverseness, choice of, in relay characteristics, 305
limiting ground-fault-current magnitude, effect of, 317
loop circuits, effect on relay adjustment, 303
misoperation prevention of single-phase directional-overcurrent relays during ground faults, 314
negative-phase-sequence ground directional relays, 330
open phases, effect of, 323, 325
overreach of instantaneous overcurrent relays, 308
polarizing ground-relay directional units, 326
reclosing, automatic, 333
restoration of service after prolonged outage, 334
single-phase vs. polyphase directional-overcurrent relays, 313
transient CT errors, 318
two vs. three relays for phase-fault protection, 311
Line protection with overcurrent relays, *see also* Overcurrent relays
Line protection with pilot relays, a-c wire-pilot relaying, 374
back-up protection, 378
CT requirements, 377
multiterminal lines, 375
sensitivity, 374
see also Pilot relaying, a-c wire-; Pilot relaying, d-c wire-
carrier-current-pilot relaying, attenuation, 378
sleet detection, 379
supervision, automatic, 378
combined phase- and directional comparison, mutual induction, effect on ground relays, 393
when to use, 392
directional comparison, electronic, 396
low-tension voltage, use of, 386
multiterminal lines, 387
sensitivity, 386
transients, effect of, 392
when to use, 385
see also Pilot relaying, carrier-current-
microwave, 396
phase comparison, back-up protection, 385
multiterminal lines, 382
sensitivity, 381
when to use, 380
see also Pilot relaying, carrier-current
reclosing, high-speed, 399
when to use pilot relaying, 373
Line trap, 100
Linear couplers, 284
Load shedding, 334, 363
Locking in, with generator differential relaying, 202
with transformer differential relaying, 251
Loss-of-excitation protection, 223
Loss-of-field protection, 223
Loss of synchronism, characteristics on *R-X* diagram, 177
derivation of relay current and voltage, 176
effect on distance relays, 181
generator protection, 218
trip-blocking relay, 364, 390
tripping relay, 361
Low-tension current for distance relays, 356

Low-tension voltage, for directional-comparison relaying, 386
for distance relaying, 148, 352
general, 145

Magnetizing-current inrush, effect on distance relays, 359
effect on transformer differential relays, 254
in parallel transformer banks, 259
Maximum torque, angle of, adjustment, 57
power relays, 52, 55
short-circuit relays, 55
Memory action, described, 83
effect of voltage-source location, 144
Mho relay, characteristics on *R-X* diagram, 81
for line protection, 340
operating characteristic, 80
Microwave-pilot relaying, *see* Pilot relaying, microwave-
Minimum pickup of directional relays, 38
Mixing transformer for wire-pilot relaying, 95
Modified-impedance relay, 77
Motor protection, field ground, 237
fire-pump, 230
loss of excitation, 237
loss of synchronism, 236
rotor overheating, 235
stator overheating, 232
stator short circuit, 230
unattended motors, 230
undervoltage, 237
Multiterminal-line protection, with a-c wire-pilot relaying, 375
with directional-comparison pilot relaying, 387
with phase-comparison pilot relaying, 382
Mutual induction, effect of, on directional-ground relays, 393
from power circuit to pilot wires, 98

Neutralizing transformers for wire-pilot circuits, 99
Normally blocked trip circuit, 109

Open phase, effect of, on directional-ground relays, 323, 325
equivalent circuits for, 323
protection of generators against, 215
Operating principles, basic, electromagnetic-attraction relays, 16
Operating principles, directional type, 24
single-quantity type, 22
induction relays:
directional type, 33
single-quantity type, 31
Operation indicator, 17
Operator vs. protective relays, 11
Opposed-voltage pilot relaying, 92, 95
Out of step, *see* Loss of synchronism
Overcurrent relays, combination of instantaneous and time delay, 49
pickup or reset, 45
time delay, 45, 46
see also Line protection with overcurrent relays
Overreach, of distance relays, 82, 350, 351
of instantaneous overcurrent relays, 308
Overtravel, defined, 48
effect of, on overcurrent-relay adjustment, 301
Overvoltage, in CT secondaries, 124
see also Generator protection
Overvoltage relays, pickup or reset, 45
time delay, 45, 46

Percentage-differential relays, description, 65
for bus protection, 284
for generator protection, 195
for transformer protection, 241
locking-in for internal faults, 202, 251
product restraint, 204
variable percent slope, 197, 203
Petersen coil, *see* Ground-fault neutralizer
Phase-comparison relaying, principle of operation, 101
see also Line protection with pilot relays
Phase-sequence filter, for wire-pilot relaying, 95
principle of operation, 188
Pickup, adjustment of, 19
defined, 17
of single-quantity relays, 45
ratio to reset of single-quantity relays, 22
ratio to reset of distance relays, 346, 351
Pilot relaying, a-c wire-, application limitations, 96, 374

Pilot relaying, circulating current, 92, 94
  general, 86
  line protection, 373, 374
  opposed voltage, 92, 95
  pilot-wire protection, 98
  pilot-wire requirements, 97
  pilot-wire supervision, 97
  remote tripping, 97
  see also Line protection with pilot relays
Pilot relaying, carrier-current-, directional comparison, 106
  general, 86
  intermittent or continuous, 109
  phase comparison, 101
  pilot described, 100
  supervision of pilot channel, 378
  see also Line protection with pilot relays
Pilot relaying, d-c wire-, basic principles, 89
  description of pilot 86
Pilot relaying, micro-wave-, description of channel, 397
  general, 86, 101, 396
  remote tripping, 398
  see also Line protection with pilot relays
Pilot relaying, principles of, ground preference, 91
  purpose of a pilot, 87
  sensitivity levels, 91, 105
  tripping and blocking pilots, 88, 90
Polarizing quantity, of a-c directional relays, 37
  of d-c directional relays, 51
  of directional units of ground relays, 151, 326
Polyphase directional relays, advantages and disadvantages, 313
  response to positive- and negative-phase-sequence volt-amps, 183
Potential device, *see* Capacitance potential device
Potential transformers, accuracy, 133
  connections, 146
  effect of fuse blowing, on distance relaying, 361
    on generator relaying, 228
  low-tension voltage for distance relays, 148
Power-balance relaying for line protection, 330
Power-rectifier-transformer protection, 271
Power swings, *see* Loss of synchronism
Power transformer and autotransformer protection, external-fault back-up, 264
  gas-accumulator and pressure relays, 261
  grounding protective relay, 263
  overcurrent relaying, 261
  percentage-differential relaying, 241
    CT accuracy requirements, 251
    CT connections, 242
    CT ratios, 250
    magnetizing inrush, effect of, 254
    parallel banks, 258
    percent slope, choice of, 252
    two-winding relay for three-winding transformer, 252
    zero-phase-sequence-current shunt, 249
  remote tripping, 263
Primary relaying defined, 4
Product restraint for generator differential relays, 204
Protective relaying, evaluation of, 12
  function of, 3
  functional characteristics of, 9
  fundamental principles of, 4

Ratings, relay, contact, 43
  current and voltage, 42
  holding coil, seal-in coil, and target, 43
Ratio correction factor, *see* Current transformers; Voltage transformers
Reactance relay, characteristic on R-X diagram, 79
  operating characteristic, 78
  partial-differential bus protection, 282
  see also Line protection with distance relays
Reclosing, automatic, affected by remote tripping, 367
  effect on blocking-terminal pilot-relaying equipment, 376
  mitigation of fault effects, 2
  of bus breakers, 292
  single-phase switching, 399
  synchronism check, 333, 366
  with distance relaying, 366
  with overcurrent relaying, 333
  with pilot relaying, 399
Rectifier-transformer protection, 271
Regulating-transformer protection, external fault back-up, 268
  in-phase type, 265
  phase-shifting type, 267

Reliability, a functional relay characteristic, 9
Remote tripping, by carrier current, 384
  by microwave, 398
  by pilot wire, 97
  for power-transformer protection, 263
  frequency-shift system, 264
Replica-type relays, for generator protection, 216
  for motor protection, 232, 235
Reset, adjustment of, 19
  defined, 17
  ratio to pickup, electromagnetic-attraction relays, 22
    induction relays, 32
  time of induction relays, 33, 49
Reversed third zone, for back-up with carrier pilot, 349
  where a line includes a transformer, 175
R-X diagram, loss-of-synchronism characteristics, 176
    appearance to distance relays, 181
  principle of, 72, 156
  short-circuit characteristics, 160
    appearance to distance relays, 167
    wye-delta transformer between relay and fault, 172
  superposition of relay and system characteristics, 157

Seal-in coil or relay, 18
Secondary excitation curves, 117
Selectivity, a functional relay characteristic, 9
Sensitivity, a functional relay characteristic, 9
  levels of blocking and tripping for pilot relaying, 91, 105, 386
Shaded-pole structure, 29
Short circuits, derivation of relay current and voltage, 160
  impedance seen by distance relays, 167
  table of currents, 164
  table of voltages, 165
  see also Faults
Single-phase switching, 399
Sleet-accumulation detection with carrier current, 379
Speed, a functional relay characteristic, 9
Split-phase relaying, combined with percentage differential, 207
  for generators, 204
Stability, system, benefited by protective relaying, 12
Step-voltage-regulator protection, 268
Supervision, automatic, of carrier-current channel, 378
  of pilot-wire circuits, 90, 97
Synchronism check with automatic reclosing, 333, 366

Target, 17
Telephone-circuit restrictions, 97
Testing, relay, field, 10
  manual, 10
Time characteristics, adjustment of, 45
  definitions, 19
  of d-c directional relays, 52
  of electromagnetic-attraction relays, 23
  of induction relays, 33, 39, 46
Time constant of d-c component, 279
Torque control, *see* Directional control
Torque production in induction relays, 26
Tower-footing resistance, 303
Transferred tripping, *see* Remote tripping
Transformer-drop compensation, for directional-comparison pilot relaying, 386
  for distance relaying, 352
Transformer protection, *see:* Electric arc-furnace-transformer protection; Grounding-transformer protection; Power-rectifier-transformer protection; Power-transformer and autotransformer protection; Regulating-transformer protection; Step-voltage-regulator protection
Transients, effect of, on directional-comparison pilot relaying, 392
  on distance relaying, 350
  on electromagnetic-attraction relays, 23
  on overcurrent relays, 308, 317, 318
Transient blocking, 392
Transient shunt, for distance relays, 351
  for overcurrent relays, 310
Transmission-line protection, with distance relaying, 340
  with overcurrent relaying, 296

Transmission-line protection, with pilot relaying, 373
Trap, line, 100
Tripping pilot, 88, 90
Tripping suppressor for transformer differential relays, 256
Tripping, undesired, vs. failure to trip, 11

Undercurrent and undervoltage relays, pickup or reset, 45
  time, 46
Universal relay torque equation, 39

Variable restraint for generator differential relays, 203
Vector conventions, 53
Vibration, of electromagnetic-attraction relays, 23
  protection for generators, 219, 225
Voltage-balance relays, 61
Voltage compensation for ground distance relays, 360
Voltage-regulator protection, 268
  see also Regulating-transformer protection; Step-voltage-regulator protection
Voltage switching for distance relays, 368
Voltage transformers, *see* Capacitance potential devices; Potential transformers

Watthour-meter structure, 30
Wire-pilot relaying, *see* Pilot relaying, a-c wire-; Pilot relaying, d-c wire-

Zero-phase-sequence-current shunt,
  current-transformer connections, 130
  used with transformer differential relays, 249